150 YEARS OF QUANTUM MANY-BODY THEORY

Series on Advances in Quantum Many-Body Theory

Edited by R. F. Bishop, C. E. Campell, J. W. Clark and S. Fantoni
(International Advisory Committee for the Series of International Conferences on Recent Progress in Many-Body Theories)

Published

Vol. 1: Proceedings of the Ninth International Conference on Recent Progress in Many-Body Theories
Edited by D. Neilson and R. F. Bishop

Vol. 3: Proceedings of the Tenth International Conference on Recent Progress in Many-Body Theories
Edited by R. F. Bishop, K. A. Gernoth, N. R. Walet and Y. Xian

Forthcoming

Vol. 2: Microscopic Approaches to the Structure of Light Nuclei
Edited by R. F. Bishop and N. R. Walet

Vol. 4: Microscopic Approaches to Quantum Liquids in Confined Geometries
Edited by E. Krotscheck and J. Navarro

Series on Advances in Quantum Many-Body Theory – Vol. 5

150 YEARS OF QUANTUM MANY-BODY THEORY

A FESTSCHRIFT IN HONOUR OF THE 65TH BIRTHDAYS OF

John W. Clark
Alpo J. Kallio
Manfred L. Ristig
Sergio Rosati

UMIST, Manchester, UK July 10 – 14, 2000

Editors

Raymond F. Bishop
Klaus A. Gernoth
Niels R. Walet

UMIST (University of Manchester Institute of Science and Technology), UK

World Scientific
Singapore • New Jersey • London • Hong Kong

Published by

World Scientific Publishing Co Pte Ltd
5 Toh Tuck Link, Singapore 596224
USA office: 27 Warren Street, Suite 401-402, Hackensack, NJ 07601
UK office: 57 Shelton Street, Covent Garden, London WC2H 9HE

British Library Cataloguing-in-Publication Data
A catalogue record for this book is available from the British Library.

150 YEARS OF QUANTUM MANY-BODY THEORY

ISBN-13 978-981-02-4730-0
ISBN-10 981-02-4730-3

CONTENTS

SERIES EDITORIAL BOARD AND OTHER COMMITTEES

FOREWORD BY THE EDITORS

In the summer of 1998 we became aware that John Clark, Alpo Kallio, Manfred Ristig, and Sergio Rosati, four of our most valued colleagues, friends, collaborators and teachers, were all of a similar age and were all approaching their 65th birthdays. Since we did not wish to let this collective occasion pass unmarked, we decided to organise a workshop in their honour at UMIST in Manchester, UK, during the period 10-14 July, 2000. Since our four honorees share a lifetime of research in quantum many-body theory and its application to diverse physical systems, the general theme and shape of the workshop were immediately obvious to us. But what to call it posed more of a problem. Eventually we settled on the eye-catching title "150 Years of Quantum Many-Body Theory". At first sight this title is somewhat startling, since the field is quite obviously not yet that old by a long chalk. Nevertheless, we justified it by somewhat arbitrarily, but it turns out rather accurately, assigning a working life of $37\frac{1}{2}$ years as many-body theorists to each of the four, starting with the award of their PhD degree.

This reasoning encapsulated our main aim in holding this meeting, namely that it should celebrate the professional careers of four of the leading quantum many-body physicists of their generation. All four started their scientific careers as nuclear (many-body) physicists. Each has made important contributions to the formal aspects of nuclear theory, and each has also progressively moved in other directions as their interests changed, and they found other fields where their methods could fruitfully be applied. The introductory article by Hermann Kümmel gives many more details of their unfolding careers.

The scientific CV's of the four honorees, which are included in this volume, also attest to the fact that their collective interests cover a remarkably wide area of theoretical physics. It was therefore only fitting that the topics both covered at the workshop and included in the Festschrift should reflect this diversity. While nuclear and subnuclear physics are perhaps emphasised most, many other fields are also represented. We are keenly aware that both the classification scheme we have adopted, and the grouping of articles within it, are somewhat arbitrary. In particular, many articles would fit comfortably under more than one heading. The interested reader should not, therefore, follow the scheme too slavishly.

The contents of this volume include articles both from among the speakers and participants at the workshop, and from colleagues who, for one reason or another, were unable to attend but nevertheless wished to contribute to the Festschrift. We thank all of the contributors for the extremely high quality of the articles. We are confident that they will live up to the standards of the remarkable men whose scientific careers they are intended collectively to honour in this Festschrift.

Raymond F. Bishop
Klaus A. Gernoth
Niels R. Walet

Manchester, UK
30 March, 2001

Scientific CVs of the Honorees

JOHN WALTER CLARK

Curriculum Vitae

Born	7 April, 1935; Lockhart, Texas, USA
Address	Department of Physics, Washington University, St. Louis, MO 63130 USA
Email	jwc@wuphys.wustl.edu

Education

1955	B.S., University of Texas, Austin
1957	M.A., University of Texas, Austin
1959	Ph.D., Washington University, St. Louis (Advisor: Eugene Feenberg)

Employment

1961	Associate Research Scientist, The Martin Company, Denver
1959	Research Associate, Washington University, St. Louis
1963-1966	Assistant Professor of Physics, Washington University
1966-1972	Associate Professor of Physics, Washington University
1972-	Professor of Physics, Washington University
1975-	Fellow, McDonnell Center for the Space Sciences, Washington University
1996-1997	Interim Chairman, Washington University
1999-	Wayman Crow Professor of Physics, Washington University

Honors and Awards

1954	Phi Beta Kappa, University of Texas
1959-1961	NSF Postdoctoral Fellow, Princeton (Advisor: Eugene Wigner)
1962-1963	NATO Postdoctoral Fellow, University of Birmingham, England and CENS Saclay, France (Advisors: Rudolf Peierls and Claude Bloch)
1965-1967	Alfred P. Sloan Foundation Fellow
1972	NATO Senior Fellow in Science, Laboratorio di Cibernetica, Naples
1972-	Fellow of the American Physical Society
1987	Eugene Feenberg Medal for Many-Body Physics

Visiting Appointments

Niels Bohr Institute, Denmark, Summer and Fall 1966, Summer 1967-71
Summer Associate, Atomic Energy Research Establishment, Harwell, UK, 1970
Guest Professor of Physics at University of Köln, Germany, 1971, 1973, 1974, 1977
Guest Professor of Physics at Åbo Akademi, Åbo, Finland, 1971-72
Aspen Center for Physics, Summer 1964, 1965, 1968, 1981
Consultant, (McDonnell-)Douglas Advanced Research Laboratories, 1965, 1969
Consultant or Summer Research Visitor, Argonne National Laboratory, 1969, 1973, 1975, 1976, 1979, 1980, 1982, 1991, 1992, 1994, 1995
Summer Visitor, Los Alamos Scientific Laboratory, 1977
Associated Western Universities Sabbatical Appointee and Laboratory Collaborator, Los Alamos National Laboratory, 1987
Scientist in Residence, Argonne National Laboratory, 1987
Summer Visitor, Theoretical Physics Institute, University of Minnesota, 1989, 91
Faculty, Spring College on Many-Body Techniques, Isfahan, Iran, 1991
Faculty, 17th Nathiagali Summer College on Physics and Contemporary Needs, Nathiagali, Pakistan, 1992
Visiting Faculty, Institute for Advanced Studies in Basic Sciences, Zanjan, Iran, Summer 1996

Service to Profession

Washington University Delegate to Argonne Universities Association, 1974-82
International Advisory Committee and Trustee, International Workshops on Condensed Matter Theories, 1987-; Local Organizer, 6th Workshop (St. Louis, 1982)
Editorial Board, Plenum/Nova Series *Condensed Matter Theories*
International Advisory Committee, International Conferences on Recent Progress in Many-Body Theories, 1981-, Chairman 1981-87.
Feenberg Medal Selection Committee, 1987-1994, Chairman 1991-94
Editorial Board, World Scientific Series *Advances in Quantum Many-Body Theory*
Co-Director, Workshop on Complex Dynamics in Neural Networks (Vietri, Italy, 1991)
Scientific Co-Director, 165th WE-Heraeus-Seminar "Theory of Spin Lattices and Lattice Gauge Models" (Bad Honnef, Germany, 1996)
Member, International Advisory Committee, European Summer School on Microscopic Quantum Many-Body Theories (Valencia, Spain, 1997)
Scientific Co-Director, WE-Heraeus-Seminar "Scientific Applications of Neural Nets" (Bad Honnef, Germany, 1998)
International Advisory Council, Institute for Advanced Studies in Basic Sciences, Zanjan, Iran, 1992-
Associate of ECT*, Trento, Italy
Scientific Co-Director of the ECT* Workshop on New Perspectives of Pairing Phenomena in Nuclear Systems (Trento, Italy, 2000)
Selection Committee for APS Wheatley Prize, Member 1998-01, Chairman 2000-01
Group Organizer, ZiF Research Year on the Sciences of Complexity (Zentrum für Interdisziplinäre Forschung, Bielefeld, 2000-2001)

Published Papers

1. J. W. Clark and E. Feenberg, Simplified treatment for strong-short range repulsions in N-particle systems. *Phys. Rev.* **113**, 388-399 (1959).
2. J. W. Clark, Effective spin-orbit potential in correlated heavy nuclei. *Ann. Phys. (NY)* **11**, 483-500 (1960).
3. J. W. Clark, The role of repulsive cores in the photonuclear effect. Can. J. Phys. **39**, 385-392 (1961).
4. J. W. Clark, D. Chakkalakal, and P. Westhaus, Simple derivation of cluster expansions for many-fermion systems. *Prog. Theor. Phys.* **34**, 726-733 (1965).
5. T.-P. Wang and J. W. Clark, Dynamical correlations and the nuclear photoeffect. *Prog. Theor. Phys.* **34**, 776-813 (1965).
6. J. W. Clark and P. Westhaus, Method of correlated basis functions. *Phys. Rev.* **141**, 833-857 (1966); *ibid.* **149**, 990 (1966).
7. J. W. Clark and T.-P. Wang, Theory of α matter. *Ann. Phys. (NY)* **40**, 127-152 (1966).
8. P. Westhaus and J. W. Clark, Λ binding to nuclear matter. *Phys. Lett.* **23**, 109-111 (1966).
9. J. W. Clark and P. Westhaus, Cluster expansions in many-fermion theory, I. *J. Math. Phys.* **9**, 131-148 (1968).
10. P. Westhaus and J. W. Clark, Cluster expansions in many-fermion theory, II. *J. Math. Phys.* **9**, 149-154 (1968).
11. G. Mueller and J. W. Clark, The problem of lambda-particle binding in nuclear matter. *Nucl. Phys.* B **7**, 227-249 (1968).
12. S.-O. Bäckman, D. A. Chakkalakal, and J. W. Clark, Numerical comparison of reaction-matrix and Jastrow methods for nuclear matter. *Nucl. Phys.* A **130**, 635-649 (1969).
13. L. Rickertsen, B. Block, J. W. Clark, and F. B. Malik, Nuclear heavy-ion-heavy-ion collisions and the intermediate-state model. *Phys. Rev. Lett.* **22**, 951-955 (1969).
14. G. Mueller and J. W. Clark. Status of Jastrow calculations of the Λ binding energy of nuclear matter. In *Proc. International Conf. on Hypernuclear Physics*, eds. A. R. Bodmer and L. G. Hyman (Argonne National Laboratory, Argonne, IL, 1969), pp. 598-618.
15. J. W. Clark and N.-C. Chao, Effect of attractive nuclear forces on the onset of ferromagnetism in neutron star matter. *Lett. Nuovo Cim.* **2**, 185-188 (1969).
16. J. W. Clark and G. Mueller, Observations on Λ overbinding in nuclear matter. *Nuovo Cim.* B **64**, 217-239 (1969).
17. J. W. Clark, Magnetic susceptibility of neutron matter. *Phys. Rev. Lett.* **23**, 1463-1466 (1969).
18. G. P. Mueller and J. W. Clark, Ground state properties of alpha matter. *Nucl. Phys.* A **155**, 561-581 (1970).
19. J. W. Clark and M. L. Ristig, Cluster expansion procedures for the correlated charge form factor. *Nuovo Cim.* A **70**, 313-322 (1970).
20. J. W. Clark and C.-H. Yang, Superfluid ground state of neutron matter and other strongly-interacting many-fermion systems. *Lett. Nuovo Cim.* **3**, 272-276

(1970); *ibid.* **2**, 379 (1970).
21. B. Block, J. W. Clark, M. D. High, R. Malmin, and F. B. Malik, Fission and the ion-ion interaction. *Ann. Phys. (NY)* **62**, 464-491 (1971).
22. M. L. Ristig, W. J. Ter Louw, and J. W. Clark, Tensor correlations in nuclear matter. *Phys. Rev.* C **3**, 1504-1513 (1971).
23. J. W. Clark, Science fiction and the finitude of man. *Washington University Magazine*, Summer, 32-37 (1971).
24. C.-H. Yang and J. W. Clark, Superfluid condensation energy of neutron matter. *Nucl. Phys.* A **174**, 49-62 (1971).
25. J. W. Clark, H. Heintzmann, W. Hillebrandt, and M. Grewing, Nuclear forces, compressibility of neutron matter and the maximum mass of neutron stars. *Astrophys. Lett.* **10**, 21-25 (1971).
26. N.-C. Chao, J. W. Clark, and C.-H. Yang, Proton superfluidity in neutron-star matter. *Nucl. Phys.* A **179**, 320-332 (1972).
27. M. Miller, C. W. Woo, J. W. Clark, and W.J. Ter Louw, Neutron-matter equations of state. *Nucl. Phys.* A **184**, 1-12 (1972).
28. M. L. Ristig, W. J. Ter Louw, and J. W. Clark, Tensor correlations in nuclear matter: three-body effects. *Phys. Rev.* C **5**, 695-706 (1972).
29. M. L. Ristig and J. W. Clark, Role of the dispersion effect in the method of correlated basis functions. *Phys. Rev.* C **5**, 1233-1237 (1972).
30. J. W. Clark and M. L. Ristig, Subsidiary conditions on nuclear many-body theories. *Phys. Rev.* C **5**, 1553-1561 (1972).
31. J. W. Clark and N.-C. Chao, On the crystallization of neutronic matter. *Nature Phys. Sci.* **236**, 37-38 (1972).
32. C.-H. Yang and J. W. Clark, Thermodynamic critical field of superconducting neutron-star matter. *Lett. Nuovo Cim.* **4**, 969-972 (1972).
33. S.-O. Bäckman, J. W. Clark, W. J. Ter Louw, D. A. Chakkalakal, and M. L. Ristig, Numerical comparison of reaction-matrix and Jastrow methods for nuclear matter: A reassessment. *Phys. Lett.* B **41**, 247-250 (1972).
34. J. W. Clark and M. L. Ristig, Embedding of the Brueckner approximation in the extended Jastrow scheme. *Phys. Rev.* C **7**, 1792-1803 (1973).
35. E. Krotscheck, J. Nitsch, M. L. Ristig, and J. W. Clark, Grand-canonical formulation of renormalized cluster expansions for the nuclear-matter problem. *Lett. Nuovo Cim.* **6**, 143-147 (1973).
36. E. Krotscheck, J. Nitsch, M. L. Ristig, and J. W. Clark, A numerical study of renormalized Jastrow theory. *Lett. Nuovo Cim.* **6**, 148-151 (1973).
37. M. L. Ristig and J. W. Clark, The Iwamoto-Yamada cluster expansion: its structure and renormalization. *Nucl. Phys.* A **199**, 351-368 (1973).
38. J. W. Clark and M. L. Ristig, The factor-cluster expansion as a framework for many-body theories. In *The Nuclear Many-Body Problem*, Vol. II, eds. F. Calogero and C. Ciofi degli Atti (Editrice Compositori, Bologna, 1973), pp. 273-298.
39. M. L. Ristig and J. W. Clark, Structure and renormalization of the factor-cluster expansion. In *The Nuclear Many-Body Problem*, Vol. II, eds. F. Calogero and Ciofi degli Atti (Editrice Compositori, Bologna, 1973), pp. 299-322.

40. J. W. Clark, Solid neutron stars—fact or fiction? In *The Nuclear Many-Body Problem*, Vol. II, F. Calogero and C. Ciofi degli Atti, eds. (Editrice Compositori, Bologna, 1973), pp. 675-704.
41. J. W. Clark, N.-C. Chao, and C.-G Källman, Would alpha matter be solid?. *Physica Fennica* **8**, 335-344 (1973).
42. J. W. Clark, The celestial timescope and the reincarnation of the universe. *Washington University Magazine*, Spring, 20-25 (1974).
43. J. W. Clark, Inside a neutron star. *Analog Science Fiction—Science Fact*, September, 61-76 (1974).
44. J. W. Clark, Review of *A Random Walk in Science. J. Biol. Psych.* **16**, 45-46 (1974).
45. J. W. Clark and M. L. Ristig, On short-range correlation effects in one- and two-nucleon ejection by light probe particles. *Phys. Lett.* B **5B**, 149-152 (1975).
46. J. W. Clark and D. G. Sandler, Evidence against solidification of a model neutron system. *Phys. Rev.* D **11**, 3365-3369 (1975).
47. N.-C. Chao and J. W. Clark, Magnetic properties of neutron-star matter. *Rev. Bras. Fis.* **5**, 169-187 (1975).
48. M. L. Ristig and J. W. Clark, A correlated model of the nuclear photoeffect. *Nuovo Cim.* A **30**, 609-631 (1975).
49. J. W. Clark, P. M. Lam, and W. J. Ter Louw, Perturbation corrections to the Jastrow energy for simple models of nuclear matter. *Nucl. Phys.* A **255, 1-12** (1975).
50. M. L. Ristig, P. M. Lam, and J. W. Clark, Condensate fraction and momentum distribution of liquid helium. *Phys. Lett.* A **55**, 101-103 (1975).
51. J. W. Clark, C.-G. Källman, C.-H. Yang, and D. A. Chakkalakal, Effect of polarization on superfluidity in low-density neutron matter. *Phys. Lett.* B **61**, 331-334 (1976).
52. M. L. Ristig and J. W. Clark, Density matrix of quantum fluids. *Phys. Rev.* B **14**, 2875-2887 (1976).
53. P. M. Lam, H. W. Jackson, M. L. Ristig, and J. W. Clark, Dynamic structure function of liquid ^{3}He. *Phys. Lett.* A **58**, 454-456 (1976).
54. D. A. Chakkalakal, C.-H. Yang, and J. W. Clark, Cluster-variational calculations for extended nuclear systems. *Nucl. Phys.* A **271**, 185-209 (1976).
55. P. M. Lam, J. W. Clark, and M. L. Ristig, Density matrix and momentum distribution of helium liquids and nuclear matter. *Phys. Rev.* B **16**, 222-230 (1977).
56. J. W. Clark, M. T. Johnson, P. M. Lam, and J. G. Zabolitzky, Numerical comparison of three theories of nuclear matter. *Nucl. Phys.* A **283**, 253-268 (1977).
57. M. L. Ristig, P. Hecking, P. M. Lam, and J. W. Clark, Pairing energy of liquid ^{4}He. *Phys. Lett.* A **63**, 94-96 (1977).
58. K. E. Kürten, M. L. Ristig, and J. W. Clark, Hypernetted-chain calculations for model nuclear matter. *Lett. Nuovo Cim.* **20**, 313-318 (1977).
59. J. W. Clark, H. Rafelski, J. Rafelski, and J. Winston, Quadrilogue on quarks. *Washington University Magazine*, Winter, 32-36 (1978).
60. J. W. Clark, P. M. Lam, J. G. Zabolitzky, and M. L. Ristig, Theoretical mo-

mentum distributions for liquid ^{3}He. *Phys. Rev.* B **17**, 1147-1151 (1978)
61. K. E. Kürten, M. L. Ristig, and J. W. Clark, Application of the method of correlated basis functions to nuclear matter. *Phys. Lett.* B **74**, 153-157 (1978).
62. J. W. Clark and M. T. Johnson, Alpha-cluster model of low-density nuclear matter. In *Clustering Aspects of Nuclear Structure and Nuclear Reactions*, Winnipeg, 1978, AIP Conf. Proc. 47, eds. W. T. H. Van Oers, J. P. Svenne, J. S. C. McKee, and W. R. Falk (American Institute of Physics, New York, 1978), p. 544-545.
63. J. W. Clark, Variational theory of nuclear matter. *Prog. Part. Nucl. Phys.* **2**, 89-199 (1979).
64. M. L. Ristig, K. E. Kürten, and J. W. Clark, Density matrix and spin-dependent correlations in normal liquid ^{3}He. *Phys. Rev.* B **19**, 3539-3551 (1979).
65. D. G. Sandler and J. W. Clark, A cluster-variational investigation of static density waves in nuclear matter. *Nucl. Phys.* A **317**, 23-36, Feenberg Memorial Issue (1979).
66. K. E. Kürten, M. L. Ristig, and J. W. Clark, A correlated-basis-functions approach to realistic nuclear matter. *Nucl. Phys.* A **317**, 87-115, Feenberg Memorial Issue (1979).
67. V. C. Aguilera-Navarro, R. Barrera, J. W. Clark, M. de Llano, and A. Plastino, Abnormal boson occupation in alpha matter. *Phys. Lett.* B **80**, 327-331 (1979).
68. J. W. Clark, L. R. Mead, E. Krotscheck, K. E. Kürten, and M. L. Ristig, Studies in the method of correlated basis functions I. *Nucl. Phys.* A **328**, 45-72 (1979).
69. E. Krotscheck and J. W. Clark, Studies in the method of correlated basis functions II. *Nucl. Phys.* A **328**, 73-103 (1979).
70. J. W. Clark, Update on the crisis in nuclear matter theory: A summary of the Trieste conference. *Nucl. Phys.* A **328**, 587-595 (1979).
71. E. Krotscheck and J. W. Clark, Studies in the method of correlated basis functions III. *Nucl. Phys.* A **333**, 77-115 (1980).
72. M. T. Johnson and J. W. Clark, Variational theory of ideal alpha matter. *KINAM, Revista de Fisica* **2**, 3-34 (1980).
73. L. R. Mead and J. W. Clark, Method of correlated basis functions for low levels of ^{16}O. *Phys. Lett.* B **90**, 331-334 (1980).
74. J. W. Clark, Variational approaches to the nuclear-matter problem: A progress report. In *The Meson Theory of Nuclear Forces and Nuclear Matter*, eds. D. Schütte, K. Holinde, and K. Bleuler (B. I. Wissenschaftsverlag, Mannheim, 1980), pp. 82-135.
75. T. J. Tarn, G. Huang, and J. W. Clark, Modelling of quantum mechanical control systems. *Mathematical Modelling* **1**, 109-121 (1980).
76. D. G. Sandler and J. W. Clark, Electric-dipole sum rule in correlated nuclear matter. *Nucl. Phys.* A **342**, 213-228 (1980).
77. J. W. Clark, E. Krotscheck, and R. M. Panoff (1980) Ground-state properties of ^{3}He and D within the method of correlated basis functions. *J. de Physique, Colloq.* **41**, C7-197-212 (1980).
78. D. G. Sandler and J. W. Clark, Stability of nuclear matter against neutral pion

condensation. *Phys. Lett.* B **100**, 213-218 (1981).

79. J. W. Clark, The correlated wave function approach to finite nuclear systems. In *The Many-Body Problem, Jastrow Correlations versus Brueckner Theory*, Springer Lecture Notes in Physics, Vol. 138, eds. R. Guardiola and J. Ros (Springer-Verlag, Berlin, 1981), pp. 184-259.
80. E. Krotscheck and J. W. Clark, Brueckner theory with Jastrow wave functions. In *The Many-Body Problem, Jastrow Correlations versus Brueckner Theory*, Springer Lecture Notes in Physics, Vol. 138, eds. R. Guardiola and J. Ros (Springer-Verlag, Berlin, 1981), pp. 356-372.
81. D. G. Sandler, N.-H. Kwong, J. W. Clark, and E. Krotscheck, New approaches to the study of collective excitations in strongly-interacting Fermi systems. In *Recent Progress in Many-Body Theories*, Springer Lecture Notes in Physics, Vol. 142, eds. J. G. Zabolitzky, M. de Llano, M. Fortes, and J. W. Clark (Springer-Verlag, Berlin, 1981), pp. 228-234.
82. E. Krotscheck, R. A. Smith, and J. W. Clark, Recent developments and future prospects in CBF theory. In *Recent Progress in Many-Body Theories*, Springer Lecture Notes in Physics, Vol. 142, eds. J. G. Zabolitzky, M. de Llano, M. Fortes, and J. W. Clark (Springer-Verlag, Berlin, 1981), pp. 270-279.
83. E. Krotscheck, R. A. Smith, J. W. Clark, and R. M. Panoff, Ground-state energetics of helium and deuterium fermion fluids. *Phys. Rev.* B **24**, 6383-6403 (1981).
84. T. J. Tarn, J. W. Clark, C. K. Ong, and G. M. Huang, Continuous-time quantum mechanical filter. In *Proc. Joint Workshop on Feedback and Synthesis of Linear and Nonlinear Systems*, eds. D. Hinrichsen and A. Isidori (Springer-Verlag, Berlin, 1982).
85. V. C. Aguilera-Navarro, R. Barrera, M. de Llano, J. W. Clark, and A. Plastino, Abnormal occupation in boson matter. *Phys. Rev.* C **25**, 560-570 (1982).
86. J. M. C. Chen, J. W. Clark, and D. G. Sandler, An extension of RPA theory to strongly-interacting systems. *Z. Phys.* A **305**, 223-229 (1982); *ibid.* **305**, 367 (1982).
87. R. M. Panoff, J. W. Clark, M. A. Lee, K. E. Schmidt, M. H. Kalos, and G. V. Chester, Variational Monte Carlo calculations for spin-aligned deuterium. *Phys. Rev. Lett.* **48**, 1675-1677 (1982).
88. J. W. Clark, Variational methods and the effective interactions. In *Proceedings of the 3rd International Conference on Nuclear Reaction Mechanisms*, Varenna, June 14-19, 1982, ed. E. Gadioli, *Ricerca Scientifica ed Educazione Permanente, Universitá degli Studi di Milano, Supplemento* **28**, 464-479 (1982).
89. J. W. Clark, Effective interactions for all seasons. In *Methods for the Many-Body Problem*, eds. J. M. C. Chen, J. W. Clark, and P. Suntharothok-Priesmeyer (Washington University, St. Louis, 1983), pp. 97-100.
90. Garng M. Huang, T. J. Tarn, and J. W. Clark, On the controllability of quantum-mechanical systems. *J. Math. Phys.* **24**, 2608-2618 (1983).
91. E. Krotscheck, J. W. Clark, and A. D. Jackson, Properties of elementary excitations in spin-polarized liquid ^{3}He. *Phys. Rev.* B **28**, 5088-5099 (1983).
92. T. J. Tarn, J. W. Clark, and G. M. Huang, Analytic controllability of quantum-mechanical systems. In *Proc. 1983 Symposium on the Mathematical Theory of*

Networks and Systems, ed. P. A. Fuhrmann (Springer-Verlag, Berlin, 1984).
93. C. K. Ong, G. M. Huang, T. J. Tarn, and J. W. Clark, Invertibility of quantum-mechanical control systems. *Mathematical Systems Theory* **17**, 335-350 (1984).
94. J. W. Clark and E. Krotscheck, Old dogs and new tricks: beyond the ground state with CBF theory. In *Recent Progress in Many-Body Theories*, Springer Lecture Notes in Physics, Vol. 198, eds. H. Kümmel and M. L. Ristig (Springer-Verlag, Berlin, 1984), 127-145.
95. J. W. Clark, J. V. Winston, and J. Rafelski, Self-organization of neural networks. *Phys. Lett.* A **102**, 207-211 (1984).
96. K. E. Kürten and J. W. Clark, Variational Monte Carlo study of spin-dependent correlations in Liquid ^{3}He. *Phys. Rev.* B **30**, 1342-1348 (1984).
97. J. W. Clark, E. Krotscheck, and B. Schwesinger, Nuclear-matter particle-hole force and a correlated RPA theory of ^{16}O. *Phys. Lett.* B **143**, 287-292 (1984).
98. M. F. Flynn, J. W. Clark, R. M. Panoff, O. Bohigas, and S. Stringari, Models and approximations for the momentum distribution in nuclear matter. *Nucl. Phys.* A **427**, 253-277 (1984).
99. J. W. Clark, C. K. Ong, T. J. Tarn, and G. M. Huang, Quantum nondemolition filters. *Mathematical Systems Theory* **18**, 33-55 (1985).
100. J. W. Clark, E. Krotscheck, and B. Schwesinger, Correlated RPA Treatment of Nuclei. *Anales de Fisica* **81**, 116-120 (1985).
101. M. F. Flynn, J. W. Clark, E. Krotscheck, R. A. Smith, and R. M. Panoff, Ground-state properties of spin-aligned deuterium. *Phys. Rev.* B **32**, 2945-2951 (1985).
102. K. E. Kürten and J. W. Clark, Variational Monte Carlo study of heavy-atom impurities in liquid ^{4}He. *Phys. Rev.* B **32**, 2952-2959 (1985).
103. J. W. Clark, J. Rafelski, and J. V. Winston, Brain without mind: Computer simulation of neural networks with modifiable neuronal interactions. *Phys. Rep.* **123(4)**, 215-273 (1985).
104. J. W. Clark, J. Cleymans, and J. Rafelski, Clustered quark matter. *Phys. Rev.* C **33**, 703-708 (1986).
105. K. E. Kürten and J. W. Clark, Chaos in neural systems. *Phys. Lett.* A **114**, 413-418 (1986).
106. K. E. Kürten, M. L. Ristig, and J. W. Clark, Variational theory of impurities in liquid ^{4}He. In *Condensed Matter Theories*, Vol. 1, ed. F. B. Malik (Plenum, New York, 1986), pp. 107-114.
107. J. W. Clark, J. M. C. Chen, E. Krotscheck, and R. A. Smith, Pairing in low-density neutron matter. In *Condensed Matter Theories*, Vol. 1, ed. F. B. Malik (Plenum, New York, 1986), pp. 313-326.
108. J. M. C. Chen, J. W. Clark, E. Krotscheck, and R. A. Smith, Nucleonic superfluidity in neutron stars: $^1S_0$ neutron pairing in the inner crust. *Nucl. Phys.* A **451**, 509-541 (1986).
109. J. W. Clark and T. J. Tarn, Quantum nondemolition filtering. In *Information, Complexity, and Control in Quantum Physics*, eds. A. Blaquiere, S. Diner, and G. Lochak (Springer-Verlag, Berlin, 1987), pp. 331-346.
110. J. W. Clark, Modeling of neural networks. In *Proceedings of the First Latin American School on Biophysics,* ACIF Series–Volume 6, eds. R. Fayad, A. M.

Rodrigues-Vargas, and G. Violini (World Scientific, Singapore, 1987), pp. 163-211.
111. K. E. Kürten and J. W. Clark, Order and chaos in neural systems. In *Condensed Matter Theories*, Vol. 2, eds. P. Vashishta, R. K. Kalia, and R. F. Bishop (Plenum, New York, 1987), pp. 83-88.
112. E. Mavrommatis, R. Davé, and J. W. Clark, The response function of the hard-sphere Fermi gas. In *Condensed Matter Theories*, Vol. 2, eds. P. Vashishta, R. K. Kalia, and R. F. Bishop (Plenum, New York, 1987), pp. 249-258.
113. R. M. Panoff and J. W. Clark, Ground-state phases of polarized deuterium species. *Phys. Rev.* B **36**, 5527-5539 (1987).
114. J. W. Clark, Statistical mechanics of neural networks. *Phys. Rep.* **158**, 91-158 (1988).
115. J. W. Clark, G. C. Littlewort, and J. Rafelski, Topology, structure, and distance in quasirandom neural networks. In *Computer Simulation in Brain Science*, ed. R. M. J. Cotterill (Cambridge University Press, Cambridge, 1988), pp. 104-118.
116. J. W. Clark, K. E. Kürten, and J. Rafelski, Access and stability of cyclic modes in quasirandom networks of threshold neurons obeying a deterministic synchronous dynamics. In *Computer Simulation in Brain Science*, ed. R. M. J. Cotterill (Cambridge University Press, Cambridge, 1988), pp. 316-344.
117. G. C. Littlewort, J. W. Clark, and J. Rafelski, Transition to cycling in neural networks. In *Computer Simulation in Brain Science*, ed. R. M. J. Cotterill (Cambridge University Press, Cambridge, 1988), pp. 345-356.
118. K. E. Kürten and J. W. Clark, Exemplification of chaotic activity in nonlinear neural models obeying a deterministic dynamics in continuous time. In *Computer Simulation in Brain Science*, ed. R. M. J. Cotterill (Cambridge University Press, Cambridge, 1988), pp. 357-371.
119. J. W. Clark, Probabilistic neural networks: In or out of equilibrium? In *Condensed Matter Theories*, Vol. 3, eds. J. S. Arponen, R. F. Bishop, and M. Manninen (Plenum, New York, 1988), pp. 381-391.
120. J. W. Clark, Whither many-body theory? – A summary of the Oulu Conference. In *Recent Progress in Many-Body Theories*, Vol. 1, eds. A. J. Kallio, E. Pajanne, and R. F. Bishop (Plenum, New York, 1988), pp. 365-381.
121. J. W. Chen, J. W. Clark, and K. E. Kürten, Modelling of neural systems in continuous time. *Mathematical and Computer Modelling* **10**, 503-513 (1988).
122. J. W. Clark and R. N. Silver, Final-state effects in inclusive quasielastic electron scattering from nuclei: clues from quantum fluids. In *Proceedings of the 5th International Conference on Nuclear Reaction Mechanisms*, Varenna, June 13-18, 1988, ed. E. Gadioli, *Ricerca Scientifica ed Educazione Permanente, Universitá degli Studi di Milano, Supplemento* **66**, 531-540 (1988).
123. J. W. Clark, Probabilistic neural networks. In *Evolution, Learning and Cognition*, ed. Y. C. Lee (World Scientific, Singapore, 1988), pp. 129-180.
124. T. J. Tarn, J. W. Clark, and G. M. Huang, Local controllability of generalized quantum mechanical systems. In *Modeling and Control of Systems in Engineering, Quantum Mechanics, Economics and Biosciences*, ed. A. Balquiere (Springer-Verlag, Berlin, 1989), pp. 161-172.

125. J. W. Clark, Microscopic calculations on spin-polarized quantum fluids. In *Spin Polarized Quantum Systems*, ed. S. Stringari (World Scientific, Singapore, 1989), pp. 97-108.
126. J. W. Clark and R. M. Panoff, Microscopic predictions for spin-aligned deuterium. In *Condensed Matter Theories*, Vol. 4, ed. J. Keller (Plenum, New York, 1989), pp. 1-15.
127. P. E. Sokol, R. N. Silver, and J. W. Clark, Momentum distributions: an overview. In *Momentum Distributions*, eds. R. N. Silver and P. E. Sokol (Plenum, New York, 1989), pp. 1-35.
128. J. W. Clark and M. L. Ristig, Overview of momentum distribution calculations. In *Momentum Distributions*, eds. R. N. Silver and P. E. Sokol (Plenum, New York, 1989), pp. 39-58.
129. M. L. Ristig and J. W. Clark, Generalized momentum distributions. In *Momentum Distributions*, eds. R. N. Silver and P. E. Sokol (Plenum, New York, 1989), pp. 365-370.
130. M. L. Ristig and J. W. Clark, Two-body density matrix of Bose fluids. *Phys. Rev.* B **40**, 4355-4368 (1989).
131. J. W. Clark, Introduction to neural networks. In *Non-linear Phenomena in Complex Systems*, ed. A. N. Proto (Elsevier, Amsterdam, 1989), pp. 1-102.
132. J. W. Clark, J.-W. Chen, and K. E. Kürten, Analog simulation of circuits in the olfactory bulb. In *Models of Brain Function*, ed. R. M. J. Cotterill (Cambridge University Press, Cambridge, 1989), pp. 327-347.
133. R. D. Davé, J. W. Clark, and R. M. Panoff, Elementary excitations of spin-aligned deuterium. *Phys. Rev.* B **41**, 757-760 (1990).
134. J. C. Witt and J. W. Clark, Experiments in artificial psychology: conditioning of asynchronous neural network models. *Mathematical Biosciences* **99**, 77-104 (1990).
135. M. L. Ristig and J. W. Clark, Two-body density matrix of a normal Fermi fluid. *Phys. Rev.* B **41**, 8811-8823 (1990).
136. J. W. Clark and M. L. Ristig, Generalized momentum distributions of quantum fluids. In *Condensed Matter Theories*, Vol. 5, ed. V. C. Aguilera-Navarro (Plenum, New York, 1990), pp. 47-60.
137. E. Mavrommatis and J. W. Clark, Correlated RPA calculations for model nuclear matter. In *Condensed Matter Theories*, Vol. 5, ed. V. C. Aguilera-Navarro (Plenum, New York, 1990), pp. 97-107.
138. J. W. Clark, Long-term behavior of neural networks. In *Relaxation in Complex Systems and Related Topics*, eds. I. Campbell and C. Giovannella (Plenum, New York, 1990), pp. 205-214.
139. M. L. Ristig and J. W. Clark, Two-body density matrix of quantum fluids. In *Recent Progress in Many-Body Theories*, Vol. 2, ed. Y. Avishai (Plenum, New York, 1990), pp. 323-329.
140. M. P. Qian, G. Gong, and J. W. Clark, Relative entropy and learning rules. *Phys. Rev.* A **43**, 1061-1070 (1991).
141. A. Dabringhaus, M. L. Ristig, and J. W. Clark, Vacuum ground and excited states of the U(1) lattice gauge Hamiltonian. *Phys. Rev.* D **43**, 1978-1990 (1991).

142. J. W. Clark and S. Gazula, Artificial neural networks that learn many-body physics. In *Condensed Matter Theories*, Vol. 6, eds. S. Fantoni and S. Rosati (Plenum, New York, 1991), pp. 1-24.
143. E. Mavrommatis and J. W. Clark, Microscopic study of the response of nuclear matter. In *First Hellenic Symposium on Theoretical Nuclear Physics*, eds. G. Anagnostatos, M. Grypeos, E. Mavrommatis, D. Bonatsos, and L. Skouras (Aristotle University of Thessaloniki, Thessaloniki, Greece, 1991), pp. 11-26.
144. J. W. Clark, Neural network modelling. *Physics in Medicine and Biology* **36**, 1259-1317 (1991).
145. J. W. Clark, S. Gazula, and H. Bohr, Teaching nuclear systematics to neural networks. In *Neural Networks: From Biology to High-Energy Physics*, eds. O. Benhar, C. Bosio, P. del Giudice, and E. Tabet (ETS Editrice, Pisa, 1991), pp. 121-133.
146. J. W. Clark, R. D. Davé, and J. M. C. Chen, Microscopic calculations of superfluid gaps. In *The Structure and Evolution of Neutron Stars*, eds. D. Pines, R. Tamagaki, and S. Tsuruta (Addison-Wesley, New York, 1992), pp. 134-147.
147. J. W. Clark, S. Gazula, and H. Bohr, Nuclear phenomenology with neural nets. In *Complex Dynamics in Neural Networks*, eds. J. G. Taylor, E. Caianiello, R. M. J. Cotterill, and J. W. Clark (Springer-Verlag, Berlin, 1992), pp. 305-322.
148. G. Senger, M. L. Ristig, C. E. Campbell, and J. W. Clark, Correlated density-matrix theory of normal quantum fluids. *Ann. Phys. (NY)* **218**, 160-196 (1992).
149. S. Gazula, J. W. Clark, and H. Bohr, Learning and prediction of nuclear stability by neural networks. *Nucl. Phys.* A **540**, 1-26 (1992).
150. J. W. Clark, S. Gazula, K. A. Gernoth, J. Hasenbein, J. Prater, and H. Bohr, Collective computation of many-body properties by neural networks. In *Recent Progress in Many-Body Theories*, Vol. 3, eds. T. L. Ainsworth, C. E. Campbell, B. E. Clements, and E. Krotscheck (Plenum, New York, 1992), pp. 371-386.
151. E. Mavrommatis, M. Petraki, J. W. Clark, and N. H. Kwong, On a response function and two-body density matrix of nuclear matter. In *2nd Hellenic Symposium on Nuclear Physics*, eds. G. R. Anagnostatos, D. Bonatsos, and E. Mavrommatis, (NCSR "Demokritos," Attiki, Greece, 1992), pp. 15-31.
152. J. W. Clark, Views from the Ivory Tower's basement – A commentary on "The clock and the cloud: chaos and order in *El diablo mundo*." *Revista de Estudios Hispánicos* **26**, 227-250 (1992).
153. J. W. Clark, Neural network models: From biology to many-body phenomenology, Lecture notes for the Spring College on Many-Body Techniques, Isfahan University of Technology, Isfahan, Iran. April 20 - May 1, 1991. In *Proceedings of the International Nathiagali Summer College on Physics and Contemporary Needs*, Vol. 17, eds. K. Yaldram and M. M. Awais (Pinstech, Islamabad, 1992), pp. 15-102. [Reprinted as Vol. 7 of *Lecture Notes of Institute for Advanced Studies in Basic Sciences* (Publication Section, IASBS, Zanjan, Iran, 1996), 79 pp.]
154. J. W. Clark and K. A. Gernoth, Teaching neural networks to do science. In *Structure: From Physics to General Systems*, Vol. 2, eds. M. Marinaro and G. Scarpetta (World Scientific, Singapore, 1992), pp. 64-77.

155. K. A. Gernoth, J. W. Clark, J. S. Prater, and H. Bohr, Neural network models of nuclear systematics. *Phys. Lett.* B **300**, 1-7 (1993).
156. J. M. C. Chen, J. W. Clark, R. D. Davé, and V. V. Khodel, Pairing gaps in nucleonic superfluids. *Nucl. Phys.* A **555**, 59-89, Migdal Memorial Issue (1993).
157. K. A. Gernoth, J. W. Clark, G. Senger, and M. L. Ristig, Excitations of the surface of liquid ^{4}He. In *Condensed Matter Theories*, Vol. 8, eds. L. Blum and F. B. Malik (Plenum, New York, 1993), pp. 195-203.
158. J. W. Clark, R. D. Davé, and J. M. C. Chen, Nucleonic superfluids. In *Condensed Matter Theories*, Vol. 8, eds. L. Blum and F. B. Malik (Plenum, New York, 1993), pp. 227-242.
159. E. Mavrommatis, M. Petraki, and J. W. Clark, Two-body density matrix of model nuclear matter. In *3rd Hellenic Symposium on Nuclear Physics*, eds. R. Vlastou, C. T. Papadopoulos, and E. N. Gazis (National Technical University of Athens, Athens, Greece, 1993), pp. 88-103.
160. J. W. Clark, E. Mavrommatis, and M. Petraki, Generalized momentum distribution of nuclear matter. *Acta. Phys. Pol.* **24**, 659-684, Janusz Dabrowski Festschrift Issue (1993).
161. K. A. Gernoth, J. W. Clark, G. Senger, and M. L. Ristig, Surface modes of liquid ^{4}He. *Phys. Rev.* B **49**, 15836-15848 (1994).
162. K. A. Gernoth and J. W. Clark, Resonant vapor modes in ^{4}He vapor-liquid systems. In *Condensed Matter Theories*, Vol. 9, eds. J. W. Clark, A. Sadiq, and K. A. Shoaib (Nova Science Publishers, Commack, NY, 1994), pp. 37-53.
163. J. W. Clark, K. A. Gernoth, and M. L. Ristig, Connectionist many-body phenomenology. In *Condensed Matter Theories*, Vol. 9, eds. J. W. Clark, A. Sadiq, and K. A. Shoaib (Nova Science Publishers, Commack, NY, 1994), pp. 519-537.
164. K. A. Gernoth and J. W. Clark, Resonant states of the ^{4}He liquid-vapor interface. *J. Low. Temp. Phys.* **96**, 153-175 (1994).
165. K. A. Gernoth, J. W. Clark, and M. L. Ristig, The surface of liquid ^{4}He at nonzero temperatures. *Z. Phys.* B **98**, 337-340 (1995).
166. K. A. Gernoth and J. W. Clark, Neural networks that learn to predict probabilities: Global models of nuclear stability and decay. *Neural Networks* **8**, 291-311 (1995).
167. K. A. Gernoth, J. W. Clark, and M. L. Ristig, Correlated density matrix theory of spatially inhomogeneous Bose fluids. In *Condensed Matter Theories*, Vol. 10, eds. M. Casas, M. de Llano, J. Navarro, and A. Polls (Nova Science Publishers, Commack, NY, 1995), pp. 91-104.
168. J. W. Clark and K. A. Gernoth, Statistical modeling of nuclear masses with neural network algorithms. In *Condensed Matter Theories*, Vol. 10, eds. M. Casas, M. de Llano, J. Navarro, and A. Polls (Nova Science Publishers, Commack, NY, 1995), pp. 317-333.
169. M. L. Ristig, G. Senger, M. Serhan, and J. W. Clark, Correlated density matrix theory of boson superfluids. *Ann. Phys. (NY)* **243**, 247-279 (1995).
170. K. A. Gernoth and J. W. Clark, A modified backpropagation algorithm for training neural networks on data with error bars. *Comp. Phys. Com.* **88**, 1-22 (1995).

171. J. W. Clark, K. A. Gernoth, and M. L. Ristig, Connectionist statistical inference. In *Recent Progress in Many-Body Theories*, Vol. 4, eds. E. Schachinger, H. Mitter, and H. Sormann (Plenum, New York, 1995), pp. 283-292.
172. J. W. Clark, Pathways to the quantum realm – Presentation of the fifth Eugene Feenberg Memorial Medal in Many-Body Physics to David M. Ceperley. In *Recent Progress in Many-Body Theories*, Vol. 4, eds. E. Schachinger, H. Mitter, and H. Sormann (Plenum, New York, 1995), pp. 449-454.
173. E. Mavrommatis, M. Petraki, and J. W. Clark, Fermi hypernetted chain evaluation of a generalized momentum distribution for model nuclear matter. *Phys. Rev.* C **51**, 1849-1858 (1995).
174. E. Mavrommatis, M. Petraki, and J. W. Clark, Study of the generalized momentum distribution of model nuclear matter. In *Advances in Nuclear Physics*, Proceedings of the 5th Symposium of Nuclear Physics, eds. C. Syros and C. Ronchi (European Commission, Luxembourg, 1995), pp. 139-157.
175. K. A. Gernoth and J. W. Clark, Neural network models of nuclear and noisy data. In *New Computing Techniques in Physics Research IV*, eds. B. Denby and D. Perret-Gallix (World Scientific, Singapore, 1995), pp. 425-430.
176. V. A. Khodel, J. W. Clark, and V. R. Shaginyan, Rearrangement of the electron Fermi surface in layered compounds. *Solid State Comm.* **96**, 353-357 (1995).
177. J. W. Clark, Control of quantum many-body dynamics: Designing quantum scissors. In *Condensed Matter Theories*, Vol. 11, eds. E. V. Ludena, P. Vashishta, and R. F. Bishop (Nova Science Publishers, Commack, NY, 1996), pp. 3-19.
178. V. A. Khodel, V. V. Khodel, and J. W. Clark, Solution of the gap equation in neutron matter. *Nucl. Phys.* A **598**, 390-417 (1996).
179. M. Petraki, E. Mavrommatis, and J. W. Clark, Study of the half-diagonal two-body density matrix of model nuclear matter. In *Advances in Nuclear Physics*, Proceedings of the 6th Hellenic Symposium on Nuclear Physics, ed. C. N. Panos (Makedonian Publications, Athens, 1996), pp. 58-71.
180. M. L. Ristig, J. W. Kim, and J. W. Clark, The Z(2) lattice gauge vacuum and the transverse Ising model: two sides of a coin. In *Theory of Spin Lattices and Lattice Gauge Models*, Springer Lecture Notes in Physics, Vol. 494, eds. J. W. Clark and M. L. Ristig (Springer-Verlag, Berlin, 1997), pp. 62-83.
181. J. W. Clark, M. L. Ristig, T. Lindenau, and M. Serhan, Bose-Einstein condensation in liquid helium: a correlated density matrix theory. In *Condensed Matter Theories*, Vol. 12, eds. J. W. Clark and P. V. Panat (Nova Science Publishers, Commack, NY, 1997), pp. 55-76.
182. J. W. Kim, M. L. Ristig, and J. W. Clark, The transverse Ising model at zero temperature. *Phys. Rev.* B **57**, 56-59 (1998).
183. V. A. Khodel, V. V. Khodel, and J. W. Clark, Universalities of triplet pairing in neutron matter. *Phys. Rev. Lett.* **81**, 3828-3831 (1998).
184. M. L. Ristig, T. Lindenau, M. Serhan, and J. W. Clark, Broken symmetries in liquid ^{4}He. In *Condensed Matter Theories*, Vol. 13, eds. J. da Providencia and F. B. Malik (Nova Science Publishers, Commack, NY, 1998), pp. 119-139.
185. J. W. Clark, V. A. Khodel, and V. V. Khodel, New perspectives on pairing in strongly interacting Fermi systems. In *Condensed Matter Theories*, Vol. 13,

eds. J. da Providencia and F. B. Malik (Nova Science Publishers, Commack, NY, 1998), pp. 363-379.

186. E. Mavrommatis, A. Dakos, K. A. Gernoth, and J. W. Clark, Calculations of nuclear halflives with neural nets. In *Condensed Matter Theories*, Vol. 13, eds. J. da Providencia and F. B. Malik (Nova Science Publishers, Commack, NY, 1998), pp. 423-438.
187. J. W. Clark, M. L. Ristig, and J. W. Kim, The transverse Ising model by CBF. *Int. J. Mod. Phys.* **B13**, 741-747 (1999).
188. M. Petraki, O. Benhar, J. W. Clark, A. Fabrocini, S. Fantoni, and E. Mavrommatis, On final-state interactions in quasi-elastic electron scattering off nuclear matter. In *Proceedings of the European Conference on Advances in Nuclear Physics and Related Areas*, Thessaloniki, Greece 8-12 July 1997, eds. D. M. Brink, M. Grypeos, and S. E. Massen (Giapouli Publishing, Thessaloniki, 1999), pp. 334-339.
189. J. W. Clark, K. A. Gernoth, S. Dittmar, and M. L. Ristig, Higher-order probabilistic perceptrons as Bayesian inference engines. *Phys. Rev.* E **59**, 6161-6174 (1999).
190. M. L. Ristig, T. Lindenau, M. Serhan, and J. W. Clark, Toward a microscopic theory of the λ transition in liquid ^{4}He. *J. Low. Temp. Phys.* **114**, 317-348 (1999).
191. J. W. Clark, Neural networks: New tools for modeling and data analysis in science. In *Scientific Applications of Neural Nets*, Springer Lecture Notes in Physics, Vol. 522, eds. J. W. Clark, T. Lindenau, and M. L. Ristig (Springer-Verlag, Berlin, 1999), pp. 1-96.
192. J. W. Clark, V. A. Khodel, and V. V. Khodel, Universalities and degeneracies of $^3P_2$ pairing. In *Condensed Matter Theories*, Vol. 14, eds. D. Ernst, I. Perakis, and S. Umar (Nova Science Publishers, Commack, NY, 2000), pp. 1-10.
193. T. Lindenau, M. L. Ristig, and J. W. Clark, Microscopic calculations on the superfluid phase of liquid helium. In *Condensed Matter Theories*, Vol. 14, eds. D. J. Ernst, I. Perakis, and S. Umar (Nova Science Publishers, Commack, NY, 2000), pp. 131-139.
194. J. W. Clark, V. V. Khodel, and V. A. Khodel. Triplet pairing: Simplicities within complexity. In *Condensed Matter Theories*, Vol. 15, eds. G. S. Anagnostatos, R. F. Bishop, and K. A. Gernoth (Nova Science Publishers, Commack, NY, 2000), pp. 1-12.
195. T. Landenau, M. L. Ristig, and J. W. Clark, Microscopic origin of the λ transition in liquid helium. In *Condensed Matter Theories*, Vol. 15, eds. G. S. Anagnostatos, R. F. Bishop, and K. A. Gernoth (Nova Science Publishers, Commack, NY, 2000), pp. 161-176.
196. E. Mavrommatis, S. Athanassopoulos, K. A. Gernoth, and J. W. Clark, Explorations of nuclear mass systematics using neural networks. In *Condensed Matter Theories*, Vol. 15, eds. G. S. Anagnostatos, R. F. Bishop, and K. A. Gernoth (Nova Science Publishers, Commack, NY, 2000), pp. 207-220.
197. D. N. Voskrensensky, V. A. Khodel, M. V. Zverev, and J. W. Clark, Rearrangement of the Fermi surface in dense neutron matter and direct Urca cooling of neutron stars. *Ap. J. Lett.* **553**, L127-130 (2000).

198. C. E. Campbell, J. W. Clark, E. Krotscheck, and L. P. Pitaevskii, Anthony Leggett: Feenberg Medalist 1999 – Condensed Matter Theory as a Test-Bed for Fundamental Quantum Mechanics. In *Advances in Quantum Many-Body Theory*, Vol. 3, eds. R. F. Bishop, K. A. Gernoth, N. R. Walet, and Y. Xian (World Scientific, Singapore, 2000), pp. 3-9.
199. V. V. Khodel, V. A. Khodel, and J. W. Clark, Triplet pairing in neutron matter. *Nucl. Phys.* A **679**, 827-867 (2001).
200. C. H. Anderson, Q. Huang, and J. W. Clark, Harmonic analysis of spiking neuronal ensembles. *Neurocomputing* **32-33**, 279-284 (2000).
201. J. W. Clark, V. A. Khodel, and M. V. Zverev, Impact of spin-isospin fluctuations on single-particle degrees of freedom in dense neutron matter. Migdal Memorial issue of *Sov. J. Nucl. Phys.* (*Yadernaya Fisica*). In press (2001).
202. E. Mavrommatis, S. Athanassopoulos, A. Dakos, K. A. Gernoth, and J. W. Clark, Statistical modeling with neural nets: Nuclear masses and halflives. In *Advances in Nuclear Physics*, Proceedings of the 9th Hellenic Symposium on Nuclear Physics. In press (2001).
203. T. J. Tarn, J. W. Clark, and D. G. Lucarelli, Controllability of quantum mechanical systems with continuous spectra. *Proceedings of the 39th IEEE Conference on Decision and Control*, December 2000. In press (2001).
204. J. W. Clark, E. Mavrommatis, S. Athanassopoulos, A. Dakos, and K. Gernoth, Statistical modeling of nuclear systematics. In *Fission Dynamics of Atomic Clusters and Nuclei*, eds. D. M. Brink, F. F. Karpechine, F. B. Malik, and J. da Providencia (World Scientific, Singapore). In press (2001).
205. J. W. Clark, V. A. Khodel, and M. V. Zverev, Abnormal occupation revisited. In *Condensed Matter Theories*, Vol. 16, eds. S. Hernandez and J. W. Clark (Nova Science Publishers, Huntington, NY). In press (2001).
206. M. J. Barber, J. W. Clark, and C. H. Anderson, Neural propagation of beliefs. Submitted to *Neural Computation* (2001).
207. V. A. Khodel, J. W. Clark, and M. V. Zverev, Superfluid phase transitions in dense neutron matter. Submitted to *Phys. Rev. Lett.* (2001).

Books

1. *Recent Progress in Many-Body Theories*, Springer Lecture Notes in Physics, Vol. 142, eds. J. G. Zabolitzky, M. de Llano, M. Fortes, and J. W. Clark (Springer-Verlag, Berlin, 1981), 479 pp.
2. *Methods for the Many-Body Problem*, eds. J. M. C. Chen, J. W. Clark, and P. Suntharothok-Priesmeyer (Washington University, St. Louis, 1983), 154 pp.
3. *Neural Network Dynamics*, eds. J. G. Taylor, E. Caianiello, R. M. J. Cotterill, and J. W. Clark (Perspectives in Neural Computing Series, Springer-Verlag, Berlin, 1992), 370 pp.
4. *Condensed Matter Theories*, Vol. 9, eds. J. W. Clark, A. Sadiq, and K. A. Shoaib (Nova Science Publishers, Commack, NY, 1994), 543 pp.
5. *Theory of Spin Lattices and Lattice Gauge Models*, Springer Lecture Notes in Physics, Vol. 494, eds. J. W. Clark and M. L. Ristig (Springer-Verlag, Berlin, 1997), 194 pp.

6. *Condensed Matter Theories*, Vol. 12, eds. J. W. Clark and P. V. Panat (Nova Science Publishers, Commack, NY, 1997), 498 pp.
7. *Scientific Applications of Neural Nets*, Springer Lecture Notes in Physics, Vol. 522, eds. J. W. Clark, T. Lindenau, and M. L. Ristig (Springer-Verlag, Berlin, 1999), 288 pp.
8. *Condensed Matter Theories*, Vol. 16, eds. S. Hernandez and J. W. Clark (Nova Science Publishers, Huntington, NY). In press (2001).

ALPO J. KALLIO

Curriculum Vitae

Born	21 February 1935; Kannus, Finland
Address	Department of Physical Sciences, P.O. Box 3000, FIN-90014 University of Oulu, Finland
Email	Alpo.Kallio@oulu.fi

Education

1954	Student Exam, Kannus Languages: Finnish, Swedish, German, English
1959	Candidate of Philosophy, Theoretical Physics University of Turku
1961	Licenciate of Philosophy, Theoretical Physics Mathematics (Laudatur) Ph.Lic. Thesis: Neutrino-Electron Scattering
1962-1964	Stipendiate at NORDITA, Copenhagen Thesis Adviser Prof. G. E. Brown
1965	Ph.D. Awarded University of Helsinki. Ph.D. Thesis: Effective Interaction on the Fermi-Surface of Finite Nuclei

Employment

1959-1962	Acting Teaching Assistant in Theoretical Physics University of Turku
1962-1964	Teaching Assistant in Theoretical Physics University of Turku
1964	Junior Research Associate Research Institute of Theoretical Physics University of Helsinki
1965-1966	Research Associate University of Minnesota School of Physics and Astronomy
1966-1967	Resident Research Associate Argonne National Laboratory
1967	Acting Associate Professor in Theoretical Physics University of Oulu
1.1.1968	Associate Professor in Theoretical Physics University of Oulu
1968	Acting Professor in Applied Mathematics University of Oulu
1.2.1969 -29.2.2000	Professor in Theoretical Physics University of Oulu

Visiting Appointments

1971	International Center for Theoretical Physics, Miramare, Trieste Nuclear Theory Course
1976-1977	Visiting professor at the State University of New York At Stony Brook

Honors and Awards

26.5.1992	Maanpuolustuskiltojen liitto ry:n Pronssinen kiltaristi (Bronze Guild Cross of the Union of Defender's of Finland Guilds)
6.12.1994	Suomen Valkoisen Ruusun Ritarikunnan I luokan Ritarimerkki (SVR R 1) (Knight, First Class, of the Order of the White Rose of Finland)
4.12.1998	Valtion Virka-ansiomerkki (State medal of Finland)

Service to Profession

1959-1961	Part-time Teacher in Mathematics and Physics in the Technical College at Turku
1968-1980	Special Lecturer (Dosent) in Nuclear Physics at The University of Helsinki
1969-1987	Member of the Board of the Computer Science Department, University of Oulu
1969-1990	Member of the Board of the Research Institute of Theoretical Physics, University of Helsinki
1973-1990	Member of the Board of NORDITA (Nordisk Institut for Teoretisk_Atomfysik), Copenhagen
1974	Organiser, High Energy Nuclear and Particle Physics Nort-Karelian Summer School.
1975	Member of the Finnish Academy of Sciences
1985-1987	Chairman of the Board of NORDITA
1987-1997	Chairman of Board of International Advisory Committee of RPMBT

Published Papers

1. A. Kallio and K. Kolltveit, An Application of the Separation Method in Shell-Model Calculation. *Nucl. Phys.* **53**, 87 (1964).
2. A. Kallio, On the Effective Nucleon-Nucleon Interaction on the Fermi Surface of Finite Nuclei. Annales Academiae Scientiarum Fennicae, Series A. VII Physica **163**, 1-35 (194).
3. A. Kallio, On the Relation Between t-Matrices in Free Scattering, Nuclear Matter and Finite Nuclei. *Phys. Lett.* **18**, 51 (1965).
4. T. Engeland and A. Kallio, An Application of the Separation Method in ^{18}O and ^{20}Ne. *Nucl. Phys.* **59**, 211 (1964).
5. A.M. Green, A. Kallio and K. Kolltveit, Odd Parity Excitations of ^{16}O with a Realistic Nucleon-Nucleon Interaction. *Phys. Lett.* **14**, 142 (1965).
6. A. Kallio and A.M. Green, Odd Parity States of ^{16}O with a Realistic Nucleon-Nucleon Interaction. *Nucl. Phys.* **84**, 161 (1966).

7. B.F. Bayman and A. Kallio, Relative-Angular-Momentum-Zero Part of Two-Nucleon Wawe Functions. *Phys. Rev.* **156**, 1121 (1967).
8. A. Kallio and B.D. Day, Accurate Calculation of the Reaction Matrix in Light Nuclei. *Phys. Lett.* B **25**, 72-74 (1967).
9. A. Kallio and E. Bycling, Solution of the Bethe-Salpeter Equation for Fermions. Annales Academiae Scientarum Fennicae, Series A **??**, ?? (1968).
10. A. Kallio and E. Byckling, Meson Spectrum in the Quark Model. *Lett. Nuovo Cim.* **Ser I**, 27 (1968).
11. A. Kallio and B.D. Day, Accurate calculation of the Reaction Matrix in Light Nuclei and Nuclear Matter. *Nucl. Phys.* A **124**, 177 (1969).
12. A.M. Green, T.K. Dahlblom, A. Kallio and M. Rho, Muon capture in ^{16}O-renormalization of the one-body operator. *Phys. Lett.* B **31**, 189 (1970).
13. P. Erkkilä, M. Saarela and A. Kallio, Muon capture rate in ^{16}O and the ground-state wave function. *Phys. Lett.* B **34**, 18 (1971).
14. C. Ciofi degli Atti and A. Kallio, Comparison of the Brueckner and Jastrow methods in electron scattering. *Phys. Lett.* B **36**, 433 (1971).
15. M. Saarela, P. Erkkilä and A. Kallio, The effect of core-excitations and SU_4-symmetry breaking upon the muon capture rate in ^{16}O. Physica Fennica **8**, 51-65 (1973).
16. A. Kallio, P. Toropainen, A.M. Green and T. Kouki, The effect of the Δ (1236) on the charge form factor of ^{3}He and ^3H. *Nucl. Phys.* A **231**, 77 (1974).
17. A.M. Green, T. Kouki, A.J. Kallio and P. Toropainen, The effect of the Δ (1236) on the trinucleon charge form factor. *Czech. J. Phys.* B **25**, 270-275 (1975).
18. O. Benhar, C. Ciofi degli Atti, A. Kallio, L. Lantto and P. Toropainen, Subsidiary conditions and variational calculations of nuclear and neutron matters. *Phys. Lett.* B **60**, 129 (1976).
19. O. Benhar, C. Ciofi degli Atti, S. Fantoni, S. Rosati, A. Kallio, L. Lantto and P. Toropainen, Lowest-order and Hyper-Netted-Chain calculations of nuclear matter *Phys. Lett.* B **64**, 395-398 (1976).
20. A. Kallio and R.A. Smith. How simple can HNC be - and still be right. *Phys. Lett.* B **68**, 315 (1977).
21. R.A. Smith, A. Kallio, M. Puoskari and P. Toropainen, The Ground State of Bose System with HNC and BBGKY equations. *Nucl. Phys.* A **328**, 186 (1979).
22. A. Kallio, P. Pietiläinen, M. Puoskari and P. Toropainen, On the Optimal Solution of Generalized HNC-Equation for Bose Quantum Liquids and Solids. *Phys. Scripta* **22**, 91 (1980).
23. A. Kallio, Fysiikan tietomme rajat (The Boundaries of our Knowledge in Physics). Academia Scientiarum Fennica, Vuosikirja - Year Book (1980).
24. M. Puoskari and A. Kallio, The effective correlation theory for liquid ^{3}He *Phys. Scripta* **24**, 601 (1981).
25. A. Kallio, The Boundaries of our Knowledge in Physics. Faravid 4. Koillissanomat Oy, Kuusamo (1981), p. 231-247.
26. A. Kallio, Merikosken kalaportaat (Fish Ladders to Merikoski). Oulu-lehti 23.3.1981.

27. A. Kallio, M. Puoskari and P. Pietiläinen, Impurity Quasiparticle as a Weak Inhomogenicy. In *Recent progress in many-body Theories*, eds. J. G. Zabolitzky, M. de LLano, M. Fortes and J. W. Clark, *Lec. Notes in Phys.*__142__ (Springer Verlag, Heidelberg, 1981), p. 235.
28. A. Kallio, P. Pietiläinen and L. Lantto, Hypernetted Chain Theory of Charged Impurity. *Phys. Scripta* **25**, 943-951 (1982).
29. M. Saarela, P. Pietiläinen and A. Kallio, Optimized free surface of liquid ^{4}He in HNC. *Phys. Rev.* B **27**, 231 (1983).
30. A. Kallio, Eriävä mielipide Seutukaavaliiton energiatyöryhmän mietintöön (Dissenting Opinion to the Report of the Energy Working Group etablished by the Suetukaavaliito (I-river harnessing)). Pohjois-Pohjanmaan Seutukaavaliitto, Julkaisusarja A:64, Oulu (1983)
31. P. Pietiläinen and A. Kallio, NHC theory of charged impurity with mixture formalism *Phys. Rev.* B **27**, 224 (1983).
32. A. Kallio, M. Puoskari, L.J. Lantto, P. Pietiläinen and V. Halonen, Linear and Non Linear Response. In *Proceedings of Recent Progress in Many-Body Theories, Altenberg 1983*, eds. H. Kummel and M. Ristig, *Lec. Notes in Phys.*__198__ (Springer-Verlag, Berlin, 1983), p. 210-218.
33. A. Kallio, P. Pietiläinen and L. Lantto, Hypernetted chain theory of charged impurity *Phys. Scripta* **23**, 943-951 (1982).
34. L. Lantto, P. Pietiläinen and A. Kallio, Variational approach to linear dielectric response *Phys. Rev.* B **26**, 5568-5576 (1982).
35. T. Chakraborty, A. Kallio, L.J.Lantto and P. Pietiläinen, Structure of Liquid Metallic Hydrogen as a Two-Component Fermi Fluid at T = 0. *Phys. Rev.* B **27**, 3061-3064 (1983).
36. M. Puoskari, A. Kallio and P. Pollari, The Temperature Dependence of the Static Structure Factor for Liquid ^{4}He below $T\lambda$. *Phys. Scripta* **29**, 378-383 (1984).
37. M. Puoskari and A. Kallio, Hypernetted Chain Theory of the Momentum Distribution for Bose Systems with Mixture Formalism. *Phys. Rev.* B **30**, 2331-2337 (1984).
38. A. Kallio, Lehtipolemiikkia kalaportaista (newspaper polemics about fish ladders). Liitto 29.9.1984, 10.10.1984.
39. A. Kallio, Physics on the Border Between the Living and the Dead. Acta Universitatis Ouluensis, Series F, Scripta Academica No. 1, University of Oulu, Oulu (1985), pp. 47-59.
40. T. Chakraborty, A. Kallio and M. Puoskari, Structure of liquid ^{4}He at low temperatures: Random-phase approximation. *Phys. Rev.* B **33**, 635-637 (1986).
41. A. Kallio, Determinism - Closing Chapter in Science. Annales Academiae Scientiarum Fennicae, Series A, VI Physica, 431, Suomalainen Tiedeakatemia, Helsinki (1986), p. 84-98.
42. A. Kallio, P. Pollari, M. Puoskari and J. Kinaret, Correlations in the Fractional Hall Effect. In *Condensed Matter Theories*, Vol. 1, ed. F. B. Mallick (Plenum, New York, 1987) p. 235-245.
43. A. Kallio and X. Xiong, Electron-hole liquid model for high-Tc superconductors. *Phys. Rev.* B **4**, 2530-2533 (1990).

44. . Kallio, Oulujoen kalaportaat taas kerran tapetilla (The fish ladders to the Oulu river revisited). Luontoleijona no 4, s. 10, (1990).
45. A. Kallio, X. Xiong and M. Alatalo, Electron-hole liquid model for high-T_c superconductors. In *Proceedings of the Sixth International Conference on Recent Progress in Many-Body Theories*, Vol. 2, ed. Y. Avishai (Plenum Press, New York, 1990), 45-54.
46. A. Kallio and X. Xiong, Electron-hole liquid model for high-T_c superconductors: Metal-insulator transition and doping behavior. *Phys. Rev.* B **43**, 5564-5575 (1991).
47. A. Kallio, Väliaikatietoa korkeanlämpötilan suprajohteista (Preset status of high temperature superconductivity). Fysiikka tänään **4**, 17-20 (1991).
48. A. Kallio and X. Xiong, Electron-hole liquid model for high-T_c superconductors. In *Condensed Matter Theories*, Vol. 6, eds. S. Fantoni and S. Rosati (Plenum Press, New York, 1990), p. 241-257.
49. A. Kallio, V. Apaja and X. Xiong, Exponent relations for superfluids with pair breaking. *Physica* C **191**, 23-31 (1992).
50. A. Kallio and X. Xiong, The spectator fermion superfluid model for high-T_c and heavy fermion superconductors: the pressure effect. *Physica* C **199**, 340-350 (1992).
51. A. Kallio, Electron tunneling and accoustic mode in high T_c-superconductors. High-Temperature Superconductivity: Physical properties, Microscopic Theory, and Mechanisms. In *Proceedings Miami Workshop on Electronic Structure and Mechanisms for High Temperature Superconductivity, 1991*, eds. J. Ashkenazi, S.E. Barnes, F. Zuo, G.C. Vezzoli and B.M. Klein (Plenum Press, New York, 1992), p. 241.
52. A. Kallio, V. Apaja and X. Xiong, Spectator fermion model for HTS and HF. In *Proceedings of Lattice effects in High-T_c superconductors, Santa Fe, 1992*, eds. Alan R. Bishop *et al.* (World Scientific, Singapore, 1992), p. 494-499.
53. V. Apaja and A. Kallio. Acoustic plasmon velocity from Raman scattering lineshapes *Physica* C **209**, 519-530 (1993).
54. A. Kallio, V. Apaja, X. Xiong and S. Pöykkö, Boson localization and universality in $YBa_2Cu_{3-x}M_xO_{7-\delta}$. *Physica* C **219**, 340-362 (1994).
55. A. Kallio, V. Apaja and S. Pöykkö, Spectator fermion binding of bosons. *Physica* C **235**, 2349-2350 (1994).
56. A. Kallio, V. Apaja and S. Pöykkö, Spectator fermion binding of bosons. In *Recent progress in Many-Body Theories*, Vol. 4, eds. E. Schachinger, H. Mitter and H. Sorman (Plenum, New York, 1995), p. 381-391.
57. A. Kallio, V. Apaja and S. Pöykkö, Spectator fermion binding of bosons. *Physica* B **210**, 472-478 (1995).
58. A. Kallio and J. Piilo, Towards exact analytic calculation of electron gas. *Czech. J. Phys.*Suppl. **46**, 2641-2642 (1996).
59. A. Kallio and K. Honkala, Hall-coefficient Scaling and Chemical Equilibrium above T_c. *Czech. J. Phys.*Suppl. **46**, 943-944 (1996).
60. A. Kallio and J. Piilo, Novel analytic calculation of electron gas properties. *Phys. Rev. Lett.* **77**, 4237-4240 (1996).
61. A. Kallio, V. Sverdlov and M. Rytivaara, Paramagnetic Meissner effect and

time reversal non-invariance from spin polarization. *Superlatt. and Microstr.* **21**, 479-484 (1997).
62. A. Kallio, V. Sverdlov and K. Honkala, Some New Aspects of High-T_c Superconductors: Chemical Equilibrium. *Superlatt. and Microstruct.* **21 Supplement A**, 111-126 (1997).
63. A. Kallio, K. Gernoth, J. Hissa and T. Häyrynen, T_c curves for $La_{2-x}M_xCuO_4$ from chemical equilibrium. *Physica* C **282**, 1617-1618 (1997).
64. A. Kallio, J. Hissa, T. Häyrynen, V. Bräysy and T. Säkkinen, Chemical equilibrium model for high-T_c and heavy fermion superconductors: the density of states *Int. J. Mod. Phys.* B **13**, 651-657 (1999).
65. A. Kallio, J. Hissa, T. Häyrynen and V. Bräysy. Upper critical field of unconventional superconductors from chemical equilibrium. *Int. J. Mod. Phys.* B **13**, 3443-3448 (1999).
66. A. Kallio, J. Hissa, T. Häyrynen and V. Bräysy, Pressure dependence of T_c from chemical equilibrium. *Int. J. Mod. Phys.* B **13**, 3532-3537 (1999).
67. A. Kallio, J. Hissa, V. Bräysy and T. Häyrynen, Upper Critical Magnetic Field H_{c2} from Chemical Equilibrium. In *The Proceeding of the First Regional Conference on Magnetic and Superconducting Materials (MSM-99), Tehran, 1999*, eds. M. Akhavan, J. Jensen and K. Kitazawa (World Scientific, Singapore, 2000).
68. A. Kallio, J. Hissa, V. Bräysy and T. Häyrynen, Pressure Dependence of T_c from Chemical Equilibrium. In *The Proceeding of the First Regional Conference on Magnetic and Superconducting Materials (MSM-99), Tehran, 1999*, eds. M. Akhavan, J. Jensen and K. Kitazawa (World Scientific, Singapore, 2000).
69. J. Hissa, A. Kallio, V. Bräysy and T. Häyrynen, Upper Critical Magnetic Field H_{c2} from Chemical Equilibrium. *Physica* B **284-288**, 1061-1062 (2000).
70. V. Bräysy, A. Kallio, J. Hissa and T. Häyrynen, Pressure Dependence of T_c from Chemical Equilibrium. *Physica* B **284-288**, 1063-1064 (2000).
71. A. Kallio, J. Hissa and V. Bräysy, Is the Heavy Fermion Specific Heat of Bosonic Origin? *Physica* C **341-348**, 451-452 (2000).
72. A. Kallio, J. Hissa, V. Bräysy and T. Häyrynen, Pressure Effect in Cuprates – Manifestation of Le Chatelier's Principle. *Physica* C **341-348**, 1949-1950 (2000).

Books

1. *Proceedings of the international conference on Recent Progress in Many-Body Theories*, eds. C. Ciofi degli Atti, A. Kallio and S. Rosati, *Nucl. Phys.* A **328**, 1,2 (1979).
2. *Recent Progress in Many-Body Theories*, Vol. 1 eds. A. J. Kallio, E. Pajanne, and R. F. Bishop (Plenum, New York, 1988).

MANFRED RISTIG

Curriculum Vitae

Born	27 January 1935, Erfurt, Germany
Address	Universität zu Köln D-50937 Köln, Germany
Email	ristig@thp.uni-koeln.de

Education

1953	Abitur, Heinrich-Mann-Oberschule in Erfurt
1953-57	Physics, Universität Jena (Prof. G. Heber)
1958-61	"Political Prisoner" Bautzen (former East Germany)
1961-64	Physics, Universität München
1964	Physics diploma, supervisors Profs. F. Bopp and G. Obermair
1964-65	Max-Planck-Institut für Physik und Astrophysik (Heisenberg Institut) München
1966	Doctoral thesis, Universität zu Köln, supervisor Prof. P. Mittelstaedt
1972	Habilitation in Theoretical Physics and venia legendi, Universität zu Köln

Employment

1965-69	Assistant, Universität zu Köln
1969-71	Research Associate, Washington University, St. Louis
1972-74	Assistant, Universität zu Köln
1974-79	Docent, apl. Professor, Universität zu Köln
1980-87	Professor (C3), Universität zu Köln
1987-	Universitätsprofessor, Universität zu Köln

Visiting Positions

1974	Visiting assistant professor, Washington University, St. Louis
1984, 85, 88, 89, 91, 93, 96	Visiting professor, Washington University, St. Louis
1997 -	Adjunct professor of Physics, Washington University, St.Louis
1984	Gordon-Godfrey visiting fellow, University of New South Wales, Sydney, Australia
1989	Visiting professor, Universidad de Granada, Spain (1989).
1988-	Vertrauensdozent der Studienstiftung des deutschen Volkes (German Scholarship Foundation)

Service to the profession

Organisiser, International Conference on Recent Progress in Many-Body Theories, Altenberg 1983
Organiser, Current Studies in Quantum Many-Body Theory, 114. WE-Heraeus-Seminar, Bad Honnef, 7-9 June 1993.
Organiser, Theory of Spin Lattices and Lattice Gauge Models, 165. WE-Heraeus Seminar, Bad Honnef, 14-16 October 1996.
Organiser, Scientific Applications of Neural Nets, 194. WE Heraeus-Seminar, Bad Honnef, 11-13 May 1998
Organiser, Microscopic Theory of Phase Transitions: Quantum versus Thermal Fluctuations, 227. WE-Heraeus-Seminar, Bad Honnef, 13-15 Dec. 1999.

Published Papers

1. M. Ristig und G. Obermair, Konsistente Temperaturentwicklungen für den Ferromagneten. *Z. Phys.* **186**, 114 (1965).
2. P. Mittelstaedt and M. Ristig, Equivalent nonlocal nucleon-nucleon potentials. *Z. Phys.* **193**, 349 (1966).
3. M. Ristig, Zur Bestimmung regulärer äquivalenter Nukleon-Nukleon Potentiale und induzierter Mehrkörperkräfte in Vielteilchensystemen. *Z. Phys.* **199**, 325 (1967).
4. M. Ristig, On the three-body contribution to the ground state energy of nuclear matter. *Z. Phys.* **232**, 279 (1970).
5. M. Ristig and S. Kistler, The method of unitary transformations in the theory of nuclear matter. *Z. Phys.* **215**, 419 (1968).
6. J. W. Clark and M. L. Ristig, Cluster-expansion procedures for the correlated charge form factor. *Nuovo Cim.* A **70**, 313 (1970).
7. M. L. Ristig, W. J. Ter Louw and J. W. Clark, Tensor correlations in nuclear matter. *Phys. Rev.* C **3**, 1504 (1971).
8. M. L. Ristig, W. J. Ter Louw and J. W. Clark, Tensor correlations in nuclear matter, three-body effects. *Phys. Rev.* C **5**, 695 (1972).
9. J. W. Clark and M. L. Ristig, Subsidiary conditions on nuclear many-body theories. *Phys. Rev.* C **5**, 1553 (1972).
10. M. L. Ristig and J. W. Clark, Role of the dispersion effect in the method of correlated basis functions. *Phys. Rev.* C **5**, 1233 (1972).
11. S. O. Bäckman *et al.*, M. L. Ristig, Numerical comparison of reaction-matrix and Jastrow methods for nuclear matter: a reassessment. *Phys. Lett.* B **41**, 247 (1972).
12. M. L. Ristig and J. W. Clark, The lwamoto-Yamada cluster expansion: its structure and renormalization. *Nucl. Phys.* A **199**, 351 (1973).
13. J. W. Clark and M. L. Ristig, Embedding of the Brueckner approximation in the extended Jastrow-scheme. *Phys. Rev.* C **7**, 1792 (1973).
14. E. Krotscheck, J. Nitsch, M. L. Ristig and J. W. Clark, Grand canonical formulation of renormalized cluster expansions for the nuclear matter problem. *Lett. Nuovo Cim.* **6**, 143 (1973).

15. E. Krotscheck, J. Nitsch, M. L. Ristig and J. W. Clark, A numerical study of renormalized Jastrow theory. *Nuovo Cim.* **6**, 148 (1973).
16. J. W. Clark and M. L. Ristig, The factor cluster expansion as a framework for many-body theories. In *The Nuclear Many-Body Problem*, Vol. 2, eds. F. Calogero and C. Ciofi degli Atti (Editrice Compository, Bologna 1973).
17. M. L. Ristig and J. W. Clark, Structure and renormalization of the factor cluster expansion. In *The Nuclear Many-Body Problem*, Vol. 2, eds. F. Calogero and C. Ciofi degli Atti (Editrice Compository, Bologna 1973).
18. J. W. Clark and M. L. Ristig, On short-range correlation effects in one- and two-nucleon ejection by light probe particles. *Phys. Lett.* B **55**, (1975).
19. M. L. Ristig and J. W. Clark, A correlated model of the nuclear photoeffect. *Nuovo Cim.* A **30**, 609 (1975).
20. E. Krotscheck and M. L. Ristig, Hypernetted-chain approximation for dense Fermi fluids. *Phys. Lett.* A **48**, 17 (1974).
21. E. Krotscheck and M. L. Ristig, Long-range Jastrow Correlations. *Nucl. Phys.* A **242**, 389 (1975).
22. M. L. Ristig, P. M. Lam and J. W. Clark, Condensate fraction and momentum distribution of liquid helium. *Phys. Lett.* A **55**, 101 (1975).
23. M. L. Ristig and J. W. Clark, Density matrix of quantum fluids. *Phys. Rev.* B **14**, 2875 (1976).
24. P. M. Lam, J. W. Clark and M. L. Ristig, Density matrix and momentum distribution of helium liquids and nuclear matter. *Phys. Rev.* B **16**, 222 (1977).
25. M. L. Ristig, Pair condensation of liquid ^{4}He. *Phys. Lett.* A **58**, 390 (1976).
26. M. L. Ristig, Pairing function of Bose fluids. *Phys. Rev.* B **18**, 1207 (1978).
27. M. L. Ristig and P. Hecking, How good is a Jastrow trial wave function for nuclear matter. *Phys. Lett.* B **65**, 405 (1976).
28. P. M. Lam, H. W. Jackson, M. L. Ristig and J. W. Clark, Dynamic structure function of liquid ^{3}He. *Phys. Lett.* A **58**, 454 (1976).
29. K. E. Kürten and M. L. Ristig, Constrained variational calculations on nuclear matter with state dependent correlations. *Phys. Lett.* B **66**, 113 (1977).
30. S. Kilic and M. L. Ristig, A generalized cell model for liquid helium and nuclear matter. *Nuovo Cim.* A **39**, 248 (1977).
31. M. L. Ristig, P. Hecking, P. M. Lam and J. W. Clark, Pairing energy of liquid ^{4}He. *Phys. Lett.* A **63**, 94 (1977).
32. K. E. Kürten, M. L. Ristig and J. W. Clark, Hypernetted-chain calculations for model nuclear matter. *Nuovo Cim.* **20**, 313 (1977).
33. J. W. Clark, P. M. Lam, J. G. Zabolitzky and M. L. Ristig, Theoretical momentum distributions for liquid ^{3}He. *Phys. Rev.* B **17**, 1147 (1978).
34. K. E. Kürten, M. L. Ristig and J. W. Clark, Application of the method of correlated basis functions to nuclear matter. *Phys. Lett.* B **74**, 153 (1978).
35. P. M. Lam and M. L. Ristig, Condensate fraction of liquid ^{4}He at low temperatures. *Phys. Lett.* A **65**, 307 (1978).
36. M. L. Ristig, Two-body density matrix of Bose fluids. *Nucl. Phys.* A **317**, 163 (1979).
37. K. E. Kürten, M. L. Ristig and J. W. Clark, A correlated-basis-functions approach to realistic nuclear matter. *Nucl. Phys.* A **317**, 87 (1979).

38. M. L. Ristig, K. E. Kürten and J. W. Clark, Density matrix and spin-dependent correlations of normal liquid ^{3}He. *Phys. Rev.* B **19**, 3539 (1979).
39. P. M. Lam and M. L. Ristig, Condensed phase of liquid ^{4}He. *Phys. Rev.* B **5**, 1960 (1979).
40. M. L. Ristig and P. M. Lam, Long-range order in Bose fluids. *Nucl. Phys.* A **328**, 267 (1979).
41. J. W. Clark, L. R. Mead, E. Krotscheck, K. E. Kürten and M. L. Ristig, Studies in the method of correlated basis functions I. *Nucl. Phys.* A **328**, 45 (1979).
42. P. M. Lam and M. L. Ristig, Hypernetted-chain calculations for the electron plasma at metallic densities. *KINAM* **1**, 407 (1979).
43. M. L. Ristig, P. M. Lam and A. Lejeune, Momentum distributions for model nuclear matter. *Phys. Lett.* B **93**, 240 (1980).
44. T. Chakraborty and M. L. Ristig, Spin-orbit correlations in nuclear matter. *Lett. Nuovo Cim.* **27**, 65 (1980).
45. M. L. Ristig and P. M. Lam, Spatial correlations in model fermi fluids. *J. Low. Temp. Phys.* **40**, 571 (1980).
46. M. L. Ristig, P. M. Lam and H. P. Nollert, Ground state fluctuations in polarized ^{3}He. *J. Phys. (Paris)* **41**, 213 (1980).
47. N. Schulz and M. L. Ristig, The paired phonon model for liquid ^{4}He at low temperatures. *Z. Phys.* B **38**, 293 (1980).
48. M. L. Ristig, Makroskopische Quanteneffekte. In *Grundlagenprobleme der modernen Physik* (Bibliographisches Institut, Mannheim, 1980).
49. M. L. Ristig and P. M. Lam, Spin-alignment in condensed atomic hydrogen. In *Recent Progress in Many-Body Theories, Lec. Notes in Phys.* **142**, 318 (1981).
50. P. M. Lam and M. L. Ristig, On the momentum distribution in Bose fluids. *Z. Phys.* B **45**, 179 (1982).
51. M. L. Ristig, Momentum distribution in quantum fluids and nuclear matter. In *From Nuclei to Particles*, International school of Physics 'Enrico Fermi' (Amsterdam, North-Holland, 1982).
52. M. L. Ristig, S. Fantoni and K. E. Kürten, Ground state properties of boson-mixtures. *Z. Phys.* B **51**, 1 (1983).
53. K. E. Kürten and M. L. Ristig, Atomic impurities in liquid helium. *Phys. Rev.* B **27**, 5479 (1983).
54. K. E. Kürten and M. L. Ristig, Paired-phonon analysis of impurities in Bose fluids. *Nuovo Cim.* D **2**, 1057 (1983).
55. M. D. Miller, M. L. Ristig and N. Schulz, Solutions of the Ornstein-Zernike equation for non-uniform systems. In *Recent Progress in Many-Body Theories*, eds. H. Kümmel and M. L. Ristig, *Lec. Notes in Phys.* **198** (Springer, Berlin 1984).
56. C. E. Campbell, M. L. Ristig, K. E. Kürten and G. Senger, Variational statistical mechanics of quantum fluids. In *Proceedings of the 17th International conference on low temperature physics*, eds. U. Eckern, A. Schmid, W. Weber and H. Wühl (Elsevier, Amsterdam, 1984), p. 1211.
57. M. L. Ristig, G. Senger, K. E. Kürten and C. E. Campbell, Variational Theory of superfluid ^{4}He at positive temperatures. In *Proceedings of the 17th International conference on low temperature physics*, eds. U. Eckern, A. Schmid, W.

Weber and H. Wühl (Elsevier, Amsterdam, 1984) p. 1209.
58. C. E. Campbell, K. E. Kürten, M. L. Ristig and G. Senger, Variational density-matrix theory of liquid ^{4}He at nonzero temperatures. *Phys. Rev.* B **30**, 3728 (1984).
59. K. E. Kürten and M. L. Ristig, Atomic and molecular hydrogen isotopes in liquid helium. *Phys. Rev.* B **31**, 1346 (1985).
60. K. E. Kürten and M. L. Ristig, Binding of ^{3}He atoms to liquid helium. *Nuovo Cim.* D **7**, 251 (1986).
61. G. Senger, M. L. Ristig, K. E. Kürten and C. E. Campbell, Helium at nonzero temperatures: A variational density matrix approach for Bose fluids. *Phys. Rev.* B **33**, 7562 (1986).
62. K. E. Kürten, M. L. Ristig and J. W. Clark, Variational theory of impurities in liquid ^{4}He. In *Condensed Matter Theories* Vol. 1, ed. F. B. Mallick (Plenum, New York, 1986), p. 107.
63. M. L. Ristig, G. Senger, K. E. Kürten and C. E. Campbell, Variational density matrix theory. In *Condensed Matter Theories*, Vol. 1, ed. F. B. Mallick (Plenum, New York, 1986), p. 153.
64. K. E. Kürten and M. L. Ristig, Boson-mixtures at non zero temperatures. In *Condensed Matter Theories*, Vol. 2, eds. O. Vashista, R. K. Kalia, and R.F. Bishop (Publisher, twon, 1987), p. 211.
65. K. E. Kürten and M. L. Ristig, Phase-separation in isotopic helium mixtures. *Phys. Rev.* B **37**, 3359 (1988).
66. M. L. Ristig, G. Senger and K. E. Kürten, Density-matrix theory of supercooled hydrogen. In *Recent Progress in Many-Body Theories*, Vol. 1, eds. A. J. Kallio, E. Pajanne and R. F. Bishop (Plenum, New York, 1988).
67. A. Dabringhaus and M. L. Ristig, The $U(1)$ lattice gauge model: a correlated many-body system. In *Condensed Matter Theories*, Vol. 4, ed. J. Keller (Plenum, New York, 1990), p. 391.
68. J. W. Clark and M. L. Ristig, Overview of momentum distribution calculations. In *Momentum Distributions*, ed. R. N. Silver (Plenum, New York, 1989).
69. M. L. Ristig and J. W. Clark, Generalized momentum distributions. In *Momentum Distributions*, ed. R. N. Silver (Plenum, New York, 1989).
70. L. Szybisz and M. L. Ristig, New method of solving the optimized paired-phonon analysis equations and stability of thin films of liquid He at $T = 0$K. *Phys. Rev.* B **40**, 4391 (1989).
71. M. L. Ristig and J. W. Clark, Two-body density matrix of a Bose fluid. *Phys. Rev.* B **40**, 4355 (1989).
72. M. L. Ristig, The structure of quantum fluids. *Found. Phys.* **19**, 1041 (1989).
73. M. L. Ristig and J. W. Clark, Two-body density matrix of a normal Fermi fluid. *Phys. Rev.* B **41**, 8811 (1990).
74. M. L. Ristig, Achievements in quantum Monte Carlo. In *Recent Progress in Many-Body Theories*, Vol. 2, ed. Y. Avishai (Plenum, New York, 1990), p. 347.
75. G. Senger and M. L. Ristig, The normal phase of a correlated Bose fluid. In *Condensed Matter Theories*, Vol. 5, ed. V. Aguilera-Navarro (Plenum, New York, 1990), p. 133.
76. J. W. Clark and M. L. Ristig, Generalized momentum distributions of quantum

fluids. In *Condensed Matter Theories*, Vol. 5, ed. V. Aguilera-Navarro (Plenum, New York, 1990), p. 47.
77. M. L. Ristig and J. W. Clark, Two-body density matrix of quantum fluids. In *Recent Progress in Many-Body Theories*, Vol. 2, ed. Y. Avishai (Plenum, New York, 1990), 323.
78. M. L. Ristig, Correlated ground state of the Hubbard model. *Z. Phys.* B **79**, 351 (1990).
79. A. Dabringhaus, M. L. Ristig and J. W. Clark, Vacuum ground and excited states of the $U(1)$ lattice gauge Hamiltonian. *Phys. Rev.* D **43**, 1978 (1991).
80. F. V. Kusmartsev and M. L. Ristig, Chiral phase states of the Hubbard Hamiltonian. *Phys. Rev.* B **44**, 535 (1991).
81. A. Dabringhaus and M. L. Ristig, The $U(1)_3$ lattice gauge vacuum. In *Condensed Matter Theories*, Vol. 6, eds. S. Fantoni and S. Rosati (Plenum, New York, 1991), p. 291.
82. F. V. Kusmartsev and M. L. Ristig, Flux phase states of the Hubbard model. *Physica* B **169**, 587 (1991).
83. M. L. Ristig and A. Dabringhaus, Uncharged states of two-dimensional lattice gauge Hamiltonians. In *Recent Progress in Many-Body Theories*, Vol. 3, ed. T. L. Ainsworth *et al.* (Plenum, New York, 1992).
84. K. A. Gernoth and M. L. Ristig, Excitation spectrum of the vapor-liquid ^{4}He interface. *Phys. Rev.* B **45**, 2969 (1992).
85. G. Senger, M. L. Ristig, C. E. Campbell and J. W. Clark, Correlated density matrix theory of normal quantum fluids. *Ann. Phys. (NY)* **218**, 160 (1992).
86. F. J. Bermejo, F. J. Monipein, M. Garcia-Hernandez, J. L. Martinez, D. Martin-Marero, A. Chalid, G. Senger, and M. L. Ristig, Collective excitations in liquid deuterium: neutron-scattering and correlated-density-matrix results. *Phys. Rev.* B **47**, 15097 (1993).
87. W. Wang and M. L. Ristig, Possible double-pairing effects in high-T_c superconductors. *Z. Phys.* B **91**, 3 (1993).
88. K. A. Gernoth, J. W. Clark, G. Senger, and M. L. Ristig, Excitations of the surface of liquid ^{4}He. In *Condensed Matter Theories*, Vol. 8, ed. L. Blum (Plenum, New York, 1993).
89. F. J. Mompeán, F. J. Bermejo, M. Garcia-Hernandez, B. Fåk, J. L. Martinez, G. Senger, and M. L. Ristig, Temperature dependence of collective excibitions in liquid deuterium studied by neutron inelastic scattering. JPCM **5**, 5743 (1993).
90. K. A. Gernoth, J. W. Clark, G. Senger, and M. L. Ristig, Surface modes of liquid ^{4}He. *Phys. Rev.* B **49**, 1583 (1994).
91. J. W. Clark, K. A. Gernoth, and M. L. Ristig, Connectionist many-body phenomenology. In *Condensed Matter Theories*, Vol. 9, eds. J. W. Clark, K. A. Shoaib, A. Sadiq (Nova Science Publ., Commack, New York, 1994).
92. K. A. Gernoth, J. W. Clark, and M. L. Ristig, Correlated Density Matrix Theory of Spatially Inhomogeneous Bose Fluids. In *Condensed Matter Theories*, Vol. 10, eds. M. Casas *et al.* (Nova Science Publ.,Commack, New York, 1995).
93. K. A. Gernoth, J. W. Clark, and M. L. Ristig, The surface of liquid ^{4}He at nonzero temperatures. *Z. Phys.* B **98**, 337 (1995).

94. M. L. Ristig, R. Pantförder, M. Serhan, and G. Senger, Correlated Density Matrix Theory of Homogeneous Bose Systems. In *Condensed Matter Theories*, Vol. 10, eds. M. Casas, M. de Llano, J. Navarro, A. Polls (Nova Science Publ., Commack, New York, 1995).
95. M. L. Ristig, G. Senger, and M. Serhan, Correlated Density Matrix Theory of Boson Superfluids. *Ann. Phys. (NY)* **243**, 247 (1995).
96. J. W. Clark, K. A. Gernoth, and M. L. Ristig, Connectionist Statistical Inference. In *Recent Progress in Many-Body Theories*, Vol. 4, eds. E. Schachinger *et al.* (Plenum, New York, 1995).
97. O. Ciftja, S. Fantoni, J. W. Kim, and M. L. Ristig, Application of the Fermi Hypernetted-Chain Theory and Effective Correlation Factor Method for Laughlin Quantum Hall States. *J. Low. Temp. Phys.* **108**, 357 (1997).
98. M. L. Ristig, J. W. Kim, and R. Mehlmann, CBF theory of Ising spins in a transverse magnetic field. In *Condensed Matter Theories*, Vol. 11, (Nova Science Publishers, Commack, 1996), p. 113.
99. M. L. Ristig and J. W. Kim, Correlated-basis function analysis of the transverse Ising model. *Phys. Rev.* B **53**, 6665 (1996).
100. J. W. Clark, M. L. Ristig, T. Lindenau, and M. Serhan, Bose-Einstein condensation in liquid helium: a correlated density matrix theory. In *Condensed Matter Theories*, Vol. 12, (Nova Science Publishers, Commack, 1997), p. 55.
101. R. Pantförder, T. Lindenau, and M. L. Ristig, Correlated one-body density matrix of boson superfluids. *J. Low. Temp. Phys.* **108**, 245 (1997).
102. M. L. Ristig, J. W. Kim, and J. W. Clark, The $Z(2)$ lattice gauge vacuum and the transverse Ising model: two sides of a coin. In *Theory of Spin Lattice and Lattice Gauge Models*, eds. J. W. Clark and M. L. Ristig, *Lec. Notes in Phys.***494** (Springer, Berlin, 1997), p. 62.
103. J. W. Kim, M. L. Ristig, and J. W. Clark, Transverse Ising model at zero temperature. *Phys. Rev.* B **57**, 56 (1998).
104. W. Wang and M. L. Ristig, Correlations and fluctuations in Josephson junction arrays. *Phys. Lett.* A **241**, 122 (1998).
105. J. W. Clark, M. L. Ristig, and J. W. Kim, The transverse Ising model by CBF. In *Recent Progress in Many-Body Theories 5*, eds. D. Neilson and R. F. Bishop, Advances in Quantum Many-Body Theory, Vol. 1 (World Scientific, 1998), p. 462.
106. M. L. Ristig, T. Lindenau, M. Serhan, and J. W. Clark, Broken symmetries in liquid helium. In *Condensed Matter Theories*, Vol. 13, (Nova Science Publishers, Commack, 1998), p. 119.
107. J. W. Clark, K. A. Gernoth, S. Dittmar, and M. L. Ristig, Higher order probabalistic perceptrons as Bayesian inference engines. *Phys. Rev.* E **59**, 6161 (1999).
108. M. L. Ristig, T. Lindenau, M. Serhan, and J. W. Clark, Toward a microscopic theory of the λ-transition in liquid ^{4}He. *J. Low. Temp. Phys.* **114**, 317 (1999).
109. T. Lindenau, M. L. Ristig, and J. W. Clark, Microscopic calculations on the superfluid phase of liquid helium. In *Condensed Matter Theories*, Vol. 14, eds. D. J. Ernst, I. E. Perakis, and A. S. Umar (Nova Science Publishers, Huntigton, 2000).

110. R. F. Bishop, D. J. J. Farnell, and M. L. Ristig, The Ising model in a transverse magnetic field: a test-bed for many-body theories. In *Condensed Matter Theories*, Vol. 14, eds. D. J. Ernst, I. E. Perakis, and A. S. Umar (Nova Science Publishers, Huntigton, 2000).
111. T. Lindenau, M. L. Ristig, and J. W. Clark, Microscopic calculations on the superfluid phase of liquid Helium. In *Condensed Matter Theories*, Vol. 14, eds. D. J. Ernst, I. E. Perakis, and A. S. Umar (Nova Science Publishers, Huntigton, 2000).
112. T. Lindenau, M. L. Ristig, J. W. Clark, Microscopic Origin of the λ-Transition in Liquid Helium. Precursors and precedents – putting the pieces of the puzzle in place. In *Condensed Matter Theories*, Vol. 15, eds. G. S. Anagnostatos *et al.* (Nova Science Publishers, Huntington, 2000).

Books

1. *Recent Progress in Many-Body Theories, Lec. Notes in Phys.* **198**, H. Kümmel and M. L. Ristig (Springer-Verlag, Berlin, 1984).
2. *Theory of Spin Lattices and Lattice Gauge Models*, J. W. Clark and M.L. Ristig, *Lec. Notes in Phys.* **494** (Springer-Verlag, Berlin, 1997).
3. *Theory of Spin Lattices and Lattice Gauge Models*, J. W. Clark and M. L. Ristig, *Lec. Notes in Phys.* **522** (Springer-Verlag, Berlin, 1999).

SERGIO ROSATI

Curriculum Vitae

Born	18 May, 1934
Address	Department of Physics, University of Pisa Via Buonarroti 2, Ed. - 56127 Pisa (Italy)
Email	rosati@galileo.pi.infn.it

Education

1952-1956	student at the University of Pisa and at Scuola Normale Superiore (Pisa)
11.1956	Laurea in Physics with 110/110, summa cum laude
11.1956	Diploma of Scuola Normale Superiore with 70/70 and Lode

Employment

1957-1976	assistant professor of Theoretical Physics
1967	libera docenza in Theoretical Physics
1976	full professor in Nuclear Physics

Honors and Awards

1957	Italian Physics Society Award for Young Italian Physicists
2000	Ordine del Cherubino award (University of Pisa) for his scientific merits and in recognition of his successful activity as a teacher and trainer of several generations of physicists.

Visiting Appointments

1957	Department of Theoretical Physics, Manchester University
1965	Columbia University, New York

Published Papers

1. L. A. Radicati and S. Rosati, On the Spin of the K_μ Meson. *Nuovo Cim.* **10**, 729 (1957).
2. L. E. Picasso and S. Rosati, On the Spin Dependence of the Λ-N interaction. *Nuovo Cim.* **11**, 711 (1959).
3. B. Barsella and S. Rosati, On the Possible Existence of Hyperfragments with Mass Number A=6. *Nuovo Cim.* **13**, 458 (1959).

4. B. Barsella and S. Rosati, On the Effect of n-p Tensor Forces in $^3H_\Lambda$. *Nuovo Cim.* **20**, 914 (1961).
5. L. Lovitch and S. Rosati, Isobaric Spin Purity in Heavy Nuclei. *Phys. Lett.* **3**, 222 (1963).
6. J. Murphy and S. Rosati, A Two-Body Method for the Bound States of a Three-Body System. *Nucl. Phys.* **63**, 625 (1965).
7. S. Rosati, On the Effect of n-p Tensor Forces in $^3H_\Lambda$, II. *Nuovo Cim.* **36**, 1351 (1965).
8. L. Lovitch and S. Rosati, A Study of the J=2 Excited Level of ^{6}Li. *Nucl. Phys.* **73**, 648 (1965).
9. L. Lovitch and S. Rosati, Reproduction of Two-Body Scattering Phase Shifts by a Phenomenological Potential. *Phys. Rev.* **140**, 408 (1965).
10. L. Lovitch and S. Rosati, Direct Numerical Integration of the Two-Body Schroedinger Equation with Tensor Forces. *Phys. Rev.* **140**, 877 (1965).
11. L. Lovitch and S. Rosati, On the Corrector Formulas for the Numerical Solution of the Schroedinger Equation for Central Fields. *Proc. Roy. Soc.* **62**, 79 (1966).
12. M. Barbi and S. Rosati, Direct Numerical Solution of the Three-Body Problem. *Phys. Rev.* **147**, 730 (1966).
13. L. Lovitch and S. Rosati, The Numerical Evaluation of the Two-Nucleon Scattering Length and Effective Range with Tensor Forces. *Proc. Phys. Soc.* **88**, 909 (1966).
14. L. Lovitch and S. Rosati,Soft Core and Two Nucleon Tensor Forces. *Nucl. Phys.* **90**, 13 (1967).
15. L. Lovitch and S. Rosati, Velocity Dependent Potentials and the Binding Energy of the Triton. *Nucl. Phys.* B **1**, 369 (1967).
16. L. Lovitch and S. Rosati, Three-Body Model of ^{6}Li, In *Proceedings of the Symposium on Light Nuclei, Few-Body Problems and Nuclear Forces*, eds. G. Paic and I. Slaus (Gordon and Breach, N.Y., 1968) p. 533.
17. B. Barsella, L. Lovitch and S. Rosati, Two- and Three-Body Model of $^6He_\Lambda$, In *Proceedings of the Symposium on Light Nuclei, Few-Body Problems and Nuclear Forces*, eds. G. Paic and I. Slaus (Gordon and Breach, N.Y., 1968) p. 591.
18. L. Lovitch and S. Rosati, The Hypernucleus $^6He_\Lambda$. *Nuovo Cim.* A **51**, 647 (1968).
19. L. Lovitch, S. Rosati and R. H. Dalitz, The Stability of $^6Li_\Lambda$. *Nuovo Cim.* A **53**, 301 (1968).
20. B. Barsella, L. Lovitch and S. Rosati, Three-Body Model of the Isobaric Triplet with $A = 6$. *Nucl. Phys.* A **111**, 638 (1968).
21. L. Panattoni and S. Rosati, Study of Hypertriton with Realistic n-p Potentials. *Nuovo Cim.* B **61**, 425 (1969).
22. L. Lovitch and S. Rosati, Binding Energies and Wave Functions of Three Bosons Interacting through Local Potentials. *Nuovo Cim.* B **63**, 335 (1969).
23. S. Fantoni and S. Rosati, Alpha-Deuteron Model of ^{6}Li. *Nucl. Phys.* A **151**, 317 (1970).
24. L. Lovitch and S. Rosati, Bound State Solution of the Two-Nucleon Schroedinger Equation with Tensor Forces. *Comp. Phys. Com.* **2**, 353 (1971).

25. S. Fantoni and S. Rosati, Expansion Procedure for Jastrow-Type Correlated Wave Functions. *Nuovo Cim.* A **10**, 145 (1972).
26. L. Lovitch and S. Rosati , The Two-Nucleon Effective Range Parameters with Tensor Forces. *Comp. Phys. Com.* **4**, 138 (1972).
27. S. Fantoni and S. Rosati, Direct Numerical Calculation of the Two-Nucleon Correlation Function for ^{4}He. *Lett. Nuovo Cim.* **5**, 931 (1972).
28. S. Fantoni and S. Rosati, Jastrow Correlations and an Irreducible Cluster Expansion for Infinite Boson or Fermion Systems. *Lett. Nuovo Cim.* **5**, 931 (1972).
29. S. Fantoni and S. Rosati, Calculation of the Two-Body Correlation Function for Fermi Systems. *Nuovo Cim.* **10**, 545 (1974).
30. E. Campani, S. Fantoni and S. Rosati, The Fermion Hypernetted Chain and the Percus-Yevick Approximation. *Nuovo Cim.* **12**, 1975 (1975).
31. S. Fantoni and S. Rosati, Microscopic Calculation for a Model Fermi Liquid. *Nuovo Cim.* **16**, 531 (1976).
32. E. Campani, S. Fantoni and S. Rosati, Unconstrained Jastrow Variational Calculation on Nuclear Matter, *Nuovo Cim.* **16**, 531 (1976).
33. S. Fantoni and S. Rosati, The Fermi-Hypernetted-Chain Method for State-Dependent Jastrow-Correlated Functions. *Nuovo Cim.* **43**, 431 (1978).
34. R. Guardiola, A. Polls, S. Fantoni and S. Rosati, The Nuclear Matter Problem with Central Spin- and Isospin-Dependent Potentials. *Nuovo Cim.* **4**, 414 (1978).
35. O. Benhar, C. Ciofi degli Atti, S. Fantoni and S. Rosati, Variational Calculation on Nuclear Matter. *Nucl. Phys.* A **328**, 127 (1979).
36. S. Fantoni and S. Rosati, Extension of the FHNC Method to Finite Systems. *Nucl. Phys.* A **328**, 478 (1979).
37. S. Fantoni and S. Rosati, FHNC Expression for the Three-Body Term of the Kinetic Energy in the Jackson-Feenberg Form. *Phys. Lett.* B **84**, 23 (1979).
38. A. Fabrocini, S. Fantoni, A. Polls and S. Rosati, Variational Approach to the Fermi Hard-Sphere System. *Nuovo Cim.* A **56**, 33 (1980).
39. S. Rosati, Some Topics in the FHNC Approach to Fermi System. In *The Meson Theory of Nuclear Forces and Nuclear Matter*, Vol. 1, eds. D. Schutte, K. Holinde and K. Bleuler (Wissenschaftsverlag, Wien/Zurich, 1980), p. 136.
40. S. Rosati and S. Fantoni, Surface and Polarization Effects in Large Fermi Systems. *Nuovo Cim.* A **58**, 327 (1980).
41. S. Rosati and R. Schiavilla, Calculation of Coulomb Energy for Spherical Nuclei. *Lett. Nuovo Cim.* A **35**, 29 (1982).
42. S. Rosati, FHNC Variational Theory for Strongly Interacting Fermi Systems. In *From Nuclei to Particles*, Proc. Int. School E. Fermi, course LXXIX, ed. A. Molinari (North Holland, Amsterdam, 1981), p. 73.
43. A. Fabrocini and S. Rosati, The Method of Interpolating Integral Equations for Quantum Fluids. - I. *Nuovo Cim.* D **1**, 567 (1982).
44. A. Fabrocini and S. Rosati, The Method of Interpolating Integral Equations for Quantum Fluids. - II. *Nuovo Cim.* D **1**, 615 (1982).
45. S. Rosati, A. Fabrocini and M. Viviani, The Interpolating Equations Method in Quantum Fluids. In *Proceedings of the Third International Conference on Recent Progress in Many-Body Theories*, 1983, eds. H. Araki *et al.*, Lect. Notes

in Phys. **198**, 406 (1984) (Springer, New-York, 1984).
46. A. Fabrocini, S. Fantoni, A. Polls and S. Rosati, Microscopic Calculation of the Excitation Spectrum of One He Impurity in Liquid He . *Phys. Rev.* B **33**, 6057 (1986).
47. S. Rosati, M. Viviani, E. Buendia, and A. Fabrocini, Long-Range and Elementary Contribution for Quantum Fluids at Zero Temperature. In *Condensed Matter Theories*, Vol. 1, ed. F. B. Malik, (Plenum Press, New York, 1986), p. 97.
48. M. Viviani and S. Rosati, Variational and Low-Density Expansion Calculations on Spin-Aligned Hydrogen System. *Anales de Fisica* A **81**, 121 (1985).
49. M. Viviani, E. Buendia, A. Fabrocini, and S. Rosati, The Method of Interpolating Integral Equations for Quantum Fluids – III. *Nuovo Cim.* D **8**, 561 (1986).
50. L. Bracci, S. Rosati and M. Viviani, Variational Calculations for Few-Body Systems with Model Central Potential. In*Proceedings of the Secondo Convegno su Problemi di Fisica Nucleare Teorica, Cortona*, eds. L. Bracci *et al.*, p. 34 (ETS, Pisa, Italy, 1987).
51. S. Fantoni, M. Viviani, E. Buendia, S. Rosati, A. Fabrocini and V. R. Pandharipande, Microscopic Calculation on Liquid ^{3}He. In *Recent Progress on Many-Body Theories* **1**, eds. A. Kallio, E. Pajanne and R.F. Bishop (Plenum Press, London, 1987), p. 205.
52. M. Viviani, E. Buendia, S. Fantoni and S. Rosati, Spin-Dependent Correlations in the Ground State of Liquid ^{3}He. *Phys. Rev.* B **38**, 4523 (1988).
53. S. Rosati and M. Viviani, Microscopic Variational Theory for Bose Fluids at Zero Temperature. In *1st International Course on Condensed Matter*, eds. D. Prosperi *et al.* (World Scientific, Singapore, 1988), p. 232.
54. A. Kievsky, S. Rosati and M. Viviani, Euler and Correlated Harmonic Oscillator Wave Functions for Three-Nucleon Systems. *Nucl. Phys.* A **501**, 503 (1989).
55. A. Kievsky, M. Viviani and S. Rosati, On the Solution of the Eigenvalue Problem for Coupled Differential Equations. In *Proceedings del Terzo Convegno su Problemi di Fisica Nucleare Teorica, Cortona 1989*, ed. L. Bracci *et al.* (ETS, Pisa, Italy, 1990), p. 105
56. S. Rosati, M. Viviani and A. Kievsky, Correlated Hyperspherical-Harmonic Expansion for Three-Nucleon Systems. *Few-Body Sys.* **9**, 1 (1990).
57. S. Rosati, M. Viviani and E. Buendia, Correlations and Momentum Distribution in the Ground State of Liquid ^{3}He. In *Condensed Matter Theories*, Vol. 5, ed. V.C. Aguilera-Navarro (Plenum Press, N. Y., 1990), p. 119.
58. A. Kievsky, M. Viviani and S. Rosati, Euler and Correlated Harmonic Oscillator Wave Function for the Trinucleon Bound-State. *Few-Body Sys.* **11**, 111 (1991).
59. A. Kievsky, M. Viviani and S. Rosati, Use of Correlated Hyperspherical Harmonic Basis for Strongly Interacting Systems. In *Condensed Matter Theories*, Vol. 6, eds. S. Fantoni and S. Rosati (Plenum Press, New York, 1991), p. 391.
60. A. Kievsky, M. Viviani and S. Rosati, Calculation on the Three-Nucleon Bound-State with Realistic Potentials. In *Proceedings of the Quarto Convegno*

su Problemi di Fisica Nucleare Teorica, Cortona 1991, eds. L. Bracci *et al.* (ETS, Pisa, Italy, 1992), p. 143.
61. A. Viviani, A. Kievsky and S. Rosati, Correlated Hyperspherical Harmonic Calculations for Three- and Four-Body Systems. *Nuovo Cim.* A **105**, 1473 (1992).
62. A. Kievsky, M. Viviani and S. Rosati, The Hyperspherical Harmonic Method Applied to ^{12}C and ^{16}O in the α-Particle Model. In: *Condensed Matter Theories*, Vol. 7, eds. A.N. Proto and J.L. Aliaga (Plenum Press, New York, 1992), p. 387.
63. M. Viviani, S. Rosati and A. Kievsky, Study of Bound and Scattering States of Few-Nucleon Systems with Correlated Basis Functions. Yad. Fiz. 56, 38 (1993) (Translation from Russian: Phys. At. Nucl. 56, 877 (1993)).
64. A. Kievsky, M. Viviani and S. Rosati, The Three-Nucleon Bound-State with Realistic Soft and Hard Core Potentials. *Nucl. Phys.* A **551**, 241 (1993).
65. A. Kievsky, M. Viviani and S. Rosati, Study of Bound and Scattering States in Three-Nucleon Systems. *Nucl. Phys.* A **577**, 511 (1994).
66. A. Kievsky, M. Viviani and S. Rosati, Variational Calculations for Scattering States in Three-Nucleon Systems. *Few-Body Sys. Suppl.* **7**, 278 (1994).
67. M. Viviani, A. Kievsky and S. Rosati, Calculation of Bound and Scattering States of Four Nucleons. In *14th International Conference in Few-Body Problems in Physics*, AIP Conference Proceedings 334, ed. F. Gross (AIP, Washington, 1994), p. 844.
68. S. Rosati, M. Viviani and A. Kievsky, Correlated Hyperspherical Harmonic Functions for Few-Nucleon Systems. In *Proceedings of the XV European Conference on Few-Body Problems in Physics, Few-Body Sys. Suppl.* **8**, 21 (1994).
69. M. Viviani, A. Kievsky and S. Rosati, Calculation of the Alpha-Particle Ground-State. Few-Body Systems **18**, 25 (1995).
70. S. Rosati, A. Kievsky and M. Viviani, Correlated Expansion Bases for Strongly Interacting Systems. In *Condensed Matter Theories*, Vol. 10, eds. M. Casas, M. de Llano, J. Navarro, and A. Polls (Plenum, New York, 1995), p. 403.
71. M. Viviani, A. Kievsky, S. Rosati and R. Schiavilla, Recent Studies on Few-Nucleon Reactions at Low Energies. In *Proceedings of the 14th International Conference on Particles and Nuclei.* eds. C. E. Carlson and J. J. Domingo (World Scientific, Singapore, 1997), p. 686.
72. D. Huber, W. Gloeckle, J. Golak, H. Kamada, A. Kievsky, S. Rosati and M. Viviani, Realistic Phase-Shift and Mixing Parameters for Elastic Neutron Deuteron Scattering: Comparison of Momentum Space and Configuration Space Methods. *Phys. Rev.* C **51**, 1100 (1995).
73. A. Kievsky, M. Viviani and S. Rosati, Cross Section, Polarization Observables and Phase Shift Parameters in p-d and n-d Elastic Scattering. *Phys. Rev.* C **52**, 15 (1995).
74. A. Kievsky, S. Rosati, W. Tornow and M. Viviani, Critical Comparison of Experimental Data and Theoretical Predictions for N-d Scattering below the breakup Threshold. *Nucl. Phys.* A **607**, 402 (1996).
75. S. Rosati, A. Kievsky, M. Viviani, L. E. Marcucci, R. Schiavilla and W. Tornow, An alternative to the Faddeev and Yakubovsky Approaches to Three and Four

Nucleon Problems. In *Proceedings of the Workshop on Electron Nucleus Scattering, EIPC, Marciana Marina*, eds. O. Benhar and A. Fabrocini (Edizioni, Pisa, 1996), p. 246.

76. W, Tornow, A. Kievsky, S. Rosati and M. Viviani, No Evidence for Large Charge-Symmetry Breaking Effects in the ^3P_J Nucleon-Nucleon Interactions. *Phys. Rev.* C **55**, 525 (1997).
77. S. Rosati, Variational Study of Four-Nucleon Scattering Processes. In *Condensed Matter Theories*, Vol. 13, eds. J. Da Providencia and F.B. Malik (Nova Science Publishers, New York, 1997), p. 119.
78. A. Kievsky, L. E. Marcucci, S. Rosati and M. Viviani, High-Precision Calculation of the Triton Ground State within the Hyperspherical Harmonics Method. *Few-Body Sys.* **22**, 1 (1997).
79. A. Kievsky, S. Rosati, M. Viviani *et al.*, The Three-Nucleon System Near the N-d Threshold. *Phys. Lett.* B **406**, 292 (1997).
80. A. Kievsky, M. Viviani and S. Rosati, N-d Scattering Above the Deuteron Breakup Threshold. *Phys. Rev.* C **56**, 2987 (1997).
81. A. Kievsky, M. Viviani, S. Rosati *et al.*, Possible 3N Force Effects in d-p Scattering at Low Energies. *Phys. Lett.* B **428**, 13 (1998).
82. T. C. Black, H. J. Karwosky, E. J. Ludwig, A. Kievsky, S. Rosati and M. Viviani, An Energy Dependent Phase Shift Analysis of Low-Energy Proton-Deuteron Elastic Scattering. In *Proceedings of the XV International Conference on Few-Body Problems in Physics Nucl. Phys.* A **631**, 680c (1998).
83. M. Viviani, A. Kievsky and S. Rosati, Calculation of Scattering Observables in Three- and Four-Nucleon Systems. In *In Proceedings of the XVII International Symposium on Innovative Computational Methods in Nuclear Many-Body Problems, Osaka*, eds. H. Horiuchi *et al.* (World Scientific, Singapore, 1998), p. 12.
84. M. Viviani, S. Rosati and A. Kievsky, Variational Calculations for Three and Four Nucleon Scattering Processes. In *Proceeding of the Settimo Convegno su Problemi di Fisica Nucleare Teorica, Cortona, Italy*, eds. A. Fabrocini *et al.* (ETS, Pisa, Italy, 1999), p. 349
85. M. Viviani, S. Rosati and A. Kievsky, Neutron-^{3}H and Proton-^{3}He Zero Energy Scattering. *Phys. Rev. Lett.* **81**, 1580 (1998).
86. R. Schiavilla, S. Rosati *et al.*, Weak Capture of Protons by Protons. *Phys. Rev.* C **58**, 1263 (1998).
87. A. Kievsky, M. Viviani, S. Rosati *et al.*, Benchmark Calculations for Polarization Observables in Three-Nucleon Scattering. *Phys. Rev.* C **58**, 3085 (1998).
88. S. Rosati and M. Viviani, Extended Hyperspherical Harmonic Expansion Bases. *Few-Body Sys.* **27**, 73 (1999).
89. H. J. Karwowski, C. R. Brune, W. H. Geist, E. J. Ludwig, K. D. Veal, M. H. Wood, A. Kievsky, S. Rosati and M. Viviani, New Physics in p-d Elastic Scattering at Low Energies. *Acta. Phys. Pol.* **30**, 1479 (1999).
90. A. Kievsky, M. Viviani and S. Rosati, Proton-Deuteron Scattering above the Deuteron Breakup. *Phys. Rev. Lett.* **82**, 3759 (1999).
91. E. A. Wulf, S. Rosati *et al.*, New photodisintegration threshold observable in ^{3}He, *Phys. Rev.* C **61**, 021601 (2000).

92. T. C. Black, H. J. Karwosky, E. J. Ludwig, A. Kievsky, S. Rosati and M. Viviani, Determination of Proton-Deuteron Scattering Lengths. *Phys. Lett.* B **471**, 103 (1999).
93. M. Viviani, A. Kievsky, L. E. Marcucci, S. Rosati and R. Schiavilla, Photo- and Electro-Disintegration of ^{3}He at Threshold and *pd* Radiative Capture. *Phys. Rev.* C **61**, 064001 (2000).
94. M. Viviani, A. Kievsky and S. Rosati, The Kohn Variational Principle for Elastic Proton-Deuteron Scattering above the Deuteron Breakup Threshold. *Few-Body Sys.*, in press.
95. S. Rosati and M. Viviani, " Hyperspherical Harmonic Methods for Strongly Interacting Systems", In "Microscopic Approaches to the Structure of Light and Medium Mass Nuclei", eds. R.F. Bishop and N. R. Walet, (World Scientific, Singapore, in press).
96. L. E. Marcucci, R. Schiavilla, M. Viviani, A. Kievsky and S. Rosati, Realistic Calculation of the hep Astrophysical Factor, *Phys. Rev. Lett.* **84**, 5959 (2000).
97. A. Kievsky, M. H. Wood, C. R. Brune, B. M. Fisher, H. J. Karwowski, D. S. Leonard, E. J. Ludwig, S. Rosati and M. Viviani, Evidence for Three-Nucleon Force Effects in p-d Elastic Scattering, *Phys. Rev.* C, in press.
98. M. Viviani, A. Kievsky, S. Rosati, E. A. George, and L. D. Knutson, Proton-^{3}He Elastic Scattering, Submitted to *Phys. Rev. Lett.*.
99. A. Kievsky, J.L. Friar, G.L. Payne, S. Rosati and M. Viviani, Phase Shifts and Mixing Parameters for Elastic Proton-Deuteron Scattering, Submitted to *Phys. Rev.* C.

Books

1. S. Rosati, *Appunti di Fisica Atomica* (Pellegrini, Pisa, 1961)
2. S. Rosati, *Fisica Generale I* (Ed. Ambrosiana, Milano, I edition 1978)
3. Proceedings of the International Conference on Recent Progress in Many-Body Theories, Trieste, Italy, 1978 eds. C. Ciofi degli Atti, A. Kallio and S. Rosati, *Nucl. Phys.* A 328 (1979),
4. S. Rosati and R. Casali, *Problemi di Fisica Generale I* (Ed. Ambrosiana, Milano, I edition 1982)
5. S. Rosati and L. Lovitch, Fisica Generale II, (Ed. Ambrosiana, Milano, I edition 1979)
6. S. Rosati and L. Lovitch, *Problemi di Fisica Generale II* (Ed. Ambrosiana, Milano, 1980)
7. *Proceedings of the Primo Convegno su Problemi di Fisica Nucleare Teorica*, Cortona, Italy, 1985, eds. L. Bracci, S. Rosati *et al.* (ETS, Pisa, Italy, 1986).
8. *Proceedings of the Secondo Convegno su Problemi di Fisica Nucleare Teorica*, Cortona, Italy, 1987, eds. L. Bracci, S. Rosati *et al.* (ETS, Pisa, Italy, 1986).
9. *Proceedings of the First International Course on Condensed Matter*, Bogota, 1986, eds. D. Prosperi, S. Rosati and G. Violini(World Scientific, Singapore, 1987).
10. *Proceedings of the Terzo Convegno su Problemi di Fisica Nucleare Teorica*, Cortona, Italy, 1989, eds. L. Bracci, S. Rosati *et al.* (ETS, Pisa, Italy, 1990).

11. *Proceedings of the Fourteenth International Workshop on Condensed Matter Theories*, Island of Elba, Italy, 1990, eds. S. Fantoni and S. Rosati (Plenum Press, New York, 1991).
12. *Proceedings of the Quarto Convegno su Problemi di Fisica Nucleare Teorica*, Cortona, Italy, 1991, eds. L. Bracci, S. Rosati *et al.* (ETS, Pisa, Italy, 1992).
13. *Proceedings of the V Convegno su Problemi di Fisica Nucleare Teorica*, Cortona, Italy, 1993, eds. A. Bonaccorso, S. Rosati *et al.* (ETS, Pisa, Italy, 1994).
14. *Proceedings of the VI Convegno su Problemi di Fisica Nucleare Teorica*, Cortona, Italy, 1995, eds. I. Bombaci, S. Rosati *et al.* (ETS, Pisa, Italy, 1996).
15. *Proceedings of the VII Convegno su Problemi di Fisica Nucleare Teorica*, Cortona, Italy, 1998, eds. A. Fabrocini, G. Pisent and S. Rosati (ETS, Pisa, Italy, 1999).

A Historical Perspective

THE MUSIC OF THE QMBT QUARTET

HERMANN G. KÜMMEL
Institut für Theoretische Physik II, Ruhr Universität Bochum
44780 Bochum, Germany
E-mail: kuemmel@hadron.tp2.ruhr-uni-bochum.de

The members of the quartet in question, of course, are our dear colleagues John Clark, Alpo Kallio, Manfred Ristig, and Sergio Rosati. It as a great honor and a great pleasure to open this meeting by talking a bit about the times I have met these four players and about their music as I have heard or seen it. Also I would like to thank Ray Bishop for the splendid idea to arrange this meeting and to carry it through.

Please forgive me if you find this talk rather subjective: it has to be, since it is very much interwoven with personal recollections and naturally I did not see all quartet members equally often. In this situation some of you may feel a bit neglected. But let me mention the little known fact that George Bernard Shaw started as a rather militant music critic. As such he was a great admirer of Mozart and Wagner, but he disliked very much Brahms. My advice then is: step into the shoes of Brahms if you feel treated badly by me.

1 First movement: Youth

It cannot be an accident that the players of this quartet are coming of age, i.e. they reach the adult age of 65, in the first year of the new millennium, which makes it quite easy to establish the year 1935 ± 1 as the year in which they opened their eyes for the first time. A competent astrologer possibly would have been able to predict from this date of birth their later fate and could tell why all four ended up as many-body theorists. Not being of this profession, I just have to guess how it all started and how they did arrive at this final goal. The only thing I am sure about is that there was a first cry, probably a rather dissonant tune, as introduction to the quartet's music. From then on they must have been quite different from most other people. They certainly wondered more than ordinary babies in a rather early stage about the strange physical laws around them: why did things never move upwards except if thrown? Why did it once per day get dark? And why did the sun rise again and darkness was gone? But these quartet players were a rather special gang of four: they never did stop wondering and looking for answers to these -and of course other- questions. In this way they did remain children for all their life, although they did grow up to full length. And apparently even the schools they were attending did not spoil their interest in looking behind the things happening around them and not taking them at their face value. No wonder, then, that they arrived at the strange and not very profitable idea to study such an esoteric field as physics.

2 Second Movement: Adulthood (almost)

Since naturally I don't know any details about this first period of their life, let me now jump over some years to the time I myself come into the picture, i.e. the time I had the first contact with the members of the quartet and/or its music. I would like to do it in alphabetical order, but as it turns out at one occasion, I have to deviate from it. To some extent the many body (MB) conferences will serve as convenient time ticks. It necessarily will be a bit of a review of 25 years of QMBT, since these four people have made essential contributions to it over the whole period.

I think I met John Clark for the first time in 1975 in St. Louis as I gave a talk at Washington University. I remember some discussion in his office, some talk with Eugene Feenberg (who made a deep impression on me as a very human being) and some dinner we had (including my family with two small kids). The second time was at the occasion of the first conference on "Recent Progress in Many-Body Theories" in Trieste in 1978.[1] At this conference John Clark gave the summary talk with the subtitle "Crisis of Nuclear Matter Theory" which was an update of a much earlier review. 22 years later it is legitimate to ask, whether or not this crisis is gone, i.e. to make an update of the update. He at that time stated that there were two problems contributing to this crisis: first, the lack of reliable MB methods and secondly the essentially unknown nucleon-nucleon (NN) forces. We can safely say that we have settled the first one with sufficient accuracy. The old Bethe-Brueckner perturbation theory has been replaced by a less perturbative version, and other methods, like the variational correlated basis function (CBF) and hyperspherical harmonics methods have superseded the rather naive older approaches. However, the second part of this crisis is not settled, as a careful look into the few nucleon problem reveals. The boundary between few and MB theories becomes rather fuzzy as the few body techniques become more and more powerful. Nowadays one is able to exactly solve the ground state problem of up to four particles via Faddeev or Faddeev-Yakubovsky methods including three-body forces, which together with the Green's function Monte Carlo (GFMC) method is an "exact" procedure. For very delicate features and small effects one must rely on rigorous methods -and this is what is needed in this context. Fortunately, Walter Glöckle as one of the leading few-body experts has the office next to mine and he keeps me well informed about what is going on: As is well known, the general forms of the various two- or three-body forces are derived from meson exchanges. Unfortunately, there always have been and still are many different versions on the market. Table 1 with data provided by Walter Glöckle[2] shows the most recent results obtained with some of these forces. In this table the first column indicates the various NN-potentials by their conventional notation. Only those with "+ TM" include three body forces. Clearly, two body forces alone do not suffice. But including three-body forces it is easy to reproduce the triton and ^3He binding energy by adjusting the cutoff parameter a bit. With some of these NN potentials then ^4He also is reproduced very well with no further adjustments, which means that four-body forces are not needed here. This seems to be very satisfactory and may even not be very surprising. But after close inspection there are several troublesome facts, which mean that the problems have not been removed. Using the NN-potentials successful in describing the ground states and

Table 1. Binding energies (in MeV) from some two plus three NN potentials for three- and four-nucleon systems in comparison with experimental data. From Ref. 2.

Potential	^3He	^3H	^{4} He
CD Bonn	-8.012	-7.272	-26.26
CD Bonn +TM	-8.464	-7.720	-29.06
AV18	-7.623	-6.924	-24.28
AV18 + TM	-8.444	-7.728	-28.36
Nijm II	-7.654	-7.012	-24.56
Nijm II +TM	-8.392	-7.720	-28.60
Exp.	-8.48	-7.72	-28.30

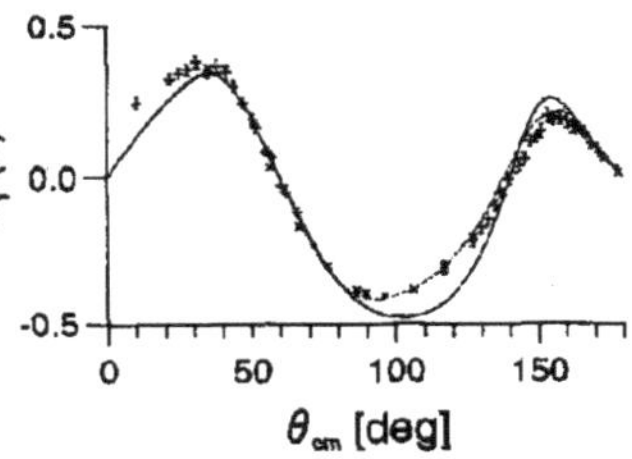

Figure 1. Deuteron vector analyzing power in elastic pd scattering at a deuteron energy of 230 MeV. Dotted/straight curves correspond to theory with/without three-body forces. From Ref. 2.

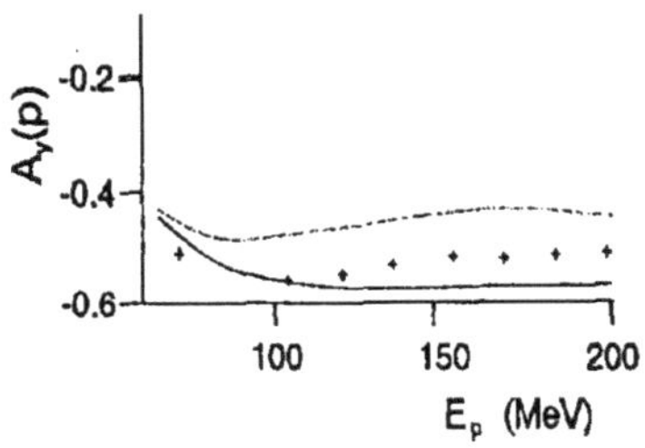

Figure 2. Proton vector analyzing power for pd scattering vs. proton energy. Curves as in Fig. 1. From Ref. 2.

computing again via the Faddeev method[2,3] the nd or pd-scattering one finds the following results: sometimes there is a surprising improvement due to three-body forces: the deuteron vector analyzing power of pd scattering of Fig.1 is a good example[2] (the curve through the experimental points includes three-body forces). But one also finds clear discrepancies, for instance in the proton vector analyzing power, see Fig. 2. for pd scattering,[2] or for the neutron vector analyzing power for nd scattering even *at very low energies*,[3] see Fig. 3. In other words: *the NN potential fitting the ground state cannot be correct.* Thus this aspect of the crisis in nuclear matter has *not* been removed. One cannot relax and use the smallness or exotic nature of these effects as an excuse. One cannot argue with their exotic nature, since already the total nd cross section has this same feature, although to a lesser degree. Admittedly, the smallness is a point one may argue about: we are willing to accept and to use the concept of a potential in low energy (atomic, solid state, fluid ..) physics, because there the corrections to this concept are *extremely* small in an absolute sense and compared to the masses involved. Here "smallness in an absolute sense" is defined by experimental inaccessibility. The mere fact that in nuclei there are experiments showing deviations forces us to treat nuclei differently. But observing the endless discussion about NN forces over the last 30

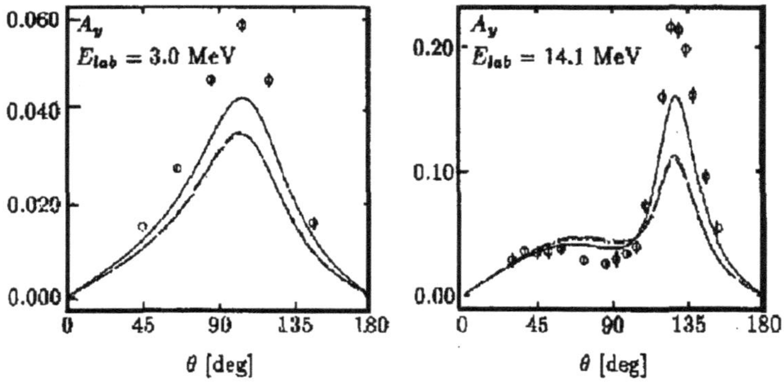

Figure 3. Neutron vector analyzing power for Nd scattering. Upper/lower curve:without/with three-body forces. From Ref. 3

years, without ever arriving at any definite conclusion, I am rather skeptical about the future. This skepticism has not been diminished by the fact that from time to time colleagues have claimed to have the final answer. I recall a rather heated discussion on three-body forces during the 1978 Trieste conference because several people rather vigorously claimed that their own and only their own version was the correct one. In a strict sense there will never be a description of nuclei as a system of nucleons. There is no NN potential even if the pions etc. could be eliminated rigorously or one could deal with quarks and gluons. There is not even an eigenvalue problem of the form $H\psi = E\psi$: for non super-renormalizable field theories there is no finite Hamiltonian (only a finite S-matrix) and all attempts to go around this problem lead at best to approximations to or simulations of reality. I personally believe that the only sound quantum field theories are lattice theories possibly of a form more sophisticated than the various forms used at present. This is a good message for the people of the MB community working on lattice models! The other extreme position would be that something is wrong with quantum mechanics -a very unlikely alternative in view of the recent successful tests and applications with the keywords Einstein-Fock-Podolsky "paradox", Bell inequality, teleportation, quantum cryptography,quantum computer etc.

Returning to the nuclear MB problem, the upshot of all this is that this part of the crisis has not been completely resolved 22 years later. After all, it is not the business of the MB theorists anymore: they have merely a service function in the sense that they can check whether a given model provided by the elementary particle theorists does or does not the job. Of course, it is still necessary and worthwhile to work on nuclear problems with the MB techniques we have.

Jumping over three more years to the MB conference in Mexico in 1981 there was a paper entitled "Recent Developments and Future Prospects in CBF Theory" (by Clark with Krotschek and Smith),[4] which is of a somewhat transitional nature, mainly announcing what later has been done with great success. I mention this also because in this paper a combination of the coupled cluster method (CCM) and CBF theory called the "correlated coupled cluster theory" is described as something with a good chance to compute high density hard core interacting systems with relatively little effort. Some years later Eckhart Krotschek complained to me that we never arrived at a collaboration using this method. It probably was a mistake not to try it. This was the consequence of another mistake I made: I don't like to admit it, but at that time I disliked the CBF method for purely irrational, that is aesthetic, reasons in spite of its successes in combination with the various hypernetted chain (HNC) approaches. I certainly have misjudged CBF. Since its birth there was permanent progress and extensions to new applications did appear. This is rather evident from the title of a talk by John Clark (together with Krotschek)[5] "Beyond the ground state with CBF theory" given at the next MB conference in Altenberg in 1983. I quote from this paper: "The focus of activity is shifting from ground state properties -where acceptable quantitative accuracy is either at hand or realizable in the near future- to elementary excitations and dynamical properties...". The claims made in this paper that CBF is a very general method were well-founded. In some respects it is more general than my beloved CCM: At that time it was hard to extract reliable physical properties beyond low energy spectra from the latter method. Also, to my knowledge even now CCM never has successfully been applied to the helium fluids and the like. At the next MB conference 1987 in Oulu John Clark was awarded the Feenberg medal and in response to this gave an overview[6] of the status of the MB theory at that time. Let me quote one of the remarks he was making, namely "the work presented at this conference is weighted far more heavily toward applications and toward the explication of mechanisms underlying observed phenomena... than was the case of previous conferences of this series, which were more concerned with formalism and with methods per se. This shift of emphasis is a reflection of the maturation of the field of MB theory....". I used to say the same in a very compact and somewhat provoking manner: what we MB theorists do now in some sense mostly is some kind of an an engineering feat. This is not meant as a degradation of our work, of course: it may and often does involve a lot of cleverness and sophistication. It typically is by no means routine. Anyway, from this time on the many body literature mostly is dealing with applications, improvements and extensions of existing methods -very many of them as an offspring of the CBF method and stimulated by the ideas put forward by John Clark. But remarkably enough, with his work on Neural Networks he has set an example of how one can start a completely new research direction in the MB environment I just described. A series of lectures on this topic he gave in Isfahan in 1991 made a lasting impression on me.

Returning to applications originating from John Clark's work I have to deviate from the alphabetical order and to introduce Manfred Ristig as the next name of the quartet: John Clark and Manfred Ristig have worked together almost all their scientific life and they seem to do so even now. They both play quite harmoniously together the same or a similar melody: it is the application of the CBF and its improvements to various systems. One of them was their extensive paper on nuclear matter published in the Nuclear Physics volume of 1979 dedicated to the memory of Eugene Feenberg.[7] I won't go into any details because I have already dwelt too much on nuclei. Suffice it to say that it did play an important role in solving the MB part of the nuclear matter crisis. From the many other papers they published

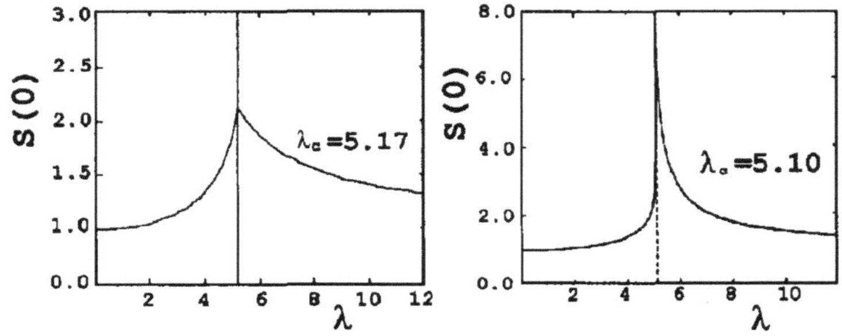

Figure 4. Static structure function at momentum zero for an Ising model as a function of the external field strength λ . Left/right figure: without/with optimization. From Ref. 8

together I have selected a fairly recent one:[8] it concerns the Ising model on a cubic lattice, with Pauli spins and nearest neighbor interaction in a transverse field. In Fig. 4 I show you the static structure function at momentum zero as a function of the field strength. The left figure is based on some older work by Ristig and Kim using variational CBF with a trial function, the right one by solving the Euler-Lagrange equations. They differ not only quantitatively but even qualitatively: only the better method reproduces a singularity which must be there. Having now arrived at Manfred Ristig I should say that I met him for the first time relatively late in spite of the geographical nearness. I refrain from giving the reason, since I would have to elaborate on the German university system with its much too hierarchical structure, or better, the atmosphere which makes this structure so ineffective. As far as I know, I became aware of Manfred Ristig as a MB theorist during the 1978 Trieste conference. Together with me he later organized the 1983 Altenberg conference. For instance we had to go together to the relevant institutions begging for money. This was not always very amusing. But Manfred did more than was his share at this occasion. I remember that I was rather reluctant in accepting the church related youth center as conference site found out by Manfred Ristig, because it was such a modest place. But just because of the church relation it was quite cheap and we could support more people -especially from the East European countries- than otherwise would have been possible. Also the place had some flair with the famous gothic church around the corner and the organ concert we could arrange. Years later several of the participants told me that they found it a most wonderful place for a conference.

Returning to MB physics there are two main research fields pursued later by Manfred Ristig and/or John Clark: the CBF applied on lattice gauge theories and the correlated density matrix approach. As a strong believer in lattice theories I would like to see the MB methods applied to realistic gauge fields. After all, these methods are more intelligent than the Monte-Carlo type applied by so many people, especially if one uses some intuition for the input. I know that in realistic theories the number of quantum fields is rather large as compared to our non-relativistic or quasi-relativistic MB problems. But can't one overcome this? Anyway, I find it important that first steps in this direction have been done using both CBF as

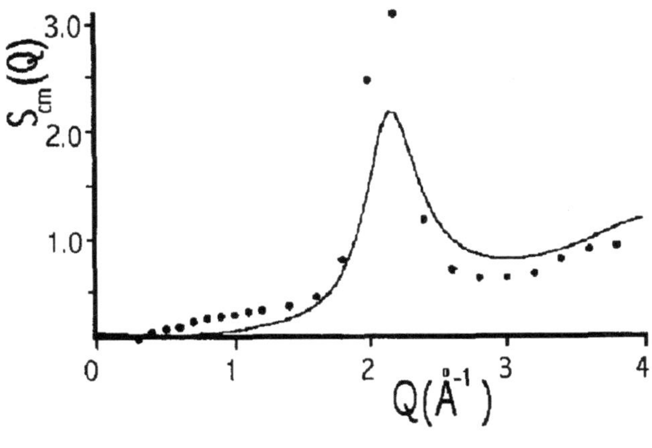

Figure 5. Neutron-scattering on deuteron fluid, first frequency moments of the center of mass dynamic structure function. From Ref. 12.

well as CCM. The CBF application to the density matrix is another line of research pursued by John Clark and Manfred Ristig. There is a review about its formal aspects in the Feenberg memorial issue of Nuclear Physics.[9] One step forward was its implementation using a variational principle for the Helmholtz free energy. Thus finite temperature features could be described.[10] And there is a whole series of other extensions and applications of the correlated density matrix, beginning with the Arad MB conference[11] in 1989. Especially for dynamical effects this method is a good choice. In this context there also was a close collaboration with C. E. Campbell and his Minneapolis group. I have selected a largely experimental paper[12] from 1993. It deals with the neutron scattering on a deuterium fluid. As far as I know, Manfred Ristig and G.Senger are the only theorists among the eight authors. They try to understand the experimental results using the correlated density matrix method. In Fig. 5 the first frequency moments of the center of mass dynamic structure function are shown as an example. Whereas the location of the peak is reproduced quite satisfactory, the absolute values do not come out well. The authors give the reasons for this. But generally one should not expect too much from whatever theory, because the excitations due to the impinging neutrons are small compared to the total energy of the system.

I met Alpo Kallio for the first time in 1966 at the Liperi Summer School in Finland, but I don't remember any more details. Shortly after that he visited me at the Max-Planck Institute in Mainz. Again I must admit that I don't know the physics we were talking about. Looking into the literature I found out that we both did old-fashioned nuclear physics with methods one nowadays would not call true MB theory, like shell model or BCS pairing calculations, to name only few. But I remember very well myself sitting with him and some other people in a garden restaurant with a beautiful view down to the Rhine river opposite Mainz, eating a local specialty called "Handkäs mit Musik" (translated literally "hand cheese with music". It is a Harz type cheese together with oil, vinegar, onions, and mustard). We had somewhat more than one glass of Rhine wine, not only because it was a good one, but also to make this extremely spicy and hot food more digestible. With a thunderstorm approaching it was a very romantic affair, although we had to run for shelter.

Let me return to the sober realities of MB theories. To my knowledge Alpo Kallio started the concert with some improvements via Padé approximation on the hypernetted chain method (presented in 1978 at the Trieste conference[13]). Later he extended the HNC to inhomogeneous systems and applied them to non-uniform physical objects. My general impression is that he likes complicated systems like those with impurities plus inhomogeneities, turning finally to the most complicated condensed matter I could imagine, the high-T_c superconductors. In spite of the immense amount of work done on this subject, these systems still seem to be very far from understood. I don't know very much about it, but on asking colleagues one obtains rather contradictory answers. Let me quote from the essay on condensed matter physics in the quite recent centennial issue of Reviews of Modern Physics by W. Kohn:[14] "While there is no doubt that the carriers are again electron pairs, there is a wide consensus, consistent with generally small isotope effects, that one or more mechanisms beyond electron-phonon coupling are at work; but there is no consensus about their nature". It is a fact that no version of a theory explains all the very many experimental results -almost 15 years after its detection. I remember very well the Oulu MB conference in 1987 where high T_c-superconductivity was a rather new and important topic and where there were as many theories as were people giving lectures about it. People seem only to agree that it is not of BCS type. The devil somehow must enjoy annoying the theorists by mixing simplicity with extreme complexity. After all, the phase transition to superconductivity per se has some elements of simplicity in it. Anyway, in this situation it is somewhat risky to enter this field, a risk Alpo Kallio did and still does take. At the next MB conference in Arad in 1989 he presented a model[15] with mobile electrons plus hole bosons, made up of electron pairs, and applied it to some high T_c compounds. In a series of papers he later has extended this model and put it on a more solid basis, and finally successfully applied it to a surprisingly large set of experimental data. Of course, there is a long way to go from first principles, or more modestly, from some microscopic picture below the fluid to this fluid-like model. I would like to quote in this context from Laughlin's Nobel lecture.[16] He first mentions that he sometimes gives to "bright but unsuspecting and innocent students" the home-work problem of deriving superfluidity from first principles. Then he continues: "There no doubt is a special place in hell being reserved for me ... for this mean trick, for the task is impossible. Superfluidity, like the fractional quantum Hall effect, is an emergent phenomenon -a low energy collective effect of huge numbers of particles *that cannot be deduced from the microscopic equations of motion in a rigorous manner.....*". Of course, this is even more true for the much more complicated high T_c superconductivity. Anyway, Alpo Kallio has demonstrated via HNC that his model is a possibility.[17] Let me present two examples of applications, selected from very many I have found. The first one[15] from the Arad conference is seen in Fig. 6. It is the transition temperature as function of the strontium content of one

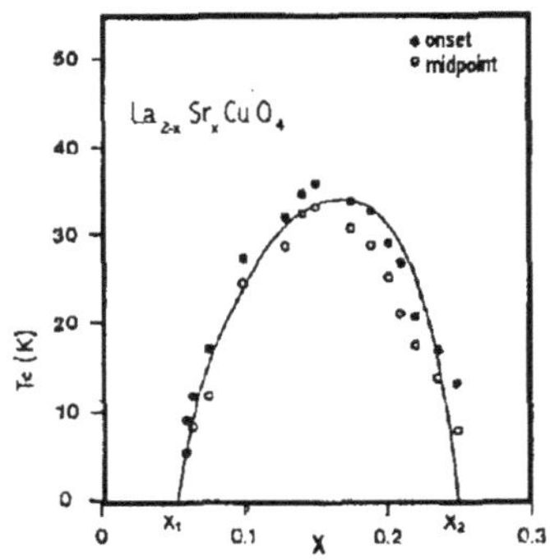

Figure 6. Transition temperature T_c as a function of the Sr content x for the $La_{2-x}Sr_xCuO_4$ compounds. From Ref. 15.

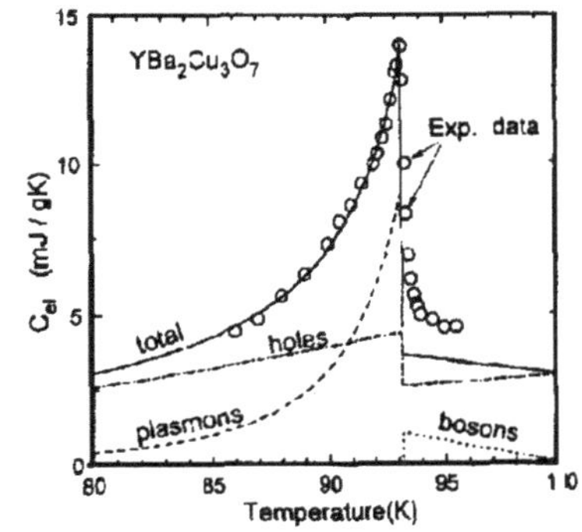

Figure 7. Mean specific heat of an YBCO compound. From Ref. 18

of those compounds. Fig. 7 from a 1994 paper[18] shows the specific heat around the transition temperature. In a paper of 1997[19] there are some more examples. All this looks very good. There is a caveat however: the theory has some adjustable parameters in it -and I am sure that there are some other experimental data which do not fit so well. But any theory I know of has this fault.

In a very general sense there are two categories of research papers: the first one embodies those papers where one believes (not always correctly) that one could have had the same ideas and done the same things, if one only would have worked hard and long enough on the problem at hand. The second category encompasses those papers, where on critical self evaluation one is sure that one never could have done the same thing, and where one just feels compelled to admire or envy the authors for their intuition, ingenuity and perseverance. Normally I would never tell anybody what I think in this respect about a given paper. Let me now make an exception since it happened a long time ago, although I remember it quite well: I did put into this second category the seminal 1975 paper by Fantoni and Rosati[20] on HNC and Fermi hypernetted chain (FHNC) with their powerful integral equations summing the terms of expectation values in terms of Jastrow type wave functions. The existence of some earlier papers (with Rosati and Ristig as authors) did and still does not diminish my high esteem of this one. Nowadays this is a widely used standard method which I therefore was forced to mention before. One easily forgets that its invention required some ingenuity and much more than routine. This now establishes the link to Sergio Rosati whom too I met at the 1978 meeting for the first time. Having expressed great respect for their HNC and FHNC work I would like to present to you a conceptually simple example, namely the ground state energy of liquid ^{3}He as function of the density[21] obtained via CBF with FHNC (Fig. 8). Although more recent methods -one of them to be presented at this meeting by Sergio Rosati himself- are at least as good as this one, this

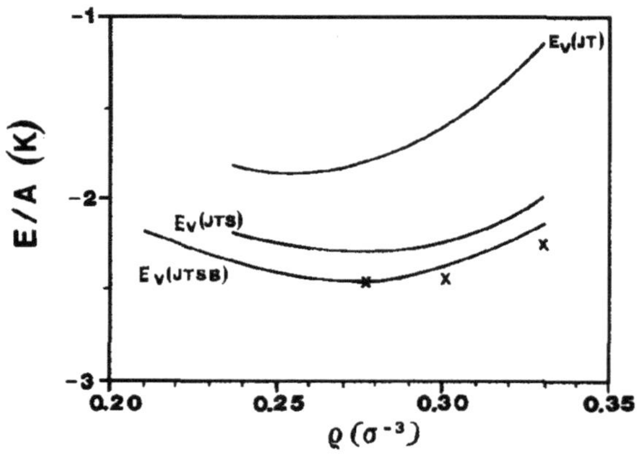

Figure 8. Ground state energy of ^4He fluid as function of density, for three different approximations, see text. From Ref. 21

example is quite instructive for two reasons. Firstly, it shows how one improves the result by introducing more complexity in the construction of the trial variational wave function. The upper/lower/lowest curve correspond to (i) inclusion of triplet correlations, (ii) triplet a plus spin correlations, and (iii) triplet plus spin correlation plus backflow. Each of the contributions occurs as a factor. It is seen that one systematically approaches the experimental values. It is true that by adding one such factor after the other the Raleigh-Ritz variational principle *must* lead to lower energies. But one observes that the second step is substantially smaller than the first one and thus one may be quite confident that one is rapidly approaching the truth -as indeed one is. Of course *certainty* (accepting some statistical errors) can be obtained only by the Green's function Monte-Carlo method. Secondly, this example also illustrates the necessity to use physical intuition and thus it demonstrates that MB theory sometimes is more than engineering work -whatever I have said before. Quite in the spirit of Laughlin's remarks again one had to make an educated guess before ever (if at all) finding a microscopic justification. Especially -but not only- in the realm of phase transitions one often has to start with some basic structure, which requires some intuition, physical insights etc., typically suggested by experimental results. From then on one may let the various MB methods do the fine tuning. There is a second very fundamental achievement by Sergio Rosati and collaborators which I have mentioned before: the correlated hyperspherical harmonics (CHH) method, dating back essentially to 1994.[22] As a variational method it slightly underbinds energies. But, let me quote from Carlson's and Schiavilla's review on few-nucleon systems[23] "the accuracy of these calculations is comparable to that achieved in "exact" calculations". Certainly, for systems with few, but more than four particles CHH has the GFMC as main competitor. However, because it is hard to extract more than just energies from GFMC one can say that CHH in some respects is superior to the former. And I believe it is the cheaper method of the two. There will be more information by Sergio Rosati and some other people in one of the later sessions.

3 Third Movement: Adulthood (real)

Traditionally any serious talk ends with a topic called "future prospects" or "outlook" and the like. On this special occasion these keywords get a special meaning: I know that some (if not all) of our four quartet members will be forced to retire from their official duties. However, there are several things which make this phase transition quite bearable. Firstly, they may look back with satisfaction to their pioneering work in MB physics. Secondly, as a theorist they have the privilege of being able to continue with research. Thirdly, the fact that in this stage of life there is no chance of a promotion anymore turns out as a great advantage: one can do research more relaxed and without any ambition with a career in mind: the only driving force is the fun one has from solving problems and finding answers to questions which nobody has found before. As seen from my age and experience this marks the real adulthood and *not* old age. Keeping the brain going will keep all four young and will shift the time where they really are getting old into a very distant future. They will not be lost to the science community!

References

1. J. W. Clark, *Nucl. Phys.* A **328**, 587 (1979).
2. W. Glöckle, H. Witala, H. Kamada, J. Golak, A. Nogga, and G. Ziemer, preprint, June 2000.
3. H. Witala, D. Hüber, and W. Glöckle, *Phys. Rev.* C **49**, 1214 (1994).
4. E. Krotschek, R. A. Smith, and J. W. Clark, in *Recent Progress in Many Body Theories*, Eds. J. G. Zabolitzky, M. de Llano, M. Fortes, and J. W. Clark (Springer, Berlin, 1981).
5. J. W. Clark and E. Krotschek, in *Recent Progress in Many Body Theories*, Eds. H. G. Kümmel and M. L. Ristig (Springer, Berlin, 1983).
6. J. W. Clark, in *Recent Progress in Many Body Theories I*, Eds. A. J. Kallio, E. Pajanne, and R. F. Bishop, (Plenum, New York, 1987).
7. K. E. Kürten, M. L. Ristig, and J. W. Clark, *Nucl. Phys.* A **317**, 87 (1979).
8. J. W. Kim, M. L. Ristig, and J. W. Clark, *Phys. Rev.* B **57**, 56 (1998).
9. M. L. Ristig, *Nucl. Phys.* A **317**, 163 (1979).
10. C. E. Campbell, K. E. Kürten, G. Senger, and M. L. Ristig, in *Condensed Matter Theories I*, Eds. F. B. Malik (Plenum, New York, 1985).
11. M. L. Ristig and J. W. Clark, in *Recent Progress in Many Body Theories II*, Eds. Y. Avishai, (Plenum, New York, 1989).
12. F. J. Bermejo, F. J. Mompeán, M. García-Hernández, J. L. Martínez, D. Martin-Marero, A. Chahid, G. Senger, and M. L. Ristig, *Phys. Rev.* B **B47**, 15097 (1993).
13. R. A. Smith, A. Kallio, M. Puoskari, and P. Toropainen, *Nucl. Phys.* A **328**, 186 (1979).
14. W. Kohn, *Rev. Mod. Phys.* **71**, S59 (1999)
15. A. Kallio, X. Xiong, and M. Alatalo, in *Recent Progress in Many Body Theories*

II, Eds. Y. Avishai (Plenum, New York, 1989).
16. R. B. Laughlin, *Rev. Mod. Phys.* **71**, 863 (1999).
17. A. Kallio, V. Apaja, and S. Pöykkö, in *Recent Progress in Many Body Theories IV*, Eds. E. Schachinger, H. Mitter, and H. Sormann (Plenum, New York, 1994).
18. A. Kallio, V. Apaja, X. Xiong, and S. Pöykkö, *Physica* C **219**, 340 (1994).
19. A. Kallio, V. Sverdlov, and K. Honkalla, *Superlattices and Microstructures* **21**, Suppl. A, 111 (1997).
20. S. Fantoni and S. Rosati, *Nuovo Cim.* A **25A**, 595 (1975) and preceding papers.
21. S. Fantoni, M. Viviani, A. Buenda, S. Rosati, A. Fabrocini and V. R. Pandharipande in *Recent Progress in Many Body Theories I*, Eds. A. J. Kallio, E. Pajanne, and R. F. Bishop, (Plenum, New York, 1987).
22. A. S. Kievsky, M. Viviani, and S. Rosati, *Nucl. Phys.* A **577**, 511 (1994).
23. J. Carlson and R. Schiavilla, *Rev. Mod. Phys.* **70**, 743 (1998).

Formal Aspects of Many-Body Theory

DIAGRAMS ARE THEORETICAL PHYSICIST'S BEST FRIENDS

JOUKO ARPONEN
Theoretical Physics Division, P.O.Box 9, FIN-00014 University of Helsinki, Finland
E-mail: Jouko.Arponen@Helsinki.FI

Quantum many-body theory and quantum field theory are among the most difficult mathematical constructs in existence. Historically it has been fortunate that these theories often allow diagrammatic treatments, which almost miraculously rationalize the calculation rules. The visualization of the underlying mathematics by using diagram expansions has led to new insight and clever summation rules allowing physically meaningful approximations to be introduced. In this paper I consider some of the basic features of diagram techniques and the conditions for their existence, starting from the simplest examples.

1 Introduction

The celebrities of this meeting, John Clark, Alpo Kallio, Manfred Ristig and Sergio Rosati, have much more in common than just their age: they have dedicated most of their scientific lives to profound and serious studies of the methods and applications of quantum many-body theory. In particular, they all have had deep impact in our understanding of various quantum liquids and the methods used in their studies. A great part of what they have done is related to various diagrammatic techniques of the many-body theory, and in particular to methods originating from Jastrow-type variational approaches, such as the hypernetted chain (HNC) approach, Fermi hypernetted chain theory (FHNC), and the correlated basis function theory (CBF). It is not, however, my intention here to review their works and achievements in theoretical physics; that will be obviously amply covered by other articles of the present volume. Instead, in this paper I will be content in presenting some personal views on diagram methods in general.

Indeed, diagram expansion methods form a rather essential tool in developing improved numerical and qualitative results from an otherwise very abstract formalism. It is amazing to see how versatile diagram expansions are and how frequently they can be applied. In what follows I try to pinpoint some of the basic reasons why diagram expansions are in general possible and feasible, and why they consequently are so abundant in theoretical physics.

2 Elementary examples

Diagrams can be associated already to very simple mathematical expressions. To start, consider the integral

$$Z(s,v) = \int_{-\infty}^{\infty} dx\, \mathrm{e}^{-\frac{1}{2}sx^2 - \frac{1}{4}vx^4}, \tag{1}$$

which may be regarded as a partition function or as a simple nontrivial Feynman path integral.

The analytic behaviour of Z in both the variables s and v is well understood, and even asymptotic approximations and accurate numerical results can easily be found.[1] Nevertheless, it is illuminating to expand the integral in a perturbation series in powers of v. We denote the "unperturbed" partition function by $Z_0(s) \equiv Z(s,0) = \Gamma\left(\frac{1}{2}\right)\left(\frac{s}{2}\right)^{-\frac{1}{2}}$, and define the unperturbed averages as

$$\langle \mathcal{O} \rangle_0 \equiv Z_0(s)^{-1} \int_{-\infty}^{\infty} dx \, \mathrm{e}^{-\frac{1}{2}sx^2} \mathcal{O}(x). \tag{2}$$

The full partition function is

$$Z/Z_0 = \sum_{n=0}^{\infty} \frac{1}{n!} \left(-\frac{1}{4}v\right)^n \langle x^{4n} \rangle_0,$$

and the nonzero moments of x are

$$\langle x^{2n} \rangle_0 = (2n-1)!! \, s^{-n}, \qquad (n > 0) \tag{3}$$

where $(2n-1)!! = (2n-1)(2n-3)\cdots \times 3 \times 1$.

Let the unperturbed "correlation function" or "propagator" be defined as

$$G_0 \equiv \underset{\smile}{xx} \equiv \langle x^2 \rangle_0 = s^{-1},$$

where s is the "energy denominator", and where the hook denotes a "contraction". It is then readily verified that the average $\langle x^{2n} \rangle_0$ can be expressed as the sum of terms containing all possible full pairwise contractions,

$$\langle x^{2n} \rangle_0 = xxxxxxx\ldots + xxxxxxx\ldots + \cdots .$$

Namely, starting from the leftmost x, it can be contracted with any of the $2n-1$ other factors; the next free x can be contracted with any of the remaining $2n-3$ factors, and so on. This produces the combinatorial factor $(2n-1)!!$.

The result can be expressed in terms of a sum of (fully contracted) diagrams, each of which contains $2n$ dots (one for each factor x), and lines connecting the dots pairwise in all possible different ways. A factor G is associated with each line. Each dot is connected to only one line, and no free dots remain in the diagrams.

The diagrams for the partition function Z (or the factor $U = Z/Z_0$) are built from those of $\langle x^{4n} \rangle_0$ for all n by grouping 4 dots into one square, or vertex, for each factor v. A typical diagram of n:th order for U is then composed of n squares for the v:s, from each of which four lines go to some other v-vertices.

For $\Re v > 0$ the function $Z(s,v)$ is an entire function of s and thus allows a convergent power series expansion. We may try to use the quartic exponential $\exp(-\frac{1}{4}vx^4)$ as the "model distribution", and define the averages with respect to this distribution. Let $Z_1(v) \equiv Z(0,v) = 2^{-\frac{1}{2}}\Gamma\left(\frac{1}{4}\right)v^{-\frac{1}{4}}$, and consider

$$\langle x^{2n} \rangle_1 = Z_1^{-1} \int_{-\infty}^{\infty} dx \, \mathrm{e}^{-\frac{1}{4}vx^4} x^{2n}.$$

The result is

$$\langle x^{2n} \rangle_1 = 2^n \frac{\Gamma\left(\frac{2n+1}{4}\right)}{\Gamma\left(\frac{1}{4}\right)} v^{-\frac{n}{2}}.$$

This expression cannot be essentially simplified.

The expansion of Z as power series of s now involves positive powers of $v^{-\frac{1}{2}}$, and the combinatorial factors are such that no diagrammatic interpretation to the terms seems possible. From this simple example it seems evident that the gaussian weight factor is essential for the diagram expansion to exist.

A slightly more complicated model would involve n variables x_a, $a = 1, \ldots, n$, and the partition function

$$Z = \int_{-\infty}^{\infty} dx_1 \cdots \int_{-\infty}^{\infty} dx_n \, \mathrm{e}^{-T[x]-V[x]},$$

where $T(x) = \frac{1}{2}\sum_{ab} t_{ab}x_a x_b$, and where V is of higher order in powers of x, as for example $V = \frac{1}{4}\sum_{abcd} v_{abcd}x_a x_b x_c x_d$. The diagram expansion looks the same as in the first example, but the lines are equipped with index pairs ab and with the factors $G_{ab} = (t^{-1})_{ab}$, and the vertices carry weights v_{abcd}.

3 Correlations

Jastrow-type variational methods for bose systems are based on trial wave functions of the form

$$\Psi_J = \prod_{i<j}^{N} f(r_{ij})\Phi_0 = \prod_{i<j}^{N} \mathrm{e}^{\frac{1}{2}u(r_{ij})}\Phi_0, \tag{4}$$

where Φ_0 is the unperturbed (constant) N-body wave function and $f = \exp(\frac{1}{2}u)$ a two-body correlation factor. By the Jackson-Feenberg energy formula[2] the total energy can be expressed as a simple functional of $u(r)$ and $g(r)$, the pair correlation function. The problem is now to express the pair correlation function in terms of the function $u(r)$. The square of the wave function is

$$\begin{aligned}|\Psi_J|^2 &= \textstyle\prod_{i<j}(1+b_{ij})\\ &= 1 + X_2 + X_3 + \cdots,\end{aligned}$$

where the bond function is $b(r) = f(r)^2 - 1 = \exp(u(r)) - 1$, and the terms X_n involve n coordinate indices (dots) and contain a number of lines between pairs representing factors b_{ij}. In this case no restriction is set to the number of lines associated with a dot. The pair correlation function is obtained, apart from normalization, by integrating over all other coordinates except two fixed dots.

Another example of similar kind is the Ursell-Mayer expansion in the theory of classical fluids. In this case one has to calculate the N-body partition function which includes the factor

$$\begin{aligned}Q_N &= \textstyle\int d^3\boldsymbol{r}_1 \cdots \int d^3\boldsymbol{r}_N \exp\left[-\beta\sum_{i<j} v(\boldsymbol{r}_{ij})\right]\\ &= \textstyle\int d^3\boldsymbol{r}_1 \cdots \int d^3\boldsymbol{r}_N \prod_{i<j}(1+b_{ij}).\end{aligned} \tag{5}$$

The bond function $b_{ij} = \exp(-\beta v_{ij}) - 1$ is fixed in the present case, but otherwise the formalism is very much like in the Jastrow theory.

4 Wick theorem

In our first example the elementary contraction $\underline{xx}$ is a moment of a probability distribution, and the higher moments are obtained by a set of rules which resemble the rules found by G. Wick[3] in the context of Green's functions in the time-dependent perturbation theory. Given a set of elementary interaction-picture operators $A_{\nu_1}(t_1)$, $A_{\nu_2}(t_2)$, ..., $A_{\nu_n}(t_n)$, which are linear and homogeneous in the one-particle creation and annihilation operators (in the second-quantized formalism), one can define the time- and normal-ordered products with the following operator expansion at zero tremperature,

$$T[A_{\nu_1}(t_1)A_{\nu_2}(t_2)A_{\nu_3}(t_3)\cdots A_{\nu_n}(t_n)] = N[A_{\nu_1}(t_1)A_{\nu_2}(t_2)A_{\nu_3}(t_3)\cdots A_{\nu_n}(t_n)]$$
$$+ \underline{N}[A_{\nu_1}(t_1)A_{\nu_2}(t_2)A_{\nu_3}(t_3)\cdots A_{\nu_n}(t_n)] + \underline{N}[A_{\nu_1}(t_1)A_{\nu_2}(t_2)A_{\nu_3}(t_3)\cdots A_{\nu_n}(t_n)]$$
$$+ \underline{N}[A_{\nu_1}(t_1)A_{\nu_2}(t_2)A_{\nu_3}(t_3)\cdots A_{\nu_n}(t_n)] + \ldots .$$

The sum goes over all terms with different contraction sets, and the theorem can be proven by induction over n. In the time-dependent perturbation theory only the fully contracted terms from the right-hand side are needed, since the averages of the others in the model state vanish.

At nonzero temperature the temperature-ordered product can be expanded in an analogous way using a generalized Wick theorem,[4] but only when averaged over the noninteracting statistical distribution $\varrho_0 = Z_0^{-1}\exp(-\beta H_0)$, and counting only the terms with full contractions (thus the normal-ordering is not a necessary concept).

The Wick theorem is used in calculating the (time- or temperature-ordered) Green's functions, such as

$$G_{AB}(t_1, t_2) = -i\langle T[SA(t_1)B(t_2)]\rangle_0 / \langle T[S]\rangle_0,$$

where S is the scattering operator.

It should be noted that in spite of apparent differences between this and the two first examples of the previous subsection, an essential feature again is the gaussian weight factor defining the zero-order distribution. In the present case the gaussian weight appears in disguise and is defined in the functional space.

5 Linked-cluster theorem

Perhaps the most important property classifying diagrams is whether they are connected or not. The calculation rules associate factorizable contributions to disconnected diagrams. In fact, the disconnected parts of diagrams cancel in such physically meaningful additive quantities as the "free energy" $F = -\ln Z$ (in the first examples) or the grand potential of the Ursell-Mayer theory. Also, in Jastrow theory, the norm of the wave function cancels the disconnected parts of the expansion for the pair correlation function $g(r)$. And in time-dependent perturbation theory, expectation values and Green functions are always calculated using only connected diagrams.

The precise form of the linked-cluster theorem varies from case to case, but the central idea is always the same. One might say that it is the linked-cluster theorem which makes classical mechanics possible, because otherwise it might be impossible to derive – starting from microscopic physics – the concepts of "extensive" and "intensive" state variables, which lie at the heart of thermodynamics and macrophysics.

6 Green's functions and partial summations

An extremely wealthy literature exists about Green's functions or n-point functions, whether derived from Feynman path integral method or from second-quantized field-theoretic approach. Once a problem has been cast into a form allowing diagrams, the human can use her power of visualization to find smarter and smarter ways to perform summations. For example, F. Dyson invented Dyson's equations for the two-point function, or the one-particle Green's function. It is based on the analysis of the diagrams contributing to G, by checking if they can be divided into two parts by cutting only one line such that the end points lie at different sides of the cut. Thus the Green's function becomes expanded in terms of the proper self-energy, which cannot be cut in this way, as

$$G = G_0 + G_0 \Sigma G_0 + \ldots = G_0 + G_0 \Sigma G, \tag{6}$$

with the simple solution $G^{-1} = G_0^{-1} - \Sigma$. This equation reduces the calculation of the Green's function into the calculation of the self-energy Σ. The expansion of Σ is more complicated, and makes it necessary to analyse the 4-point function, or the two-body Green's function. In fact, by diagram methods one can usually derive a hierarchy of equations for the n-point functions, connecting a lower-order function to higher-order functions, *ad infinitum.*

Analogous diagram analysis can be made in the Jastrow HNC and Ursell-Mayer theories by recognizing certain characteristic diagram structures (*e.g.*, articulation points, nodal points, composite and elementary graphs) based on which infinite subsummations can easily be done. A.D. Jackson and coworkers[5] have studied the connections between the parquet summation technique in the Green function theory and the JHNC technique in the variational formulation, as applied to liquid ^{4}He. The idea in the parquet summation technique is to sum planar Feynman diagrams with crossing symmetry in the s-, t-, and u-channels for the two-particle vertex Γ. The equations for the diagram structures in the case of a bose liquid are as follows,

$$\begin{aligned} L &= (V + C)G_{pp}(V + C) + (V + C)G_{pp}L, \\ C &= (V + L)G_{ph}(V + L) + (V + L)G_{ph}C, \\ \Gamma &= V + L + C; \end{aligned} \tag{7}$$

here V is the interaction potential, C and L represent the infinite sums of chain-type (rings) and ladder-type diagrams, respectively. Iteration of the equations leads to "ladders of chains of ladders of ...", creating intertwined (approximately) planar structures. The practical difficulty in solving these equations even in a suitably chosen local approximation should be respected in spite of their deceptively simple form. The results in these and subsequent refined calculations have been both qualitatively and numerically very encouraging.

A tacit hope in these and similar expansions is that the lower-order Green's functions may become analytically and numerically well approximated even if the higher-order functions are more crudely approximated, or if the hierarchy is abruptly terminated. There are guidelines which can be used in constructing such approximations. For example, in many cases certain sum rules can be proven to hold. If an approximation can take as many of these and other similar consistency requirements into account, the results may be expected to be both reliable and highly illuminating.

7 Coupled cluster method

Coupled-cluster or $\exp S$ theory[6] has a quite different algebraic origin, and is not immediately related to Feynman diagram techniques. Nevertheless, it can be interpreted in terms of partially time-integrated Feynman diagrams, or Goldstone diagrams. In fact, since the method exploits to high extent the disentanglement of the energy denominators of usual Goldstone diagrams, we might rather name the diagrams in this case coupled-cluster diagrams.

This method is based on the observation that the ground-state wave function of the many-body system is of the exponential form $|\Psi\rangle = \mathrm{e}^S|\Phi\rangle$ (up to a normalization factor), where $|\Phi\rangle$ is the model state, and S a creation operator expressible in the form $S = \sum_{J\neq 0} S_J C_J^\dagger$, in which the $C_J^\dagger$ are a complete set of configuration creation operators corresponding to the many-body configurations J, and the S_J are linked-cluster many-body amplitudes. In the normal coupled-cluster method (NCCM) one writes the ground-state energy eigenvalue problem in the form of the "equations for S_n",[6,7] *i.e.*, $\langle C_J^\dagger\Phi|\mathrm{e}^{-S}H\mathrm{e}^S\Phi\rangle = E_0\delta(J,0)$, where H is the Hamiltonian. The ingenious feature in this formalism is that the similarity transformation automatically guarantees the linked-cluster theorem and the size-extensivity property without extra requirements. Applications of NCCM range from quantum chemistry to elementary particle physics, and a rich experience of its approximation methods has evolved.

The extended coupled cluster method (ECCM)[8] introduces additional cluster amplitudes, and a consistent development of the formalism leads to a canonical structure in the space of cluster amplitude functions. The coupled cluster methods are basically defined in a non-hermitean form, and therefore also in the ECCM one can introduce two non-hermitean canonical coordinate sets, which are complex conjugate to each other. In the first case we have the set $\{\sigma_J, \tilde{\sigma}_J\}$, in the second the set $\{\tilde{\sigma}_J^*, \sigma_J^*\}$.[9] In each of the cases the expectations of observables can be expressed as functions of these variables, *e.g.*, $\langle \mathcal{O}\rangle \equiv \bar{\mathcal{O}}[\sigma,\tilde{\sigma}] \equiv \langle\Phi|\mathrm{e}^{S^\dagger}\mathcal{O}\mathrm{e}^S|\Phi\rangle/\langle\Phi|\mathrm{e}^{S^\dagger}\mathrm{e}^S|\Phi\rangle$. The connection between the original cluster operators $S, S^\dagger$ and the amplitudes $\{\sigma,\tilde{\sigma}\}$ or $\{\tilde{\sigma}^*,\sigma^*\}$ is explained in the references given above, and will not be spelled out here.

In ECCM the expectation of a product of operators can be given in terms of the expectations of the operators and their first derivatives by the help of a second-rank tensor X,[8] (see also Ref. 10 where the formalism is extended to fermion systems by introducing anticommuting cluster coordinates)

$$\langle AB\rangle = \langle A\rangle \star \langle B\rangle \equiv \bar{A}\bar{B} + X(\mathrm{d}\bar{A}, \mathrm{d}\bar{B}). \tag{8}$$

The precise form of the tensor depends on which set of variables are used as the basic coordinates in the ECCM phase space. In all cases, the elements of which the tensor X is made have diagrammatic interpretations in terms of well-defined substructures of coupled-cluster diagrams. A particularly interesting form is obtained in the hermitized form of the ECCM,[9] regarding as free variables either the set $\{\sigma, \sigma^*\}$ or the set $\{\tilde{\sigma}^*, \tilde{\sigma}\}$. In the previous case we get

$$X = \frac{\partial}{\partial\sigma}(\sigma + \sigma L\sigma)\frac{\partial}{\partial\sigma} + \frac{\partial}{\partial\sigma}KQ^{-1}K^{\dagger}\frac{\partial}{\partial\sigma^*} + \frac{\partial}{\partial\sigma^*}\sigma^* Q\sigma\frac{\partial}{\partial\sigma} + \frac{\partial}{\partial\sigma^*}(\sigma^* + \sigma^* L^*\sigma^*)\frac{\partial}{\partial\sigma^*},$$

and in the latter

$$X = \frac{\partial}{\partial\tilde{\sigma}}L\frac{\partial}{\partial\tilde{\sigma}} + \frac{\partial}{\partial\tilde{\sigma}}LQ^{-1}L^*\frac{\partial}{\partial\tilde{\sigma}^*} + \frac{\partial}{\partial\tilde{\sigma}^*}Q\frac{\partial}{\partial\tilde{\sigma}} + \frac{\partial}{\partial\tilde{\sigma}^*}L^*\frac{\partial}{\partial\tilde{\sigma}^*}.$$

The diagrammatically interpretable functions (with matrix indices) L, K, Q appearing in these formulae are fully explained in the original references.

Although the ECCM formalism has not yet matured to the degree of the better established NCCM, it has features which are fascinating and to my understanding new among other diagrammatically derived methods. First of all, it emphasizes the geometric phase-space picture, where all the basic variables (cluster amplitudes) are coordinates on a manifold. Secondly, the picture is that of classical canonical formalism, *i.e.*, the quantum problem is reduced to a mean-field theory on a certain symplectic manifold. This is a consequence of the fact that the ECCM diagrams are formally tree diagrams where the branches are multiparticle on-energy-shell propagators.

8 Comments

In present days the brute-force numeric simulations on a discretized lattice are by far the most popular method used in both gauge field theories and in non-relativistic many-body theory. Without underestimating the efficiency of such methods (and see, *e.g.*, the criticism in Ref.[11]) it may be fair to say that they lack the beauty and inspiration of the analytical studies, wherever such are available. As an example, the method of canonical transformations has provided very elegant and effortless qualitative explanations for the superfluidity of bose and fermi liquids (the Bogoliubov and Bogoliubov-Valatin transformations, the BCS theory). The diagrammatic Green's function method can easily account for such qualitative features and extend the theory to encompass the main effects of interactions and correlations (for example, the Gorkov and Eliashberg equations for weak- and strong-coupling superconductivity). Such qualitative understanding is not equally easy to obtain in numerical lattice simulations, which otherwise are very robust and safe in principle.

Of course there are serious pitfalls also in using diagrammatic methods. Inconsiderate infinite summations may easily lead to divergent results even though the individual diagrams were finite, as in renormalizable theories. Such problems stem from the closeness of the diagrammatic method to ordinary perturbation theory which almost always is divergent at the interesting physical coupling strength (*c.f.* our first example). The nature of things is such that the physically interesting additive quantities, like the the ground-state energy or the free energy of a many-body system, are nonanalytic functions of the coupling strength, and one has to live with that! Therefore the partial infinite summations of diagrams must be chosen with understanding and careful consideration.

In conclusion, it may be safe to say that most of our qualitative understanding of the properties of many-body systems stem from analytic and diagrammatic methods. Therefore, when studying a new problem, it may always be a good idea to think about diagrams.

References

1. C. M. Bender and T. T. Wu, *Phys. Rev.* **184**, 1231 (1969); B. Simon, *Ann. Phys. (NY)* **58**, 76 (1970).
2. E. Feenberg, *Theory of quantum liquids* (Academic Press, New York, 1969); G. Ripka, *Phys. Rep.* **56**, 1 (1979).
3. G. C. Wick, *Phys. Rev.* **80**, 265 (1950).
4. T. Matsubara, *Prog. Theor. Phys.* **14**, 351 (1955); M. Gaudin, *Nucl. Phys.* **15**, 89 (1960).
5. A. D. Jackson, A. Lande, and R. A. Smith, *Phys. Rep.* **86**, 55 (1982); E. Krotscheck, G.-X. Qian, and W. Kohn, *Phys. Rev.* B **36**, 4245 (1985).
6. F. Coester, *Nucl. Phys.* **7**, 421 (1958); F. Coester and H. Kümmel, *Nucl. Phys.* **17**, 477 (1960).
7. H. G. Kümmel, K. H. Lührmann, and J. G. Zabolitzky, *Phys. Rep.* **36**, 1 (1978).
8. J. S. Arponen, R. F. Bishop, E. Pajanne, *Phys. Rev.* A **36**, 2519 (1987); J. S. Arponen and R. F. Bishop, *Phys. Rev.* A **207**, 171 (1993).
9. Jouko Arponen, *Phys. Rev.* A **55**, 2686 (1997).
10. Jouko S. Arponen, *Theor. Chim. Acta* **80**, 149 (1991).
11. K. G. Wilson, *Nucl. Phys.* B **17**, 82 (1990).

FOURTH ORDER ALGORITHMS FOR SOLVING DIVERSE MANY-BODY PROBLEMS

SIU A. CHIN
Center for Theoretical Physics, Department of Physics, Texas A&M University, College Station, TX 77843
E-mail: s-a-chin@tamu.edu

I show that the method of factorizing the operator $e^{\epsilon(T+V)}$ to fourth order with purely positive coefficients yields excellent ***symplectic*** algorithms for solving classical dynamical problems, ***unitary*** algorithms for solving quantum dynamical problems, ***norm preserving*** algorithms for solving stochastic many-body problems and fourth order Diffusion Monte Carlo algorithms for solving quantum many-body problems.

1 Introduction

A generic evolution equation of the form

$$\frac{\partial w}{\partial t} = (T+V)w, \tag{1}$$

where T and V are non-commuting operators, occurs in many problems of physics. This equation can be solved iteratively via

$$w(t+\epsilon) = e^{\epsilon(T+V)}w(t), \tag{2}$$

provided that one has a suitable approximation for the small time evolution operator $e^{\epsilon(T+V)}$. While first and second order factorizations such as

$$e^{\epsilon(T+V)} \approx \mathcal{T}^{(1)} = e^{\epsilon T}e^{\epsilon V}, \tag{3}$$

$$\approx \mathcal{T}_A^{(2)} = e^{\frac{1}{2}\epsilon V}e^{\epsilon T}e^{\frac{1}{2}\epsilon V}, \tag{4}$$

$$\approx \mathcal{T}_B^{(2)} = e^{\frac{1}{2}\epsilon T}e^{\epsilon V}e^{\frac{1}{2}\epsilon T}, \tag{5}$$

are well known, it is only recently that one has learned how to decompose the small time evolution operator to fourth order with purely *positive* coefficients.[1,2] This has given rise to an entirely new class of algorithms for solving diverse physical problems. In this work, I will summarize some results of these new algorithms in solving the classical dynamical problem,[2,3] the time-dependent Schrödinger equation,[4] the many-body Fokker-Planck equation,[5] and the many-body Schrödinger equation in imaginary time.[6] I will begin by explaining the novel characters of these fourth order factorizations.

2 Fourth Order Factorizations

It has been known for some time, in the context of symplectic integrators,[7] that the short time evolution operator can be factorized to arbitrarily high order in the form[8–10]

$$e^{\epsilon(T+V)} = \prod_i e^{a_i\epsilon T}e^{b_i\epsilon V}, \tag{6}$$

with coefficients $\{a_i, b_i\}$ determined by the required order of accuracy. For example, if $\mathcal{T}^{(2)}$ denotes either $\mathcal{T}_A^{(2)}$ or $\mathcal{T}_B^{(2)}$, then the evolution operator can be factorized to fourth order via

$$\mathcal{T}_{FR}^{(4)}(\epsilon) = \mathcal{T}^{(2)}(\tilde{\epsilon})\mathcal{T}^{(2)}(-s\tilde{\epsilon})\mathcal{T}^{(2)}(\tilde{\epsilon}) \tag{7}$$

where $s = 2^{1/3}$ is chosen to cancel the third order error term in $\mathcal{T}^{(2)}$ and $\tilde{\epsilon} = \epsilon/(2-s)$ rescales the sum of forward-backward-forward time steps back to ϵ. This standard fourth order Ruth-Forest[11] scheme, has been independently derived many times in the context of symplectic integrators.[12,13] The above derivation was first published by Creutz and Gocksch[8] in 1989. Suzuki[9] and Yoshida[10] independently published the same constructions in 1990.

Notice that the middle time step in (7) is negative. This is not accidental. Suzuki[14] has proved that, beyond second order, any factorization of the form (6) must produce some negative coefficients in the set $\{a_i, b_i\}$, corresponding to some negative time steps. While negative time steps are not detrimental in solving the time-dependent classical or quantum mechanical problem, it is impossible to do any Monte Carlo simulations in which the kinetic energy operator is the diffusion kernel simulated by Gaussian random walks. Having a negative time step means that one must simulate the diffusion process backward in time, which is impossible.

The essence of Suzuki's proof is to note that since

$$e^{\frac{1}{2}\epsilon V}e^{\epsilon T}e^{\frac{1}{2}\epsilon V} = e^{\epsilon C}, \tag{8}$$

with

$$C = \epsilon(T+V) - \frac{1}{12}\epsilon^3[T,[V,T]] + \frac{1}{24}\epsilon^3[V,[T,V]] + O(\epsilon^5) \tag{9}$$

in order to obtain a fourth order algorithm one must eliminate third order error terms involving double commutators $[T,[V,T]]$ and $[V,[T,V]]$. With purely positive coefficients a_i and b_i, one can eliminate either one or the other, but not both. Thus to obtain a fourth order factorization with only positive coefficients, one must retain one of the two double commutators. In order for the factorization to be useful, the retained commutators must be "calculable". We will designated the retained commutator as $[V,[T,V]]$.

Recently, Suzuki[1] and Chin[2] have derive a number of such fourth order factorization schemes. The two schemes derived by both Suzuki and Chin, using different methods, are:

$$\mathcal{T}_A^{(4)} \equiv e^{\epsilon\frac{1}{6}V}e^{\epsilon\frac{1}{2}T}e^{\epsilon\frac{2}{3}\widetilde{V}}e^{\epsilon\frac{1}{2}T}e^{\epsilon\frac{1}{6}V}, \tag{10}$$

with $\widetilde{V}$ given by

$$\widetilde{V} = V + \frac{1}{48}\epsilon^2[V,[T,V]], \tag{11}$$

and

$$\mathcal{T}_B^{(4)} \equiv e^{\epsilon\frac{1}{2}(1-\frac{1}{\sqrt{3}})T}e^{\epsilon\frac{1}{2}\bar{V}}e^{\epsilon\frac{1}{\sqrt{3}}T}e^{\epsilon\frac{1}{2}\bar{V}}e^{\epsilon\frac{1}{2}(1-\frac{1}{\sqrt{3}})T}, \tag{12}$$

with $\bar{V}$ given by

$$\bar{V} = V + \frac{1}{24}(2-\sqrt{3})\epsilon^2[V,[T,V]]. \tag{13}$$

In addition, Chin[2,4] has derived factorization scheme C,

$$\mathcal{T}_C^{(4)} \equiv e^{\epsilon\frac{1}{6}T} e^{\epsilon\frac{3}{8}V} e^{\epsilon\frac{1}{3}T} e^{\epsilon\frac{1}{4}\tilde{V}} e^{\epsilon\frac{1}{3}T} e^{\epsilon\frac{3}{8}V} e^{\epsilon\frac{1}{6}T}, \tag{14}$$

which minimizes the appearance of V, and scheme D

$$\mathcal{T}_D^{(4)} \equiv e^{\epsilon\frac{1}{8}\tilde{V}} e^{\epsilon\frac{1}{3}T} e^{\epsilon\frac{3}{8}V} e^{\epsilon\frac{1}{3}T} e^{\epsilon\frac{3}{8}V} e^{\epsilon\frac{1}{3}T} e^{\epsilon\frac{1}{8}\tilde{V}}. \tag{15}$$

which minimizes the appearance of T. Each of these factorization schemes can be translated into an algorithm for solving the evolution equation (1) depending on the specific form of the operators T and V.

3 Solving Classical Dynamical Problems

In classical mechanics, the equation of motion for any dynamical variable w without explicit time-dependence is given by the Poisson bracket

$$\frac{dw}{dt} = \{w, h\} = \sum_i \Big[\frac{\partial h}{\partial p_i}\frac{\partial}{\partial q_i} - \frac{\partial h}{\partial q_i}\frac{\partial}{\partial p_i}\Big] w, \tag{16}$$

where

$$h = \frac{1}{2}\sum_i p_i^2 + v(q_i) \tag{17}$$

is the Hamiltonian function. This equation can be integrated to the form (2), where T and V are now first order differential operators

$$T = \sum_i \frac{\partial h}{\partial p_i}\frac{\partial}{\partial q_i} = \sum_i p_i \frac{\partial}{\partial q_i}, \qquad V = \sum_i -\frac{\partial h}{\partial q_i}\frac{\partial}{\partial p_i} = \sum_i F_i \frac{\partial}{\partial p_i}, \tag{18}$$

with force $F_i = -\partial v/\partial q_i$. The exponentiated operators, $e^{\epsilon T}$, $e^{\epsilon V}$, are then displacement operators which displace q_i and p_i respectively forward in time via

$$q_i \to q_i + \epsilon p_i \quad \text{and} \quad p_i \to p_i + \epsilon F_i. \tag{19}$$

Every decomposition of the evolution operator $e^{\epsilon(T+V)}$ produces a sequence of p_i and q_i displacements which constitutes a *symplectic* algorithm for evolving the system forward in time. First and second order factorizations (3) and (4) produce the well known leap-frog and velocity-Verlet algorithm respectively.

For T and V defined by (18), the double commutator

$$[V, [T, V]] = 2 f_j \frac{\partial F_i}{\partial q_j}\frac{\partial}{\partial p_i} = \nabla_i |\mathbf{F}|^2 \frac{\partial}{\partial p_i} \tag{20}$$

is again a V-like operator. Thus the occurrence of $\tilde{V}$ corresponds to replacing the original force F_i by an effective force

$$\tilde{F}_i = F_i + \frac{1}{48}\epsilon^2 \nabla_i |\mathbf{F}|^2. \tag{21}$$

The case for $\bar{V}$ is similar. These new symplectic algorithms all required evaluating the gradient of the force in additional to the force.

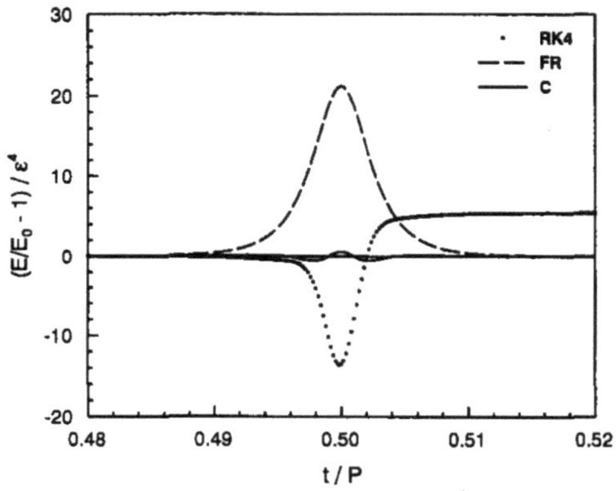

Figure 1. The normalized energy deviation of a particle in a Keplerian orbit, which measures the step-size independent energy error coefficient. P is the period of the elliptical orbit and ϵ is the time step size. RK4, FR, and C denote results for the 4th order Runge-Kutta, Forest-Ruth, and Chin's C algorithm respectively. The maximum deviations for algorithm FR and C are 21 and 0.27 respectively.

To gauge their effectiveness, we compare algorithms by solving the two dimensional Keppler problem

$$\frac{d^2\mathbf{q}}{dt^2} = -\frac{\mathbf{q}}{q^3}, \tag{22}$$

with initial conditions $\mathbf{q}_0 = (10, 0)$ and $\mathbf{p}_0 = (0, 1/10)$. The resulting highly eccentric (e=0.9) orbit provides a non-trivial testing ground for trajectory integration. Fig. 1 compares the normalized, step-size independent energy error coefficient for the 4th order Runge-Kutta (RK4), Forest-Ruth (FR) and Chin's C algorithm over one period of the orbit. The error coefficient for the two symplectic algorithms are substantial only near mid period when the particle is at its closest approach to the attractive center. For symplectic algorithms energy is conserved over one period. Its average energy error is bounded and constant in time. In contrast, the 4th order Runge-Kutta energy error function is an irreversible, step-like function over one period. Each successive period will increase the error by the same amount resulting in a linearly rising, staircase-like error function in time. The maximum error in Chin's algorithm C is smaller than that of the FR algorithm by a factor of 80. Algorithm A and B's maximum errors are 7 and 11 times as large as algorithm C.[2]

Energy conservation does not directly measure how well the orbit is determined. When the time step is not too small, a very noticeable error is that the orbit precesses. It is convenient to monitor this precession by measuring the rotation of the Laplace-Runge-Lenz vector:

$$\mathbf{A} = \mathbf{p} \times \mathbf{L} - \hat{\mathbf{r}}. \tag{23}$$

When the orbit is exact, the LRL vector is constant, pointing along the semi-major axis of the orbit. When the orbit precesses, the LRL vector rotates correspondingly.

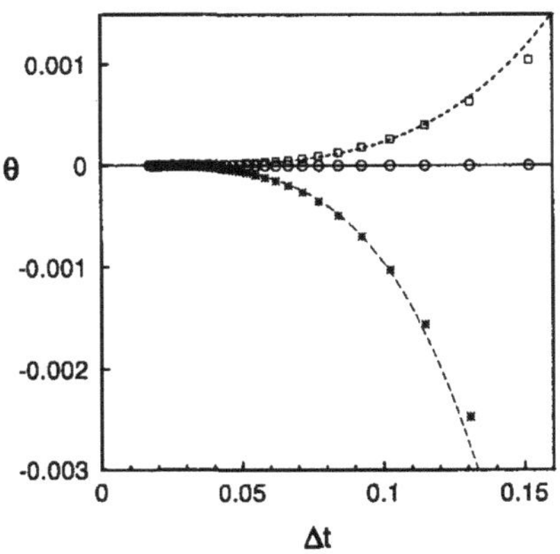

Figure 2. The angle of rotation of the Laplace-Runge-Lenz vector after one period of the Keplerian orbit as a function of the time step size for various algorithms. The squares, asterisks, and circles are results of the Runge-Kutta, Ruth-Forest and Chin's C algorithms respectively. The fitted lines are single power of Δt^4 with approximate coefficients 2.3, 9.5, and 0.004 respectively.

Fig. 2 shows the amount of rotation per period as a function of the time step size used. For a more detailed discussion, see Ref. 3.

4 Solving the Time-Dependent Schrödinger Equation

A quantum state is evolved forward in time by the Schrödinger evolution operator

$$e^{\epsilon H} = e^{\epsilon(T+V)}, \tag{24}$$

where $\epsilon = -i\Delta t$ and

$$T = -\frac{1}{2}\sum_i \nabla_i^2, \qquad V = V(\mathbf{r}_i), \tag{25}$$

are the usual kinetic and potential energy operators in units such that $\hbar = 1 = m$. Since $e^{\epsilon V}$ is diagonal in position space and $e^{\epsilon T}$ is diagonal in momentum space, any decomposition of $e^{\epsilon(T+V)}$ into products of $e^{\epsilon T}$ and $e^{\epsilon V}$ will yield a *unitary* algorithm for evolving the state forward in time by shuffling the wave function back and forth between real and Fourier space. Every occurrence of $e^{\epsilon T}$ thus requires two Fast Fourier Transforms (FFTs); one direct transform to momentum space where $e^{\epsilon T}$ becomes simple multiplications and one inverse transform to return to position space. (See detailed discussion by Takahashi and Ikeda.[15])

The advantage of this operator approach is that higher order algorithms can be constructed easily, as in the case of symplectic algorithms. However, any factorization of the form (6) must produce algorithms which evolve the system backward in time for some intermediate time steps. While this is not detrimental in solving the quantum evolution problem, it has been shown that these negative time step algorithms converge only for very small values of Δt and is far from optimal.[4] As we will show below, the factorization of the Schrödinger evolution operator to 4th order with purely *positive* time steps yielded algorithms with excellent convergent properties at relatively large values of Δt. To transcribe the factorization scheme described in Section 2, we note that in this case, the double commutator is

$$[V,[T,V]] = \sum_i |\nabla_i V|^2, \tag{26}$$

which modifies the original potential energy V, to the effective potential operator

$$\widetilde{V} = V - \frac{1}{48}\Delta t^2 \sum_i |\nabla_i V|^2. \tag{27}$$

To gauge the effectiveness of these fourth order, positive time step algorithms, we test them on a one dimensional scattering problem where a Gaussian wave pocket

$$\psi_0(x) = \frac{1}{(2\pi\sigma^2)^{1/4}} \exp\left[ik_0 x - \frac{(x-x_0)^2}{4\sigma^2}\right], \tag{28}$$

is impinged on a smooth sech-square potential given by

$$V(x) = V_0 \,\mathrm{sech}^2(x) \tag{29}$$

with $V_0 = 48.2$. Fig. 3 shows the resulting transmission coefficient for various algorithms as a function of the time step size Δt at an incident energy of $E_0 = V_0$.

The second order results (4), denoted by asterisks, can be accurately fitted by $T_0 - 0.36\Delta t^2$ for $\Delta t < 0.1$, demonstrating its quadratic convergence. This algorithm requires two FFT's, which is more economical than the alternative second algorithm corresponding to (5). The Ruth-Forest (7) results can also be well fitted by $T_0 - 74\Delta t^4$ over the same range. However, the range of convergence of the RF algorithm is not substantially greater than that of the second order algorithm. In comparison, the four 4th order algorithms with positive coefficients are distinctly superior. Whereas the fourth order error coefficient of the Ruth-Forest algorithm is 74, the corresponding coefficients for algorithms A, B, C and D are -1.07, -0.38, 0.14 and 0.14 respectively. Algorithm C and D yield identical results. Algorithm D's error coefficient is more than 500 times smaller than that of RF, and can achieve the same accuracy by using step sizes nearly 5 times as large. The comparison with second order results is even more favorable; the step size can be 10-15 times as large. Note that scheme A, remarkably, only requires 4 FFTs. To compare the computational effort involve, we timed each algorithm for 160 iterations on a Pentium II 450 MH processor using a Fortran compiler. The second order algorithm took $T_2 = 5.33$s. Relative to this time, the time required by algorithms RF, A, B, C, D are respectively, $2.98T_2$, $2.22T_2$, $3.37T_2$, $3.97T_2$, and $3.26T_2$ respectively, which roughly scale with the number of FFTs used in each algorithm. While algorithm D has the smallest error coefficient, algorithm A is also notable in that it is roughly 1/3 faster than RF but converges at time steps nearly 10 times as large.

5 Solving the Fokker-Planck Equation

The Fokker-Planck equation

$$\frac{\partial}{\partial t}P(\mathbf{x},t) = LP(\mathbf{x},t) \equiv \left[\frac{1}{2}D_{ij}\partial_i\partial_j - \partial_i G_i(\mathbf{x})\right]P(\mathbf{x},t), \tag{30}$$

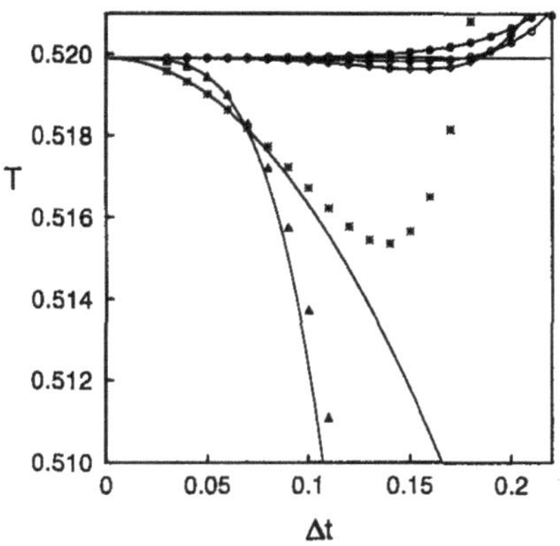

Figure 3. The transmission coefficient T as a function of time step size for various algorithms. The asterisks are second order results, (4). The solid triangles are 4th order results corresponding to the Ruth-Forest splitting scheme with negative coefficients, (7). The hollow diamonds and circles are results of algorithm A (10) and B (12) respectively. The filled circles are identical results produced by algorithms C (14), and D (15).

is used to describe a variety of physical and chemical processes.[16] While the equation is of intrinsic physical interest, it is also important computationally as the basis for developing the Langevin algorithm ,[17–20] and as the backbone of the Diffusion Monte Carlo algorithm. Even when the diffusion matrix D_{ij} is position independent, it is difficult to derive Monte Carlo algorithms for solving the Fokker-Planck equation beyond second order.[17–21] Fourth order algorithms are only now possible with the discovery of fourth order factorizations schemes described in Section 2.

The Fokker-Planck equation can be integrated to give

$$P(\mathbf{x}, t+\epsilon) = \mathrm{e}^{\epsilon L} P(\mathbf{x}, t), \tag{31}$$

which again can be solved by factorizing the short time Fokker-Planck evolution operator $\mathrm{e}^{\epsilon L} = \mathrm{e}^{\epsilon(T+D)}$ into solvable parts. While this method has no difficulty in dealing with a general but constant diffusion matrix D_{ij}, it is sufficient for us to consider the case $D_{ij} = \delta_{ij}$. Thus L consists of the two operators

$$T = \frac{1}{2}\partial_i\partial_i \qquad \text{and} \qquad D = -\partial_i G_i(\mathbf{x}), \tag{32}$$

where implied summations over repeated indices.

When the operator $\mathrm{e}^{\epsilon T}$ acts on $P(\mathbf{x}, t)$, it evolves it forward in time according to the *diffusion* equation

$$\frac{\partial}{\partial t} P(\mathbf{x}, t) = \frac{1}{2}\partial_i\partial_i P(\mathbf{x}, t). \tag{33}$$

If $\{x_i\}$ is a set of points distributed according to $P(\mathbf{x}, t)$, then ϵ time later the distribution can be exactly simulated by random walking each point according to

$$x_i' = x_i + \sqrt{\epsilon}\,\xi_i, \tag{34}$$

where $\{\xi_i\}$ is a set of Gaussian random numbers with zero mean and unit variance. When $e^{\epsilon D}$ acts on $P(\mathbf{x},t)$, it evolves the distribution forward in time according to the *continuity* equation

$$\frac{\partial}{\partial t}P(\mathbf{x},t) = -\partial_i[G_i(\mathbf{x})P(\mathbf{x},t)], \tag{35}$$

where $G_i(\mathbf{x})P(\mathbf{x},t) = J_i(\mathbf{x})$ is the probability current density with velocity field $G_i(\mathbf{x})$. The continuity equation can also be exactly simulated by setting

$$x_i' = x_i(\epsilon), \tag{36}$$

where $x_i(\epsilon)$ is the exact trajectory determined by

$$\frac{d\mathbf{x}}{dt} = \mathbf{G}(\mathbf{x}), \tag{37}$$

with initial condition $x_i(0) = x_i$. Hence if $e^{\epsilon(T+D)}$ can be factorized into products of operators $e^{\epsilon T}$ and $e^{\epsilon D}$, then there is a well-defined sequence of updating according to (34) and (36) which will evolve the distribution forward in time. For example, the second order factorization,

$$e^{\frac{1}{2}\epsilon T}e^{\epsilon D}e^{\frac{1}{2}\epsilon T} = \exp[\epsilon(T+D) + O(\epsilon^3)\cdots], \tag{38}$$

leads to the following second order Langevin algorithm[21]

$$\begin{aligned} y_i &= x_i + \xi_i\sqrt{\epsilon/2}, \\ x_i' &= y_i(\epsilon) + \xi_i'\sqrt{\epsilon/2}, \end{aligned} \tag{39}$$

where ξ_i and ξ_i' are independent sets of zero mean, unit variance Gaussian random numbers. For a second order algorithm, it is sufficient to solve the trajectory $y_i(\epsilon)$ correctly to second order in ϵ, *e.g.*, via a second order Runge-Kutta algorithm:

$$y_i(\epsilon) = y_i + \epsilon G_i\left(\mathbf{y} + \frac{1}{2}\epsilon\mathbf{G}(\mathbf{y})\right). \tag{40}$$

Since $e^{\epsilon T}$ cannot be simulated if ϵ were negative, Suzuki's proof[14] guarantees that there cannot be any fourth order Langevin algorithms consisting only of the operators $e^{\epsilon T}$ and $e^{\epsilon D}$. To derive a fourth order algorithm, the positive coefficient factorization schemes of Section 2 are indispensable. The form of the operators T and D, as given in (32), dictates that one should keep only the commutator $[D,[T,D]]$, which is at most a second order differential operator. Since the velocity field $\mathbf{G}$ is usually given in terms of a potential function $V(\mathbf{x})$,

$$G_i(\mathbf{x}) = -\partial_i V(\mathbf{x}), \tag{41}$$

the double commutator has the form

$$[D,[T,D]] = \partial_i\partial_j f_{i,j} + \partial_i v_i, \tag{42}$$

where

$$\begin{aligned} f_{i,j} &\equiv V_{i,j,k}V_k - 2V_{i,k}V_{j,k} \\ v_i &\equiv \frac{1}{2}\left(2V_{i,j,k}V_{j,k} + V_{i,j}V_{j,k,k} - V_{i,j,k,k}V_j\right). \end{aligned} \tag{43}$$

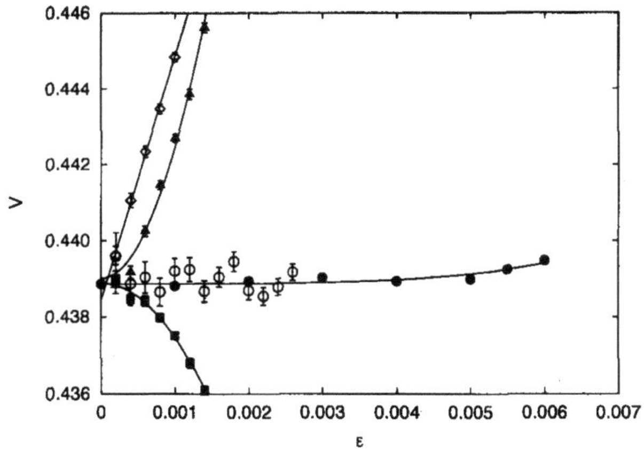

Figure 4. The convergence of fourth order Langevin algorithms for simulating the Brownian dynamics of 121 interacting colloidal particles in two dimensions. The equilibrium potential energy per particle is plotted as a function of the time step size ϵ used. Open diamonds are results using the first order Langevin algorithm. Solid triangles and solid squares denote results of the two second order algorithms. Open circles give results of our fourth order Langevin algorithm using the standard 4th order Runge-Kutta algorithm for determining the particle trajectory. The solid circles give results with improved trajectory determination as discussed in the text.

The indices on V indicate corresponding partial derivatives. Since the action of the operator D requires the solution of the particle's trajectory, we must minimize its occurrence. This dictates that we should only use scheme B to factorize

$$\begin{aligned}\exp\left[\epsilon\,(T+D)\right] &= \exp\left[\frac{\epsilon}{2}\left(1-\frac{1}{\sqrt{3}}\right)T\right]\exp\left(\frac{\epsilon}{2}D\right)\exp\left(\frac{\epsilon}{\sqrt{3}}\tilde{T}\right)\\ &\times\exp\left(\frac{\epsilon}{2}D\right)\exp\left[\frac{\epsilon}{2}\left(1-\frac{1}{\sqrt{3}}\right)T\right]+O(\epsilon^5),\end{aligned} \tag{44}$$

where we have included the double commutator in $\tilde{T}$

$$\tilde{T}=T+\frac{\epsilon^2}{24}(2\sqrt{3}-3)[D,[T,D]]. \tag{45}$$

To obtain a fourth order algorithm, we must simulate this new term

$$\exp\left(\frac{\epsilon}{\sqrt{3}}\tilde{T}\right)=\exp\left[\frac{\epsilon}{\sqrt{3}}T+\frac{\epsilon^3}{24}\left(2-\sqrt{3}\right)\left(\partial_i\partial_j f_{i,j}+\partial_i v_i\right)\right] \tag{46}$$

correctly to 4th order. How this can be done is a technical advance whose detail can be found in Ref. 5. It is suffice to quote that the entire factorization scheme

(44) can be simulated by setting

$$w_i = x_i + \xi_i\sqrt{\frac{\epsilon}{2}\left(1-\frac{1}{\sqrt{3}}\right)}, \qquad y_i = w_i(\epsilon/2) + \xi_i'\sqrt{\frac{\epsilon}{2\sqrt{3}}},$$
$$z_i = y_i - \frac{\epsilon^3}{24}\left(2-\sqrt{3}\right)v_i(\mathbf{y}) + \sqrt{\frac{\epsilon}{2\sqrt{3}}}\left[\delta_{i,j} + \frac{1}{2}\left(\frac{1}{\sqrt{3}}-\frac{1}{2}\right)\epsilon^2 f_{i,j}(\mathbf{y})\right]\xi_j'',$$
$$x_i' = z_i(\epsilon/2) + \xi_i'''\sqrt{\frac{\epsilon}{2}\left(1-\frac{1}{\sqrt{3}}\right)}, \tag{47}$$

where ξ_i to ξ_i''' are four sets of independent Gaussian random numbers with zero mean and unit variance.

To demonstrate the effectiveness of this 4th order Langevin algorithm, we used it to simulate the Brownian dynamics of 121 colloidal particles in two dimensions interacting via a pair wise strongly repulsive Yukawa potential

$$V(r) = \frac{V_0}{r}\exp[-\lambda(r-1)], \tag{48}$$

with $\lambda = 8$. This system has been described and simulated extensively via second order algorithms by Branka and Heyes.[22] In Fig. 4. we show the convergence of the potential energy at one parameter setting as a function of the time step-size used. We have also included results of a linear and two quadratic Langevin algorithms for comparison.

When our fourth order Langevin algorithm is implemented by using the standard fourth order Runge-Kutta algorithm to solve the trajectory equation (37) we obtained results as shown by open circles in Fig. 4. The variance increases abruptly at around $\epsilon = 0.0022$ and the algorithm becomes unstable at larger ϵ's. The problem can be traced to the failure of the Runge-Kutta algorithm in solving for the trajectory equation at large time steps. To compute the trajectory more accurately, we monitor the square of the difference between the results of the fourth order Runge-Kutta and the embedded second order algorithm (40). If this difference exceeds some tolerance, such as 0.01, we recompute the trajectory more accurately by applying our trajectory algorithm twice at half the time step size. This incurs only a very small overhead, even at $\Delta t = 0.004$ the trajectory recomputation required only about an additional 6% of the time. With this improvement, our fourth order Langevin algorithm give results as shown as solid circles in Fig. 4. The step-size dependence of the fourth order algorithm is very flat, and produces the converged results of lower order algorithms at step-sizes nearly 50 times as large.

6 Solving Quantum Many-Body Problems

The basic idea of the Diffusion Monte Carlo (DMC) algorithm is to solve for the ground state of the Hamiltonian H by evolving the imaginary time Schrödinger equation

$$-\frac{\partial}{\partial t}\psi(\mathbf{x},t) = (H-E)\psi(\mathbf{x},t) = \Big[-\frac{1}{2}\nabla^2 + V(\mathbf{x}) - E\Big]\psi(\mathbf{x},t) \tag{49}$$

to large time.[23–25] The constant E is added to stabilize the population of configurations used in Monte Carlo simulations. Here, $\mathbf{x}$ and ∇^2 denote the coordinate and the Laplacian of the N-particle system. In order for the algorithm to be practical, capable of handling rapidly varying potentials, it is essential to implement important sampling as suggested by Kalos.[26] This means that instead of solving for $\psi(\mathbf{x})$, one evolves the product wave function $\rho(\mathbf{x}) = \phi(\mathbf{x})\psi(\mathbf{x})$ according to[24,25]

$$
\begin{aligned}
-\frac{\partial}{\partial t}\rho(\mathbf{x},t) &= \phi(\mathbf{x})H\phi^{-1}(\mathbf{x})\rho(\mathbf{x},t), \\
&= -\frac{1}{2}\nabla^2\rho(\mathbf{x},t) + \nabla_i\Big[G_i(\mathbf{x})\rho(\mathbf{x},t)\Big] + [E_L(\mathbf{x}) - E]\rho(\mathbf{x},t), \qquad (50)\\
&= [T + D + E_L - E]\rho(\mathbf{x},t) = [L + E_L - E]\rho(\mathbf{x},t), \qquad (51)
\end{aligned}
$$

where

$$E_L(\mathbf{x}) = \phi(\mathbf{x})^{-1}H\phi(\mathbf{x}), \qquad (52)$$

is the *local energy* function. The Fokker-Planck operator L appears naturally with the *drift velocity*

$$G_i(\mathbf{x}) = \phi(\mathbf{x})^{-1}\nabla_i\phi(\mathbf{x}) = -\nabla_i S(\mathbf{x}) \qquad (53)$$

defined by the trial ground state wave function $\phi(\mathbf{x}) = \exp[-S(\mathbf{x})]$. (Here, it is conventional to define the operators T, D and L as the negative of those defined in the last section. The negative sign is then canceled everywhere by the negative prefactor $-\epsilon$.)

Once again the formal operator equation

$$\rho(\mathbf{x}, t+\epsilon) = \mathrm{e}^{-\epsilon(L+E_L-E)}\rho(\mathbf{x},t) \qquad (54)$$

can be solved by factorizing the above small time evolution operator. The effect of $\mathrm{e}^{-\epsilon L}$ acting on $\rho(\mathbf{x},t)$ has been described in the last section. The operator $\mathrm{e}^{-\epsilon(E_L-E)}$ evolves $\rho(\mathbf{x},t)$ forward in time according to

$$\rho(\mathbf{x}, t+\epsilon) = \mathrm{e}^{-\epsilon(E_L(\mathbf{x})-E)}\rho(\mathbf{x},t), \qquad (55)$$

which can be simulated by updating the *weight* W_k associated with the configuration $\mathbf{x}_k$ by

$$W_k' = \mathrm{e}^{-\epsilon(E_L(\mathbf{x}_k)-E)}W_k. \qquad (56)$$

E will be chosen to keep the weights near unity. The value that can do this is the exact ground state energy. There are various methods of keeping track of weights, the original and the simplest method[27] is to replicate the configuration $\mathbf{x}_i$ on the average $\mathrm{e}^{-\epsilon[E_L(\mathbf{x}_i)-E]}$ times.

A first order factorization of

$$\mathrm{e}^{-\epsilon(T+D+E_L)} = \mathrm{e}^{-\epsilon E_L}\mathrm{e}^{-\epsilon T}\mathrm{e}^{-\epsilon D} + o(\epsilon^2), \qquad (57)$$

leads to the following first order Diffusion Monte Carlo algorithm

$$x_i' = x_i + \epsilon G_i(\mathbf{x}) + \xi_i\sqrt{\epsilon}, \qquad (58)$$

where we have solved $x_i(\epsilon)$ to first order. The updated position $\mathbf{x}'$ is then weighted (or replicated) by the factor

$$W' = e^{-\epsilon(E_L(\mathbf{x}')-E)}. \tag{59}$$

To decompose $e^{-\epsilon(T+D+E_L)} = e^{-\epsilon(L+E_L)}$ to fourth order, one possibility is to keep the Langevin operator L intact. In this case, the double commutator

$$[E_L,[L,E_L]] = [E_L,[T,E_L]] = (\partial_i E_L)(\partial_i E_L), \tag{60}$$

is the square of the gradient of the local energy, which is a manageable coordinate function. Since the Langevin operator is complicated to simulate, we must choose a fourth order factorization of $e^{-\epsilon(L+E_L)}$ which minimizes the appearance of L. We are thus forced to use factorization B in the form

$$e^{-\epsilon(L+E_L)} = e^{-\frac{1}{6}\epsilon E_L} e^{-\frac{1}{2}\epsilon L} e^{-\frac{2}{3}\epsilon \tilde{E}_L} e^{-\frac{1}{2}\epsilon L} e^{-\frac{1}{6}\epsilon E_L} + o(\epsilon^5), \tag{61}$$

with $\tilde{E}_L$ given by

$$\tilde{E}_L = E_L + \frac{1}{48}\epsilon^2 [E_L,[L,E_L]] = E_L + \frac{1}{48}\epsilon^2 |\nabla E_L|^2. \tag{62}$$

If the local energy $E_L(\mathbf{x})$ is a smooth function, close to the exact ground state, then the double commutator correction will be also be small. If $\mathbf{x}_0$ is the initial configuration, $\mathbf{x}_{1/2}$ is the Langevin evolved configuration a time step $\epsilon/2$ later, and $\mathbf{x}_1$ the Langevin evolved configuration a time step $\epsilon/2$ later still, then the weight of the final configuration $\mathbf{x}_1$ is

$$W_1 = e^{-\epsilon\left[\frac{1}{6}E_L(\mathbf{x}_1)+\frac{2}{3}\tilde{E}_L(\mathbf{x}_{1/2})+\frac{1}{6}E_L(\mathbf{x}_0)-E\right]}. \tag{63}$$

Eq. (61) is our basic 4th order DMC algorithm and will be referred to as DMC4.

To demonstrate the practical effectiveness of DMC4, we use it to solve for the ground state energy of bulk liquid helium described by the many-body Hamiltonian

$$H = \sum_i -\frac{\hbar^2}{2m}\nabla_i^2 + \sum_{i<j} V(r_{ij}). \tag{64}$$

We use the Aziz *et al*[28] potential with $\hbar^2/m = 12.12$ Å^{2}K. Instead of the usual McMillan trial function, we use a trial function of the form,

$$\phi(\mathbf{x}) = \prod_{i<j} \exp\{-\ln(2)\exp[-(r_{ij}-c_0)/d_0]\}. \tag{65}$$

With $c_0 = 2.8$ Å and $d_0 = 0.48$ Å, this trial function gives a slightly better energy of 5.886(5) K/particle. Since the standard calculation details[29] are well known, we will just describe the results as summarized in Fig. 5. Again, the convergence of our 4th order algorithm is clearly quartic. The extrapolated values are -7.114(2) K for our 4th order algorithm and -7.111(2) K for the second order algorithm DMC2a. Both are in agreement with Boronat and Casulleras's[30] second order DMC result of -7.121(10) K. If the Langevin operator L is not kept intact, more complicated, but faster fourth order DMC algorithms are possible. These are discussed in detail in Ref. 6.

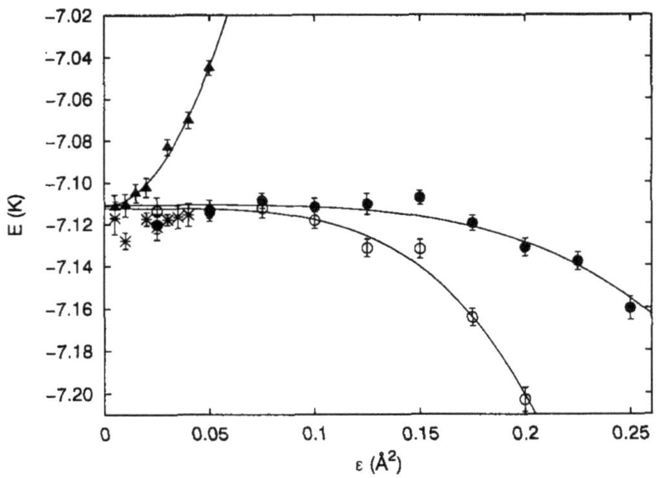

Figure 5. The time step convergence of the ground state energy per particle for bulk liquid helium in a 128 particle simulation. The solid circles are the result of our 4th order algorithm DMC4. For comparison, we also show as triangles, results from a second order DMC algorithm. The asterisks and open circles are alternative fourth order DMC algorithms described in Ref. 6. The lines are least square fits to the data.

7 Conclusions

In this work, I have demonstrated the initial success of a new class of fourth order algorithms for solving diverse many-body problems. They are all characterized by having purely positive time steps at all intermediate time steps. In cases compared, these new algorithms are orders of magnitude better than their negative time step counterparts. For the fourth order Langevin and Diffusion Monte Carlo algorithms, their flat convergence curves have virtually eliminated the systematic step size error. These algorithms should be more extensively tested in large scale quantum and molecular dynamic simulations.

Acknowledgments

The results described in this work were obtained in collaboration with D. Kidwell, C.-R. Chen and H. Forbert. This work was supported, in part, by the National Science Foundation grant PHY-9870054.

References

1. M. Suzuki, in *Computer Simulation Studies in Condensed Matter Physics VIII*, Eds. D. P. Landau, K. K. Mon, and H.-B. Shüttler (Springer, Berlin, 1996).
2. S. A. Chin, *Phys. Lett.* A **226**, 344 (1997).
3. S. A. Chin and D. W. Kidwell, *Phys. Rev.* E **62**, 8746 (2000).
4. S. A. Chin and C.-R. Chen, *J. Chem. Phys.*, in press, physics/0012017.

5. H. A. Forbert and S. A. Chin, *Phys. Rev.* E, in press, nucl-th/0006087
6. H. A. Forbert and S. A. Chin, submitted to *Phys. Rev.* B, nucl-th/0009068.
7. H. Yoshida, *Celest. Mech.* **56**, 27 (1993).
8. M. Creutz and A. Gocksch, *Phys. Rev. Lett.* **63**, 9 (1989).
9. M. Suzuki, *Phys. Lett.* A **146**, 319 (1990); **165**, 387 (1992).
10. H. Yoshida, *Phys. Lett.* A **150**, 262 (1990).
11. E. Forest and R. D. Ruth, *Physica* D **43**, 105 (1990).
12. M. Campostrini and P. Rossi, *Nucl. Phys.* B **329**, 753 (1990).
13. J. Candy and W. Rozmus, *J. Comp. Phys.* **92**, 230 (1991).
14. M. Suzuki, *J. Math. Phys.* **32**, 400 (1991).
15. K. Takahashi and K. Ikeda, *J. Chem. Phys.* **99**, 8680 (1993)
16. H. Risken, *The Fokker-Planck Equation, Methods of Solution and Applications*, 2nd edition, (Springer, New York, 1989).
17. E. Helfand, *Bell Syst. Tech. J.* **58**, 2289 (1979).
18. I. Drummond *et. al.*, *Nucl. Phys.* B **220**, 119 (1983).
19. A. Ukawa and M. Fukugita, *Phys. Rev. Lett.* **55**, 1854 (1985).
20. S. A. Chin, *Nucl. Phys.* B **9**, 498 (1989).
21. S. A. Chin, *Phys. Rev.* A **42**, 6991 (1990).
22. A. Branka and D. Heyes, *Phys. Rev.* E **60**, 2381 (1999).
23. J. B. Anderson, *J. Chem. Phys.* **63**, 1499 (1975).
24. P. Reynolds, D. Ceperley, B. Alder, and W. Lester, *J. Chem. Phys.* **77**, 5593 (1982).
25. J. Moskowitz, K. Schmidt, M. Lee, and M. Kalos, *J. Chem. Phys.* **77**, 349 (1982).
26. M. H. Kalos, D. Levesque, and L. Verlet, *Phys. Rev.* A **9**, 2178 (1974).
27. M. H. Kalos, *Phys. Rev.* **128**, 1791 (1962).
28. R. A. Aziz *et al.*, *J. Chem. Phys.* **70**, 4330 (1979).
29. M. H. Kalos, M. A. Lee, P. A. Withlock, G. V. Chester, *Phys. Rev.* B **24**, 115 (1981).
30. J. Boronat and J. Casulleras, *Phys. Rev.* B **49**, 8920 (1994).

RELATIVISTIC QUANTUM DYNAMICS OF MANY-BODY SYSTEMS

F. COESTER
Physics Division, Argonne National Laboratory, Argonne, IL 60439, USA
E-mail: coester@anl.gov

W. N. POLYZOU
Department of Physics and Astronomy, University of Iowa, Iowa City IA 52242, USA
E-mail: polyzou@uiowa.edu

Relativistic quantum dynamics requires a unitary representation of the Poincaré group on the Hilbert space of states. The Dynamics of many-body systems must satisfy cluster separability requirements. In this paper we formulate an abstract framework of four-dimensional Euclidean Green's functions that can be used to construct relativistic quantum dynamics of N-particle systems consistent with these requirements. This approach should be useful in bridging the gap between few-body dynamics based on phenomenological mass operators and on quantum field theory.

1 Introduction

The superposition principle and the space-time symmetry are realized in relativistic quantum mechanics by a Hilbert space of states with a unitary representation of the inhomogeneous Lorentz group (Poincaré group).[1] Various representations of single particle states are well known.[2] Since the components of the four-momentum p are constrained by the mass, $p^2 = -m^2$, there is a choice of convenient independent momentum variables in the wave function: For instance the components $\vec{p}$ orthogonal to some fixed time-like vector, or the components $p^+, p_\perp$ orthogonal to some null-vector. For particles with spin the functions representing state vectors are functions of spin variables undergoing Wigner rotations. All these representations are equivalent. States of noninteracting particles are represented by tensor products of single particle states.

In quantum mechanics the Hilbert space of state vectors is the same for free and interacting particles. The interactions are implemented by modifications of the Poincaré generators. Following Bakamjian and Thomas[3] this has been done modifying the mass operator, leaving the spin operator independent of interactions. The Poincaré generators obtain as functions of kinematic components of the four momentum operator and canonically conjugate positions, the mass operator, the spin operator. The choice of these kinematic components determines the "form of dynamics".[4] The principal difficulty in this approach is the realization of cluster separability.[5] The properties of any isolated cluster of particles should not depend on the presence or absence of other clusters. The solution involves the recursive construction of appropriate many-body interactions in the many-body mass operators.[6] There are no theorems defining minimal many-body interactions.

Alternatively, single-particle states can be represented by equivalence classes of covariant functions of the four-momentum with the positive semi-definite inner product measure $d\mu(p) := d^4p\delta(p^2+m^2)\theta(p^0)$. For particles with spin the functions representing state vectors are Lorentz covariant functions of spinor indices with a semi-definite inner product measure. Starting from kinematically covariant functions representing multi-particle states, interactions can be introduced modifying the semi-definite inner product measure.[7]

For free fields (free particles) Poincaré generators are obtained by integration of the energy-momentum tensor over a three-dimensional manifold in the Minkowski space.[8] Interactions in the energy-momentum tensor require local commutation relations and infinitely many degrees of freedom. The action of these Poincaré generators on Fock-space vectors produces linear functionals over the Fock space, not vectors in Fock-space. The Hilbert spaces of free and interacting fields are necessarily inequivalent.

Minkowski-Green's functions are defined by vacuum expectation values of time-ordered products of local renormalized Heisenberg fields. Using assumed spectral properties of the intermediate states and the asymptotic properties of the field operators there are simple relations to observable bound-state masses, scattering amplitudes and form factors. The principal problem is to establish a quantitative relation of "approximations" to a theory of local operators.

A central feature in the formulation of relativistic quantum theory is the absence of finite-dimensional unitary representations of the Lorentz group, $O(1,3)$. This is the reason the representation of states by Lorentz covariant functions requires a semidefinite inner product measure. However, the Lorentz group is related, by complexification, to the orthogonal group in four dimensions, $O(4)$ which does have finite dimensional unitary representations. The unitary representations of the real Euclidean group $E(4)$ together with invariant Green's operators are useful in the formulation of Poincaré invariant dynamics. This connection has been exploited extensively in quantum field theory. The equivalence of the Wightman axioms with the Osterwalder-Schrader axioms[9] establishes that Euclidean Green's functions (Schwinger functions) satisfying the Osterwalder-Schrader axioms imply the existence of field operators. In the context of Lagrangean field theory the Schwinger functions are obtained as moments of the functional measure defined by the Lagrangean. In the Euclidean formulation the locality axiom is independent of the axioms which establish unitary Poincaré representations with the appropriate spectral properties. This feature is essential for the formulation of Poincaré covariant dynamics finite many-body systems.

The purpose of this paper is to explore the formulation of relativistic quantum dynamics based on Euclidean invariant Green's functions. Ordinary quantum mechanics provides some heuristic indications. The Hilbert space, $\mathcal{H}$ of states of N particles is independent of the dynamics. It is the same for free and interacting particles. The dynamics is specified by the invariant Casimir Hamiltonian, $\mathsf{h} := H - \vec{P}^2/2M$ or the resolvent operator $G(\mathsf{E}) := 1/(\mathsf{h} - i\mathsf{E})$. Interactions are introduced by invariant additions to the Casimir Hamiltonian $\mathsf{h} = \mathsf{h}_0 + V$, or $G(\mathsf{E})^{-1} = G_0(\mathsf{E})^{-1} + V$. While the approach explored here should be applicable to many qualitatively different physical systems and illuminate relations to quantum field theory, the focus of this exploratory study is limited to systems with a fixed number of particles, for instance nucleons. The abstract framework of the dynamics to be explored is formulated in Section 2. Section 3 provides a realization for single particles. Two-body interactions are formulated in Section 4. The many-body dynamics with the realization of cluster separability is discussed in Sec. 5.

2 The Abstract Framework

2.1 The Auxiliary Hilbert Space.

Since there are no finite dimensional unitary representations of the Lorentz group $SO(1,3)$ we assume the representatives of physical states to be a subset of a larger space, subject to the following assumptions.

H1. Physical states are a linear subset of vectors Ψ in an auxiliary Hilbert space $\mathcal{H}_a$ with the norm $||\Psi||_a^2 = \langle \Psi, \Psi \rangle$.

H2. There is a unitary representation $U(\mathcal{R}, a)$ of the Euclidean group $E(4)$ with $\mathcal{R} \in O(4)$ on the Hilbert space $\mathcal{H}_a$. The self-adjoint generators of $E(4)$ are denoted by P^μ and $\mathsf{J}^{\mu\nu} = -\mathsf{J}^{\nu\mu}$.

H3. There is a self-adjoint, unitary operator Θ, on $\mathcal{H}_a$, which is invariant under the 3-dimensional Euclidean subgroup

$$[\Theta, \mathsf{P}^k] = 0 \,, \qquad [\Theta, \mathsf{J}^{ik}] = 0 \,, \tag{1}$$

and satisfies

$$\Theta \mathsf{P}^0 = -\mathsf{P}^0 \Theta \,, \qquad \Theta \mathsf{J}^{0k} = -\mathsf{J}^{0k} \Theta \,. \tag{2}$$

The two Casimir operators $\vec{j}_\pm^{\,2}$ of $O(4)$ are functions of the generators $\mathsf{J}^{\mu\nu}$,[10]

$$j_\pm^k := \frac{1}{\sqrt{2}} \left(\sum_{\mu<\nu} \epsilon^{0k\mu\nu} \mathsf{J}^{\mu\nu} \pm \mathsf{J}^{0k} \right) . \tag{3}$$

The spectra of the Casimir operators are $\sigma(\vec{j}_\pm^{\,2}) = s_\pm(s_\pm + 1)$ with non-negative integer or half-odd integer values of $s_\pm$.

The operators

$$P^0 := i\mathsf{P}^0 \,, \qquad P^k := \mathsf{P}^k \,, \qquad J^{0k} := i\mathsf{J}^{0k} \,, \qquad J^{ik} := \mathsf{J}^{ik} \tag{4}$$

satisfy the Poincaré Lie algebra, and the spectrum of P^2 is the real line, $-\infty < \sigma(P^2) < \infty$. The operators

$$\exp(iP^0 t) \equiv \exp(-\mathsf{P}^0 t) \qquad \text{and} \qquad \exp(iJ^{0k}\chi) \equiv \exp(-\mathsf{J}^{0k}\chi) \tag{5}$$

are self-adjoint. Together with the unitary representations of the 3-dimensional Euclidean group they define a non-unitary representation of the Poincaré group. The inner product $(\Psi, \Psi) := \langle \Theta\Psi, \Psi \rangle \equiv \langle \Psi, \Theta\Psi \rangle$ defines a pseudo-Hilbert space[11] of the vectors in $\mathcal{H}_a$. The inner product (Ψ, Ψ) of the pseudo-Hilbert space is Poincaré invariant, $(U(\Lambda, a)\Psi, U(\Lambda, a)\Psi) = (\Psi, \Psi)$.

2.2 Green's Operators and the Physical Hilbert Space.

The representation of physical states and the dynamical properties of the system are specified by a Green's operator, G, with the following properties.

G1. The Green's operator G is a bounded normal operator on $\mathcal{H}_a$ with an inverse defined on a dense set.

G2. The Green's operator G commutes with P^μ and $\mathsf{J}^{\mu\nu}$ and hence with P^μ, $J^{\mu\nu}$.

G3. The operator ΘG is Hermitean, $\Theta G = G^\dagger \Theta$.

G4. There is a Poincaré invariant linear manifold $\mathcal{S}$ of vectors $\Psi \in \mathcal{H}_a$ that all $\Psi \in \mathcal{S}$ satisfy the inequalities

$$0 \leq (e^{-i\mathsf{P}^0\tau}\Psi, Ge^{-i\mathsf{P}^0\tau}\Psi) \equiv (\Psi, Ge^{-2P^0\tau}\Psi) \leq (\Psi, G\Psi)\,, \quad \forall \tau \geq 0\,. \tag{6}$$

By assumption G4 the inner product $(\Psi, G\Psi)$ of vectors in this manifold is positive semi-definite. Physical states are represented by equivalence classes of vectors. Two vectors Ψ_a and Ψ_b are equivalent, $\Psi_a \sim \Psi_b$, if and only if

$$||\Psi_a - \Psi_b||^2 := ([\Psi_a - \Psi_b], G[\Psi_a - \Psi_b)]) = 0\,. \tag{7}$$

The physical Hilbert space $\mathcal{H}$ is equipped with a unitary representation of the Poincaré group. Single-particle states Ψ_M of mass M, elementary or composite, satisfy $(P^2 + M^2)\Psi_M \sim 0$.

2.3 Perturbations of Green's Operators

A perturbation $G_0 \to G$ may be defined by

$$G^{-1} := G_0^{-1} + \mathcal{U}\,, \tag{8}$$

where $\mathcal{U}$ is an $E(4)$ invariant, pseudo-Hermitean operator with domain $D(\mathcal{U}) \supset D(G_0^{-1})$ By assumption $\mathcal{U}$ is bounded relative to G_0^{-1},

$$||\mathcal{U}\Psi||_a \leq a||G_0^{-1}\Psi||_a + b||\Psi||_a\,, \tag{9}$$

with $0 \leq a < 1$ and $0 \leq b$. The operators $\mathcal{U}G_0$, $G_0\mathcal{U}$ and are bounded with a bound less than 1. It follows that $G^{-1}G_0$, and G_0G^{-1} are bounded operators with bounded inverses.

3 Realization for Single Particles

The auxiliary Hilbert space $\mathcal{H}_a$ of a spin-zero single particle is realized by square integrable functions $\Psi(\mathsf{x})$ with $\mathsf{x} := \{\mathsf{x}^0, \mathsf{x}^1, \mathsf{x}^2, \mathsf{x}^3\}$ with the inner product

$$\langle \Psi, \Psi \rangle = \int d^4\mathsf{x} |\Psi(\mathsf{x})|^2\,. \tag{10}$$

Schwartz functions $f(\mathsf{x})$ are dense in this Hilbert space. The involution operator Θ is defined by

$$\Theta\Psi(\mathsf{x}) := \Psi(-\mathsf{x}^0, \vec{\mathsf{x}})\,. \tag{11}$$

The self-adjoint generators of the real Euclidean group $E(4)$ are

$$\mathsf{P}^{\mu} := \frac{1}{i}\frac{\partial}{\partial \mathsf{x}^{\mu}} , \qquad \mathsf{J}^{\mu\nu} := \frac{1}{i}\left(\mathsf{x}^{\mu}\frac{\partial}{\partial \mathsf{x}^{\nu}} - \mathsf{x}^{\nu}\frac{\partial}{\partial \mathsf{x}^{\mu}}\right) . \tag{12}$$

The associated Poincaré generators are then defined by Eq. (4). The Green's operator is represented by the Green's function $G(\mathsf{x}-\mathsf{x}')$

$$\begin{aligned} G(\mathsf{x}-\mathsf{x}') &:= \left(\frac{1}{2\pi}\right)^4 \int d^4\mathsf{p}\, \frac{\exp[i\mathsf{p}(\mathsf{x}-\mathsf{x}')]}{\mathsf{p}^2+m^2} \\ &= \left(\frac{1}{2\pi}\right)^3 \int d^3\vec{p}\, \frac{\exp\left(i\vec{p}(\vec{x}-\vec{x}') - \omega(p)|\mathsf{x}'^0-\mathsf{x}^0|\right)}{2\omega(p)} , \end{aligned} \tag{13}$$

where $\omega(p) := \sqrt{m^2+\vec{p}^2}$. Schwartz functions $f(\mathsf{x})$ with support restricted to positive values of x^0 represent a linear manifold of vectors in $\mathcal{H}_a$ which satisfies the the requirement G4. With the support restriction it follows that

$$(f, Gf) = \langle \tilde{f}, \tilde{f}\rangle := \int \frac{d^3p}{2\omega(p)} |\tilde{f}(\vec{p})|^2 = \int d^4\mathsf{x}\, |\tilde{f}(\mathsf{x})|^2 , \tag{14}$$

where

$$\begin{aligned} \tilde{f}(\vec{p}) &:= (2\pi)^{-\frac{3}{2}} \int d^4\mathsf{x} e^{-i\vec{p}\cdot\vec{x}} e^{-\omega(p)\mathsf{x}^0} f(\mathsf{x}) , \\ \tilde{f}(\mathsf{x}) &:= (2\pi)^{-\frac{3}{2}} \int d^3p\, e^{i\vec{p}\cdot\vec{x}-\omega(p)\mathsf{x}^0} \theta(\mathsf{x}^0)\tilde{f}(\vec{p}) . \end{aligned} \tag{15}$$

The Hilbert space $\mathcal{H}$ is constructed by the usual procedure of moding out zero-norm vectors and adding Cauchy sequences. The equivalence classes of Schwartz functions are dense in $\mathcal{H}$. Two functions f_1 and f_2 are equivalent, $f_1 \sim f_2$, if nd only if

$$\|f_1 - f_2\| = 0 . \tag{16}$$

It follows that two functions f_1 and f_2 are in the same equivalence class, $f_1 \sim f_2$, iff $\tilde{f}_1 = \tilde{f}_2$.

Since

$$\theta(\mathsf{x}'^0)\Theta G(\mathsf{x}'-\mathsf{x})\theta(\mathsf{x}^0) = \theta(\mathsf{x}'^0)\theta(\mathsf{x}^0)\left(\frac{1}{2\pi}\right)^3 \int d^3p\, \frac{e^{-\omega(p)(\mathsf{x}'^0+\mathsf{x}^0)} e^{i\vec{p}\cdot(\vec{x}'-\vec{x})}}{2\omega(p)} , \tag{17}$$

it follows that the inner product $(f_a, G\, f_b)$ of functions satisfying the support condition is manifestly Lorentz invariant,

$$(f_b, G\, f_a) = \int \frac{d^3p}{\omega(p)}\, \tilde{f}_b(\vec{p})^* \tilde{f}_a(\vec{p}) = \int d^4p\, \delta(p^2+m^2)\, \theta(p^0)\, f_b(p)^* f_a(p) \tag{18}$$

with

$$f(p) := \int d^4\mathsf{x} \exp\left(-i\vec{p}\cdot\vec{\mathsf{x}} - p^0\mathsf{x}^0\right) f(\mathsf{x}) . \tag{19}$$

The time evolution evolution $f_a(t) := e^{-iP^0t} f_a$ is given explicitly by

$$(f_b, G\, e^{-iP^0t} f_a) = \langle \tilde{f}_b, \tilde{f}_a(t)\rangle , \tag{20}$$

with $\tilde{f}_a(t,p) := e^{-i\omega(p)t}\tilde{f}_a(\vec{p})$.

For a single spin-1/2 particle the Green's function is

$$G(\mathsf{x}-\mathsf{y}) := \frac{1}{(2\pi)^4}\int d^4\mathsf{p}e^{i\mathsf{p}\cdot(\mathsf{x}-\mathsf{y})}\left(\frac{\mathsf{p}\cdot\gamma_e + m}{\mathsf{p}^2+m^2}\right) , \tag{21}$$

where the spinor matrices $\gamma_e := i\beta, \beta\vec{\alpha}$ with $\vec{\alpha} := \gamma_5\vec{\sigma}$ satisfy

$$\tfrac{1}{2}\{\gamma_{e\mu}, \gamma_{e\nu}\} = -\delta_{\mu\nu} . \tag{22}$$

The involution operator Θ must also act on the spinor indices,

$$(\Theta f)(\mathsf{x}) := \beta f(-\mathsf{x}^0, \vec{\mathsf{x}}). \tag{23}$$

As in the case of spin 0 it is easy to verify positivity of the inner product (f, Gf) for Schwartz functions with support restricted to positive values of x^0.

$$(f,Gf) := \int d^4\mathsf{x}d^4\mathsf{y}f^\dagger(\mathsf{x})\Theta G(\mathsf{x}-\mathsf{y})f(\mathsf{y}) = \int d^4\mathsf{x}d^4\mathsf{y}f^\dagger(\mathsf{x})\beta G(-\mathsf{x}^0-\mathsf{y}^0,\vec{\mathsf{x}}-\vec{\mathsf{y}})f(\mathsf{y})$$
$$= \int d^3p\tilde{f}^\dagger(p)\frac{\omega(p)+\vec{\alpha}\cdot\vec{p}+\beta m}{2\omega(p)}\tilde{f}(p) = \int d^4\mathsf{x}\tilde{f}^\dagger(\mathsf{x})\tilde{f}(\mathsf{x}) \geq 0 , \tag{24}$$

where

$$\tilde{f}(\mathsf{x}) := (2\pi)^{-\frac{3}{2}}\int d^3p\; e^{i\vec{p}\cdot\vec{x}-\omega(p)\mathsf{x}^0}\theta(\mathsf{x}^0)\;\frac{\omega(p)+\vec{\alpha}\cdot\vec{p}+\beta m}{2\omega(p)}\;\tilde{f}(\vec{p}) . \tag{25}$$

4 Two-Body Dynamics

The auxiliary Hilbert space $\mathcal{H}_a$ is the tensor product of the single-particle auxiliary Hilbert spaces. The involution operator Θ is the tensor product of single particle involution operators. Schwartz functions $f(\mathsf{x}_1,\mathsf{x}_2) \equiv f(\mathsf{X},\mathsf{x})$ with $\mathsf{X} := \frac{1}{2}(\mathsf{x}_1+\mathsf{x}_2)$ and $\mathsf{x} := \mathsf{x}_1 - \mathsf{x}_2$ are dense in this Hilbert space. For spin 1/2 particles these functions depend on spinor variables as well. The $E(4)$ generators are additive,

$$\mathsf{P}^\mu = \frac{1}{i}\sum_{n=1}^{2}\frac{\partial}{\partial\mathsf{x}_n^\mu}, \qquad \mathsf{J}^{\mu\nu} = \frac{1}{i}\sum_{n=1}^{2}\left(\mathsf{x}_n^\mu\frac{\partial}{\partial\mathsf{x}_n^\nu} - \mathsf{x}_n^\nu\frac{\partial}{\partial\mathsf{x}_n^\mu} + \tfrac{1}{4}[\gamma_{ne}^\mu, \gamma_{ne}^\nu]\right) . \tag{26}$$

In general Green's operators are realized by $E(4)$ invariant tempered distributions $G(\mathsf{x}_1,\mathsf{x}_2;\mathsf{x}_2',\mathsf{x}_1')$. For free particles the Green's function is the product of single particle Green's functions, Interactions are added according to Eq. (8). A simple example of a nucleon-nucleon interaction is of the form

$$\mathcal{U}\Psi(\mathsf{X},\mathsf{x}) = \Big\{\mathsf{V}_S(\mathsf{x}) + \gamma_5^{(1)}\gamma_5^{(2)}\mathsf{V}_P(\mathsf{x}) + \sum_{\mu,\nu}[\gamma_{e\mu}^{(1)},\gamma_{e\nu}^{(1)}][\gamma_{e\mu}^{(2)},\gamma_{e\nu}^{(2)}]\mathsf{V}_T(\mathsf{x})$$
$$+ (\gamma_e^{(1)}\cdot\gamma_e^{(2)})\mathsf{V}_V(\mathsf{x}) + (\gamma_5^{(1)}\gamma_e^{(1)}\cdot\gamma_e^{(2)}\gamma_5^{(2)})\mathsf{V}_A(\mathsf{x})\Big\}\Psi(\mathsf{X},\mathsf{x}) . \tag{27}$$

Scattering wave functions are obtained by the weak time limits

$$(\Psi_b, G\Omega_{a\pm}) = \lim_{t=\pm\infty}(\Psi_b, Ge^{iP^0t}\tilde{\Psi}_a(t)), \tag{28}$$

where $\tilde{\Psi}_a(t) := \tilde{\Psi}_a^{(1)}(t) \times \tilde{\Psi}_a^{(2)}(t)$ is the tensor product of single particle states.[12] The S-matrix in this limit reads

$$S_{ba} = \lim_{t'=+\infty} \lim_{t=-\infty} \left(\tilde{\Psi}_b(t'), G e^{iP^0(t-t')} \tilde{\Psi}_a(t) \right) . \tag{29}$$

5 Many-Body Dynamics

The auxiliary Hilbert space $\mathcal{H}_a$ is the N-fold tensor product of the single-particle auxiliary Hilbert spaces. The involution operator Θ is the outer product of single particle involution operators. Schwartz functions, $f(\mathsf{x}_1, \ldots, \mathsf{x}_N)$, of Npoints and N spinor indices are dense in this Hilbert space. The $E(4)$ generators are additive,

$$\mathsf{P}^\mu = \frac{1}{i} \sum_{n=1}^{N} \frac{\partial}{\partial \mathsf{x}_n^\mu}, \qquad \mathsf{J}^{\mu\nu} = \frac{1}{i} \sum_{n=1}^{N} \left(\mathsf{x}_n^\mu \frac{\partial}{\partial \mathsf{x}_n^\nu} - \mathsf{x}_n^\nu \frac{\partial}{\partial \mathsf{x}_n^\mu} + \tfrac{1}{4}[\gamma_{ne}^\mu, \gamma_{ne}^\nu] \right) . \tag{30}$$

In general Green's operators are realized by $E(4)$ invariant tempered distributions $G_N(\mathsf{x}_1, \ldots \mathsf{x}_N; \mathsf{y}_N, \ldots, \mathsf{y}_1)$. The cluster properties can be conveniently realized using formal annihilation operators $a(\mathsf{x})$, $b(\mathsf{x})$ and creation operators $a^\dagger(\mathsf{x})$, $b^\dagger(\mathsf{x})$ which satisfy the commutation relations[13]

$$\{a(\mathsf{x}), a^\dagger(\mathsf{y})\} = \delta^{(4)}(\mathsf{x} - \mathsf{y}) , \qquad \{b(\mathsf{x}), b^\dagger(\mathsf{y})\} = \delta^{(4)}(\mathsf{x} - \mathsf{y}) ,$$
$$\{a(\mathsf{x}), b(\mathsf{y})\} = \{a(\mathsf{x}), b^\dagger(\mathsf{y})\} = 0 . \tag{31}$$

The Green's function of N free particles is related to the single-particle Green's function $G(\mathsf{x} - \mathsf{y})$ by the expression

$$G_{0N}(\mathsf{x}_1, \ldots, \mathsf{x}_n; \mathsf{y}_n, \ldots, \mathsf{y}_1) = \langle 0 | a(\mathsf{x}_1) \cdots a(\mathsf{x}_N) b(\mathsf{y}_N) \cdots b(\mathsf{y}_1) e^{\mathbf{S}_1} | 0 \rangle , \tag{32}$$

where

$$\mathbf{S}_1 := \int d^4\mathsf{x} \int d^4\mathsf{y} a^\dagger(\mathsf{x}) b^\dagger(\mathsf{y}) G(\mathsf{x} - \mathsf{y}), \tag{33}$$

and $|0\rangle$ is the cyclic vector that is annihilated by $a(\mathsf{x})$, $b(\mathsf{x})$. With the definition

$$\psi(\mathsf{x}) := e^{-\mathbf{S}_1} a(\mathsf{x}) e^{\mathbf{S}_1} = a(\mathsf{x}) - \int d^{(4)}\mathsf{y} G(\mathsf{x} - \mathsf{y}) b^\dagger(\mathsf{y}) ,$$
$$\bar{\psi}(\mathsf{x}) := e^{-\mathbf{S}_1} b(\mathsf{x}) e^{\mathbf{S}_1} = b(\mathsf{x}) + \int d^{(4)}\mathsf{y} a^\dagger(y) G(\mathsf{y} - \mathsf{x}) \tag{34}$$

it follows that

$$G_{0N} = \langle 0 | \psi(\mathsf{x}_1) \cdots \psi(\mathsf{x}_N) \bar{\psi}(\mathsf{y}_N) \cdots \bar{\psi}(\mathsf{y}_1) | 0 \rangle . \tag{35}$$

The cluster structure of the general N-particle Green's functions is realized by the expression[14]

$$G_N(\mathsf{x}_1, \ldots, \mathsf{x}_N; \mathsf{y}_N, \ldots, \mathsf{y}_1) = \langle 0 | a(\mathsf{x}_1) \cdots a(\mathsf{x}_N) b(\mathsf{y}_N) \cdots b(\mathsf{y}_1) \exp\left(\sum_n \mathbf{S}_n \right) | 0 \rangle , \tag{36}$$

if the functions $S_n(\mathrm{x}_1, \ldots, \mathrm{x}_n; \mathrm{y}_n, \ldots, \mathrm{y}_1)$ in

$$\mathbf{S_n} := \frac{1}{n!^2} \int d^4\mathrm{x}_1 \ldots \int d^4\mathrm{y}_1 \ldots \\ \times\ a^\dagger(\mathrm{x}_n) \cdots a^\dagger(\mathrm{x}_1) b^\dagger(\mathrm{y}_1) \cdots b^\dagger(\mathrm{y}_n) S_n(\mathrm{x}_1, \ldots, \mathrm{x}_n; \mathrm{y}_n, \ldots, \mathrm{y}_1) \tag{37}$$

vanish for separation of the points into widely separated clusters. The cluster structure of the the Green's operator is realized imposing this structure on the inverse Green's operator,

$$\mathbf{G}^{-1} = \exp\left[\int d^4\mathrm{x} \int d^4\mathrm{y} a^\dagger(\mathrm{x}) b^\dagger(\mathrm{y}) S^{-1}(\mathrm{x}-\mathrm{y}) + \sum_{n\geq 2} \mathbf{U}_n\right] . \tag{38}$$

The interaction operators $\mathbf{U}_n$ for different n are independent. In particular we may assume $\mathbf{U}_n \equiv 0$ for $n > 2$.

$$\mathbf{U}_2 = \tfrac{1}{4} \int d^4\mathrm{x}_1 \int d^4\mathrm{x}_2 \int d^4\mathrm{y}_1 \int d^4\mathrm{y}_2 a^\dagger(\mathrm{x}_1) a^\dagger(\mathrm{x}_2) b^\dagger(\mathrm{y}_2) b^\dagger(\mathrm{y}_1) U(\mathrm{x}_1, \mathrm{x}_2; \mathrm{y}_2, \mathrm{y}_1) , \tag{39}$$

where $U(\mathrm{x}_1, \mathrm{x}_2; \mathrm{y}_2, \mathrm{y}_1)$ may for instance be given by Eq. (27).

The full Green's function satisfying cluster separability is then given for any N by

$$G(\mathrm{x}_1, \ldots, \mathrm{x}_N; \mathrm{y}_N, \ldots, \mathrm{y}_1) = \langle 0| \psi(\mathrm{x}_1) \cdots \psi(\mathrm{x}_n) \bar{\psi}(\mathrm{y}_n) \cdots \bar{\psi}(\mathrm{y}_1) e^{\mathbf{A}} |0\rangle , \tag{40}$$

with

$$\mathbf{A} := \tfrac{1}{4} \int d^4\mathrm{x}_1 \int d^4\mathrm{x}_2 \int d^4\mathrm{y}_1 \int d^4\mathrm{y}_2 \psi(\mathrm{x}_1) \psi(\mathrm{x}_2) \bar{\psi}^\dagger(\mathrm{y}_2) \bar{\psi}^\dagger(\mathrm{y}_1) U(\mathrm{x}_1, \mathrm{x}_2; \mathrm{y}_2, \mathrm{y}_1) . \tag{41}$$

6 Conclusions

Kinematic Poincaré covariance of state vectors of many-body systems requires a dynamically determined semi-definite inner product measure. An effective realization is based on an auxiliary Hilbert space endowed with a unitary representation of the four-dimensional Euclidean group. A self-adjoint unitary involution operator provides a Poincaré invariant indefinite inner product. The Euclidean invariant Green's operator specifies the Poincaré invariant semi-definite inner product of the subspace of physical states. In this framework two-body Green's operators are sufficient to determine many-body Green's functions satisfying cluster separability.

Acknowledgments

This work was supported in part by the Department of Energy, Nuclear Physics Division, under contracts W-31-109-ENG-38 and DE-FG02-86ER40286.

References

1. E. P. Wigner, *Ann. Math.* **40**, 141 (1939).
2. R. Haag, *Local Quantum Physics* (Springer Verlag (1992) Chap. I.3 ;
   S. Weinberg, *The Quantum Theory of Fields* (Cambr. Univ. Press, 1995) Chap. 2.
3. B. Bakamjian and L. H. Thomas, *Phys. Rev.* **92**, 1300 (1953)
4. P. A. M. Dirac, *Rev. Mod. Phys.* **21**, 392 (1949)
5. L. L. Foldy, *Phys. Rev.* **122**, 275 (1961)
   F. Coester, *Helv. Phys. Acta* **38**, 7 (1965)
6. S. N. Sokolov, *Dokl. Akad. Nauk. S.S.S.R.* **233**, 575 (1977); *Theor. Math. Phys.* **36**, 193 (1978)
   F. Coester and W. N. Polyzou, *Phys. Rev.* D **26**, 1348 (1982)
7. W. N. Polyzou, *Phys. Rev.* D **32**, 995 (1985)
8. J. Schwinger, *Phys. Rev.* **127**, 324 (1964)
9. K. Osterwalder and R. Schrader, *Comm. Math. Phys.* **31**, 83 (1973); *ibid.* **42**, 281 (1975).
10. S. S. Schweber, *Relativistic Quantum Field Theory* (Harper and Row, 1961) p.33.
11. N. N. Bogolubov, A. A. Logunov, A. I. Oksak , L. T. Todorov, *General Principles of Quantum Field Theory* (Kluwer Acad. Pub., 1987) Chapter 10.
12. W. Brenig and R. Haag, *Fort. der Phys.* **7**, 183 (1959) [translation in *Quantum Scattering Theory* Ed. M. Ross (Indiana Univ. Press, 1963) p. 13]
13. F. Coester, *Phys. Rev.* **95**, 1318 (1954); J. G. Valatin, *Proc. Roy. Soc.* **229**, 221 (1955)
14. F. Coester, *Nucl. Phys.* **8**, 421 (1958)

ELASTIC N-BODY TO N-BODY SCATTERING IN THE HYPERSPHERICAL REPRESENTATION

R. YA. KEZERASHVILI

New York City Technical College, The City University of New York, 300 Jay Street, Brooklyn, NY 10201-2983, USA

It is shown that in the hyperspherical representation terms of the three- and four-body scattering functions corresponding to elastic $3 \to 3$ and $4 \to 4$ scattering have $\varrho^{-5/2}$ and ϱ^{-4} asymptotic behavior in the 6- and 9-dimensional configuration space, respectively. Nonlinear first-order system of differential equations and the optical theorem for such processes are given.

1 Introduction

Scattering theory is a subject with a long and complicated history and collision processes have become exceedingly important for physics in the last century. They have proved, in fact, to be the most powerful means to obtain information on the structure of atoms, nuclei, nucleons and elementary particles and on the nature of the interparticle interactions.

Usually, in a scattering experiment an initial state always made up with stable two particles, but it is observed that many final states include more than two particles. Scattering experiments in which three or more particles from accelerator beams are made to collide with each other are not feasible in the near future. In spite of this fact the theoretical study of the three or more particle scattering is important. One of the important applications of $3 \to 3$ scattering amplitude is a calculation of the third virial coefficients. The $3 \to 3$ and $4 \to 4$ scattering amplitudes are of great interest in chemical reactions study. The knowledge of the continuum state wave function of the three and four particles is necessary for solving many problems in nuclear physics. A direct way to find out the continuum state wave function of N particle system is a study of the elastic N-body to N-body scattering.

Great progress has been made in the study of three- and four-body scattering using the technique of Faddeev integral equations and many important results are obtained in the framework of the formal scattering theory.[1-7] The development of the theory of three- and four-particle scattering in configuration space has allowed for the determination of the asymptotic behavior of the three- and four-particle scattering function, both for short-range and Coulomb interactions.[5,7]

The three- and four-particle scattering wave functions were also studied in the framework of the hyperspherical functions method.[8-12] This approach yields to expand the partial waves method developed for the two body scattering to the N-body scattering. The hyperspherical functions method leads to introduction of the three- and four-body scattering phases, and allows one to develop a partial waves method for the $3 \to 3$ and $4 \to 4$ scattering.

2 The Asymptotics of the $N \to N$ Scattering Wave Function in the Hyperspherical Basis

Here we are discussing the asymptotic behavior of the three- and four-particle wave functions describing elastic $3 \to 3$ and $4 \to 4$ scattering, using the hyperspherical function method. Faddeev integral equations for the three-body system, Faddeev-Yakubovsky integral equations for the four-body system, and the derivation of the differential equations in configuration space for components of the three- and four-particle wave function allow to begin the study of the asymptotic form of these functions.

Let us consider a system of N pairwise interacting, non-relativistic spinless particles, participating in an elastic process with N particles before and after collision. We introduce the Jacobi coordinates in the center-of-mass, defined by

$$\vec{x}_i = \frac{1}{\sqrt{l(l+1)}}\left[\sum_{j=1}^{l} m_j\vec{r}_j - l\vec{r}_{l+1}\right], \qquad l = 1, 2, ..., N, \tag{1}$$

where m_j and $\vec{r}_j$ are the particles masses and vector positions. The Jacobi momenta $\vec{p}_j$ are conjugated to $\vec{x}_j$ in the center-of-mass system. The wave function of N unbound particle is expanded in terms of the hyperspherical functions

$$\Psi(\varrho, \kappa) = \varrho^{-\frac{3N-4}{2}} \sum_{K,\lambda} u_K^\lambda(\varrho, \kappa)\Phi_K^\lambda(\Omega_\varrho)[\Phi_K^\lambda(\Omega_\kappa)]^*. \tag{2}$$

In Eq. (2), for the sake of simplicity, we denote by λ the totality of quantum numbers on which the N-body hyperspherical functions depend and the integer K is the global momentum in the $3(N-1)$-dimensional configuration space, which is the analog of angular momentum in case of $N = 2$. The functions $\Phi_K^\lambda(\Omega_\varrho)$ and $\Phi_K^\lambda(\Omega_\kappa)$ are called "hyperspherical functions" and they are the eigenfunctions of the angular part of the $3(N-1)$-dimensional Laplace operator in configuration and momentum space, respectively, with eigenvalue $L_N(L_N+1)$, where $L_N = K + 3(N-2)/2$. In Eq. (2) we introduce the hyperspherical coordinates in $3(N-1)$-dimensional configuration space, given by the hyperradius $\varrho^2 = \sum x_l^2$ and a set of angles Ω_ϱ, which define the direction of the vector $\vec{\varrho}$ in $3(N-1)$-dimensional space. $\vec{\kappa}$ and Ω_κ are the set of conjugated hyperspherical coordinates in momentum space.

Introducing the expansion (2) in the Schrödinger equation for N unbound particles, multiplying by $[\Phi_K^\lambda(\Omega_\varrho)]^*\ \Phi_K^{\lambda'}(\Omega_\kappa)$, and integrating over $d\Omega_\varrho d\Omega_\kappa$ results in a system of coupled differential equations for the hyperradial functions $u_K^{\lambda'}(\varrho, \kappa)$,

$$\frac{d^2u_K^\lambda(\varrho,\kappa)}{d\varrho^2} + \left[\kappa^2 - \frac{L_N(L_N+1)}{\varrho^2}\right] u_K^\lambda(\varrho,\kappa) = \sum_{K'\ \lambda'} V_{KK'\lambda\lambda'}(\varrho)\, u_{K'}^{\lambda'}(\varrho,\kappa), \tag{3}$$

where

$$V_{KK'\lambda\lambda'}(\varrho) = \frac{2}{\hbar^2}\int [\Phi_K^\lambda(\Omega_\varrho)]^* \left(\sum_{i\prec j} V_{ij}\right) \Phi_{K'}^{\lambda'}(\Omega_\kappa) d\Omega_\varrho d\Omega_\kappa \tag{4}$$

is the N-particle effective potential energy defined by the two-particle potentials V_{ij}.

2.1 Elastic 3 → 3 Scattering.

Following Refs. 1, 2, 6 the asymptotic behavior of the wave function describing the processes 3 → 3, in which all three particles are free before the reaction and no two-particles bound states are formed, can be represented in the form

$$\Psi_3(\vec{\varrho} \mid \vec{\kappa}) = \Phi_3(\vec{\varrho} \mid \vec{\kappa}) + \sum_{\alpha=1}^{3} U^s(\vec{x}_{1\alpha}, \vec{x}_{2\alpha} \mid \vec{p}_{1\alpha}, \vec{p}_{2\alpha})$$
$$+ \sum_{\alpha \succ \beta}^{3} U^d(\vec{x}_{1\alpha}, \vec{x}_{2\alpha} \mid \vec{p}_{1\beta}, \vec{p}_{2\beta}) + U_{33}(\vec{\varrho} \mid \vec{\kappa}), \quad (5)$$

where $\alpha = (1,2)3$, $(2,3)1$ and $(3,1)2$, $\Phi_3(\vec{\varrho} \mid \vec{\kappa}) = \exp[i(\vec{x}_1 \cdot \vec{p}_1 + \vec{x}_2 \cdot \vec{p}_2)]$, $U_{33}(\vec{\varrho} \mid \vec{\kappa})$ is a six-dimensional spherical wave and has the asymptotic behavior $\varrho^{-5/2}$. The functions $U^s(\vec{x}_1, \vec{x}_2 \mid \vec{p}_1, \vec{p}_2)$ and $U^d(\vec{x}_1, \vec{x}_2 \mid \vec{p}_1, \vec{p}_2)$ are associated with the single and double scattering processes. They can be expressed explicitly in terms of the two-particle t-matrix as

$$U^s(\vec{\varrho} \mid \vec{\kappa}) = -\int d\vec{p}^{\,\prime}_{1\alpha} \exp[i(\vec{x}_{1\alpha} \cdot \vec{p}^{\,\prime}_{1\alpha} + \vec{x}_{2\alpha} \cdot \vec{p}_{2\alpha})] \frac{t^{\pm}_{\alpha}(\vec{p}^{\,\prime}_{1\alpha}, \vec{p}_{1\alpha})}{p'^2_{1\alpha} - p^2_{1\alpha} \pm i0}, \quad (6)$$

$$U^d(\vec{\varrho} \mid \vec{\kappa}) = \int d\vec{p}^{\,\prime}_{1\alpha} d\vec{p}^{\,\prime}_{2\alpha} \exp[i(\vec{x}_{1\alpha} \cdot \vec{p}^{\,\prime}_{1\alpha} + \vec{x}_{2\alpha} \cdot \vec{p}^{\,\prime}_{2\alpha})]$$
$$\times \frac{t^{\pm}_{\alpha}(\vec{p}^{\,\prime}_{1\alpha}, \vec{k}^{\prime}_{\alpha\beta}; \kappa^2 - p'^2_{2\alpha} \pm i0)}{p'^2_{1\alpha} - (\kappa^2 - p'^2_{2\alpha} \pm i0)} \frac{t^{\pm}_{\alpha}(\vec{k}_{\beta\alpha}, \vec{p}_{\beta})}{k'^2_{\beta\alpha} - p^2_{1\beta} \pm i0}[1 - \zeta(q^2_{\alpha}), \quad (7)$$

where t_α represents the single scattering two-body t-matrix, $\vec{k}^{\prime}_{\alpha\beta}$ and $\vec{k}_{\beta\alpha}$ are the same momenta as defined by Eq. (2.2) of Ref. 2 and $\zeta(p^2_{2\alpha})$ is a cutoff function.

Following Ref. 13, let us now expand the plane wave in six-dimensional space in terms of hyperspherical functions

$$\Phi_3(\vec{\varrho} \mid \vec{\kappa}) = \frac{(2\pi)^2}{(\kappa\varrho)^2} \sum_{K \; l_1 l_2 \; m_1 m_2} i^K J_{K+2}(\varrho\kappa) \Phi^{\lambda}_K(\Omega_\varrho)[\Phi^{\lambda}_K(\Omega_\kappa)]^*, \quad (8)$$

where $J_{K+2}(\varrho\kappa)$ is the regular Bessel function, $\lambda \equiv l_1 l_2 m_1 m_2$ and l_1,l_2,m_1, m_2 are the orbital momenta with their projections associated with the Jacobi coordinates $\vec{x}_1$ and $\vec{x}_2$. $\Phi^{\lambda}_K(\Omega_\varrho)$ and $\Phi^{\lambda}_K(\Omega_\kappa)$ are the hyperspherical functions in the nine-dimensional configuration and momentum spaces, respectively. If we take into account the asymptotic behavior of the Bessel function

$$J_{K+2}(\varrho\kappa) \lim_{\varrho \to \infty} \sqrt{\frac{2}{\pi\kappa}} \varrho^{-1/2} \sin[\kappa\varrho - \frac{1}{2}(2K+5)], \quad (9)$$

substitute the expansion (8) in Eqs. (6) and (7), and expand these functions in terms of hyperspherical functions after multiplication by $[\Phi^{\lambda}_K(\Omega_\varrho)]^* \Phi^{\lambda}_K(\Omega_\kappa)$ and integration over $d\Omega_\varrho d\Omega_\kappa$ we can conclude that single and double scattering terms for the process of 3 → 3 scattering give $\varrho^{-5/2}$ dependence in the asymptotic region.[13]

2.2 Elastic 4 → 4 Scattering.

Following Ref. 14, we consider a system of four non-relativistic particles involved in a 4 → 4 elastic process when there are four free particles before and after the reaction. The wave function $\Psi_4(\vec{\varrho} \mid \vec{\kappa})$ corresponding to four free particles in the initial state and describing the process 4 → 4, is determined by the formula[7]

$$\begin{aligned}\Psi_4(\vec{\varrho} \mid \vec{\kappa}) &= \Phi_4(\vec{\varrho} \mid \vec{\kappa}) + U_2 + U_3 \\ &+ \sum_{\alpha a} \sum_{b \neq a} \sum_{b \neq \gamma} \sum_{\sigma \subset b} U^{ab}_{\alpha\gamma\beta\sigma}(\vec{\varrho} \mid \vec{\kappa}) + U_{44}(\vec{\varrho} \mid \vec{\kappa}),\end{aligned} \tag{10}$$

where $\sigma \subset b$ means that the partition σ is obtained from the partition b by dividing its subsystems into parts. The first term $\Phi_4(\vec{\varrho} \mid \vec{\kappa})$ in Eq. (10) is a plane wave in the nine-dimensional space. This function satisfies the Schrödinger equation, which describes a noninteracting four-particle system. From the scattering theory point of view, this is the incoming wave for the processes 4 → 4. The second and third terms in Eq. (10) have a clear physical meaning. They are expressed in terms of the wave functions of two- and three-particle systems, the coordinate asymptotics of which are well known,[6] and correspond to two-body collisions of the particles within the subsystems of the partitions of the four particles into two groups. Those terms dominate asymptotically and decrease more slowly than ϱ^{-4}. We shall clearly assume that the two- and three-particle subsystems do not have bound states. This means that the corresponding two- and three-particle subsystems operators have only a continuous spectrum. The last term U_{44} in Eq. (10) is proportional to the nine-dimensional spherical wave and asymptotically behaves as ϱ^{-4}. As has been shown in Ref. 7, in those sections of the configuration space, where the formally constricted scattering amplitude becomes infinite, the asymptotic behavior of the wave function represented by the fourth term can be described by Fresnel integrals and the Fresnel type double integral. Below we will show that the fourth term represented in Eq. (10) in the hyperspherical representation asymptotically goes as ϱ^{-4} in the nine-dimensional configuration space of the center-of-mass system.

Following Ref. 7, let us split the fourth term of Eq. (10) into four terms,

$$\begin{aligned}U^{ab}_{\alpha\gamma\beta\sigma}(\vec{\varrho} \mid \vec{\kappa}) &= \delta_{\alpha\gamma}\delta_{\beta\sigma}M_{\alpha\beta\sigma}(\vec{\varrho} \mid \vec{\kappa}) + \delta_{\alpha\gamma}M^{b}_{\alpha\beta\sigma}(\vec{\varrho} \mid \vec{\kappa}) \\ &+ \delta_{\beta\sigma}M^{a}_{\alpha\gamma\beta\sigma}(\vec{\varrho} \mid \vec{\kappa}) + M^{ab}_{\alpha\gamma\beta\sigma}(\vec{\varrho} \mid \vec{\kappa}),\end{aligned} \tag{11}$$

where, for example,

$$M_{\alpha\beta\sigma}(\vec{\varrho} \mid \vec{\kappa}) = -\int d\vec{\kappa}' \, \Phi_4(\vec{\varrho} \mid \vec{\kappa}) \frac{T_{\alpha\beta\sigma}(\vec{\kappa}', \vec{\kappa}; \kappa^2 + i0)}{\kappa'^2 - \kappa^2 - i0}, \tag{12}$$

and the kernels $T_{\alpha\beta\sigma}$ can be expressed in terms of the two-particle scattering matrix. All other terms $M^{b}_{\alpha\beta\sigma}(\vec{\varrho} \mid \vec{\kappa})$, $M^{a}_{\alpha\gamma\beta\sigma}(\vec{\varrho} \mid \vec{\kappa})$ and $M^{ab}_{\alpha\gamma\beta\sigma}(\vec{\varrho} \mid \vec{\kappa})$ in Eq. (11) have the same structure as Eq. (12) and their kernels $T^{b}_{\alpha\beta\sigma}$, $T^{a}_{\alpha\gamma\beta\sigma}$, and $T^{ab}_{\alpha\gamma\beta\sigma}$ of the operators on the right-hand side can be expressed explicitly in terms of the two-particle scattering matrix. The properties of these kernels were investigated in Ref. 7. Here we use the same notations as in Ref. 7 and we do not write the corresponding expressions, in order to reduce the length of the equations.

To investigate the asymptotic behavior of these terms let us expand the function $M_{\alpha\beta\sigma}(\vec{\varrho} \mid \vec{\kappa})$ and the plane wave in nine-dimensional space in terms of the hyperspherical functions[14]

$$M_{\alpha\beta\sigma}(\vec{\varrho} \mid \vec{\kappa}) = \sum_{\substack{\mu\mu' K' K\, l_1 l_2 l_3 \\ m_1 m_2 m_3}} M^{\lambda}_{\mu\mu' KK'} \Phi^{\lambda}_{\mu K}(\Omega_\varrho)[\Phi^{\lambda}_{\mu K}(\Omega_\kappa)]^*, \tag{13}$$

$$\Phi_4(\vec{\varrho} \mid \vec{\kappa}) = \frac{(2\pi)^{9/2}}{(\kappa\varrho)^{7/2}} \sum_{K\;\; l_1 l_2 l_3\; m_1 m_2 m_3} i^K J_{K+7/2}(\kappa\varrho)\Phi^{\lambda}_{K}(\Omega_\varrho)[\Phi^{\lambda}_{K}(\Omega_\kappa)]^*, \tag{14}$$

where $J_{\mu+7/2}(\kappa\varrho)$ is the regular Bessel function, $\lambda \equiv l_1 l_2 l_3 m_1 m_2 m_3$ and $l_1 m_1$, $l_2 m_2$ and $l_3 m_3$ are the orbital momenta with their projections associated to the Jacobi coordinates x_1, x_2, x_3, and $\Phi_K^{l_1 l_2 l_3 m_1 m_2 m_3}(\Omega)$ are the hyperspherical functions, which are the eigenfunctions of the angular part of the nine-dimensional Laplace operator in configuration and momentum spaces, respectively.

By introducing Eqs. (13) and (14) in Eq. (12), multiplying with $[\Phi^{\lambda}_{\mu K}(\Omega_\varrho)]^*$ $\Phi^{\lambda}_{\mu K}(\Omega_\kappa)$, integrating over $d\Omega_\varrho d\Omega_\kappa$, and taking into account the asymptotic behavior of the Bessel function $J_{K+7/2}(\kappa\varrho)$, we obtain that Eq. (12) has a ϱ^{-4} dependence in the asymptotic region. Using the same procedure for the $M^b_{\alpha\beta\sigma}$, $M^a_{\alpha\gamma\beta\sigma}$ and $M^{ab}_{\alpha\gamma\beta\sigma}$ we can finally conclude that in the hyperspherical representation all terms of the four-particle wave function asymptotically go as ϱ^{-4} in the center-of -mass system of the nine-dimensional configuration space.[14]

3 Equations for Three- and Four-Body in the Continuum Spectrum.

The well-known method of phase functions developed by Calogero[15] for the two-body problem has been generalized to three- and four-body systems in Refs. 9 and 11. Let us write Eq. (2) in the matrix form

$$\frac{d^2 u(\varrho,\kappa)}{d\varrho^2} + \left[\kappa^2 - \frac{L^2}{\varrho^2}\right] u(\varrho,\kappa) = V(\varrho)\, u(\varrho,\kappa), \tag{15}$$

where L^2 is a diagonal matrix with elements $(K+3/2)(K+5/2)$ in the case of three particles, while in the case of four particles the elements are $(K+3)(K+4)$. In Eq. (15) $u(\varrho,\kappa)$ is a column matrix consisting of the hyperradial functions $u^{\alpha}_{K}(\varrho,\kappa)$, while $V(\varrho)$ is a square matrix composed of the effective potential energies obtained by means of averaging the nucleon-nucleon interaction over the three-body or four-body hyperspherical functions, respectively.

Let us solve the matrix Eq. (15) in the following form

$$u(\varrho,\kappa) = \{J(\varrho\kappa)U^{-1}(n;\varrho) - N(\varrho\kappa)U^{-1}(n;\varrho)T(\varrho)\}A(\varrho), \tag{16}$$

where $J(\varrho\kappa)$ and $N(\varrho\kappa)$ are diagonal matrices composed of the Bessel ($J_\nu(\varrho\kappa)$, $J_{\nu+2}(\varrho\kappa)$, $J_{\nu+4}(\varrho\kappa), ...$) and Neumann ($N_\nu(\varrho\kappa)$, $N_{\nu+2}(\varrho\kappa)$, $N_{\nu+4}(\varrho\kappa), ...$) functions ($\nu = K+2$ in case of three particles and $\nu = K + 7/5$ in case of four particles), and $A(\varrho)$ is the amplitude function representing a column matrix. In Eq. (16)

the inverse matrix $U^{-1}(n;\varrho)$ is obtained from the orthogonal matrix of nth order composed according to the following recurrent procedure:

$$U(n;\varrho) = \begin{pmatrix} 1 & 0 \\ 0 & U(n-1;\varrho) \end{pmatrix} O^T(n;\varrho), \tag{17}$$

where $O^T(n;\varrho)$ is the matrix obtained by transposing the matrix $O(n)$ of Ref. 16, with elements $\sin\varepsilon_\nu(\varrho)$ and $\cos\varepsilon_\nu(\varrho)$. $T(\varrho)$ is a diagonal matrix with elements $\tan\delta_\nu(\varrho)$, $\tan\delta_{\nu+2}(\varrho)$, $\tan\delta_\nu(\varrho)$, ... We shall call the functions $\delta_\nu(\varrho)$ and $\varepsilon_\nu(\varrho)$ introduced in the matrices $T(\varrho)$ and $U(n;\varrho)$ the phase functions and mixing functions, respectively.

We impose the following condition on the derivative,

$$\frac{du(\varrho,\kappa)}{d\varrho} = \frac{dJ(\varrho\kappa)}{d\varrho}U^{-1}(n;\varrho) - \frac{dN(\varrho\kappa)}{d\varrho}U^{-1}(n;\varrho)T(\varrho)\}A(\varrho). \tag{18}$$

Substituting Eq. (16) into Eq. (15) and taking into account the relation (18), after several simple transformations we obtain the following equations for the phase functions matrix $T(\varrho)$ and amplitude function $A(\varrho)$:

$$\begin{aligned} \frac{dT(\varrho)}{d\varrho} &= -T(\varrho)\frac{dU^{-1}(n;\varrho)}{d\varrho}U^{-1}(n;\varrho) + \frac{dU^{-1}(n;\varrho)}{d\varrho}U^{-1}(n;\varrho)T(\varrho) \\ &\quad - \frac{\pi\varrho}{2}\{U(n;\varrho)J(\varrho\kappa) - T(\varrho)U(n;\varrho)N(\varrho\kappa)\} \\ &\quad \times V(\varrho)\{J(\varrho\kappa)U^{-1}(n;\varrho) - N(\varrho\kappa)U^{-1}(n;\varrho)T(\varrho)\}, \end{aligned} \tag{19}$$

$$\begin{aligned} A(\varrho) &= U(n;\varrho)\exp\{-\frac{1}{2}\pi\int dr r N(r\kappa)V(r)[J(r\kappa)U^{-1}(n;r) \\ &\quad -N(r\kappa)U^{-1}(n;r)T(r)]U(n;r)\}A_0, \end{aligned} \tag{20}$$

where A_0 is an amplitude column matrix determined by the choice of the boundary conditions. The nonlinear first-order differential matrix, Eq. (19), makes it possible to determine the phase functions $\delta_\nu(\varrho)$ and mixing functions $\varepsilon_\nu(\varrho)$. Having solved Eqs. (19) and (20) and using Eqs. (16) and (2) we can find the three- or four-particle wave functions in continuum spectrum using the arbitrary harmonics number approximation. It is easy to see that three- or four-body functions obtained asymptotically at $\varrho \to \infty$ behave as $\varrho^{-5/2}$ and ϱ^{-4} in the six- and nine-dimensional configuration space, respectively. So these functions will have the same asymptotic behavior as we discussed in Sec. 2. The phase functions $\delta_\nu(\varrho)$ and mixing functions $\varepsilon_\nu(\varrho)$ have a clear physical meaning. In Ref. 9 have shown that the asymptotic values of the phase functions and mixing functions as $\varrho \to \infty$ directly give the scattering phases and mixing parameters for the elastic $3 \to 3$ and $4 \to 4$ scattering

4 Optical Theorem

For a two-body elastic scattering there is a relationship between a total cross section and an imaginary part of a forward scattering amplitude. In Ref. 8 the corresponding relationship for the total cross and the imaginary part of the forward scattering amplitude of elastic $3 \to 3$ scattering was obtained. The total cross section for the elastic $3 \to 3$ forward scattering is

$$\sigma = \sqrt{\frac{2}{\pi}} \frac{(4\pi)^2}{\kappa^{5/2}} \sum_{l_1 l_2 L} \frac{1}{\langle l_1 l_2 00 \mid 0L \rangle^2 (2l_1+1)(2l_2+1)} \times \mathrm{Im} \int_0^{\pi/2} F^{l_1 l_2 L}_{\vec{p_1}=\vec{p_1}';\vec{p_2}=\vec{p_2}'}(\Omega_\kappa, \Omega_{\kappa'}) \sin^2\beta \cos^2\beta d\beta. \quad (21)$$

The formula (21) connects the imaginary part of the partial amplitude for the forward $3 \to 3$ elastic scattering, $F^{l_1 l_2 L}_{\vec{p_1}=\vec{p_1}';\vec{p_2}=\vec{p_2}'}(\Omega_\kappa, \Omega_{\kappa'})$, with the total cross section. The three-body forward scattering means $\vec{p_1} = \vec{p_1}'$ and $\vec{p_2} = \vec{p_2}'$. The relation (21) is much more complex than the corresponding relation for the two-body scattering case. In the latter case forward scattering denotes $k = k'$ (k and k' are the relative momentum of the two particles before and after scattering, respectively) and the forward scattering amplitude depends on the total energy in the c.m. frame only. In the three-body elastic forward scattering case we require the equality of the two pairs of Jacobi vectors $\vec{p_1} = \vec{p_1}'$ and $\vec{p_2} = \vec{p_2}'$. Therefore the amplitude depends not only on the total energy, but also on the angle β which gives the distribution of the energy between the subsystems (1,2) and (3). This fact gives rise to the integration over the β angle in Eq. (21). Please note that Eq. (21) can easily be reformulated for elastic $4 \to 4$ scattering.

References

1. J. Nuttal, *J. Math. Phys.* **12**, 1896 (1971).
2. S. P. Merkuriev, *Theor. Math. Phys.* (USSR) **8**, 798 (1971).
3. R. G. Newton, *Ann. Phys. (NY)* **74**, 324 (1972).
4. S. P. Merkuriev, *Nucl. Phys.* A **233**, 395 (1974).
5. S. P. Merkuriev, *Ann. Phys. (NY)* **130**, 395 (1980).
6. S. P. Merkuriev and L. D. Faddeev, *Quantum Theory of Scattering of Few-Particle Systems* (Nauka, Moscow, 1985) [in Russian].
7. S. L. Yakovlev, *Theor. Math. Phys.* (USSR) 56, 673 (1983).
8. R. I. Jibuti and R. Ya. Kezerashvili,*Czech. J. Phys.* B **30**, 1090 (1980).
9. R. Ya. Kezerashvili, *Yad. Fiz.* **38**, 491 (1983) [*Sov. J. Nucl. Phys.* **38**, 293 (1983)].
10. M. Fabre de la Ripelle, *Ann. Phys. (NY)* **147**, 281 (1983).
11. R. I. Jibuti and R. Ya. Kezerashvili, *Nucl. Phys.* A **437**, 687 (1985).
12. M. Fabre de la Ripelle, H. Fiedeldey and S. A. Sofianos, *Phys. Rev.* C **38**, 449 (1988).
13. R. Ya. Kezerashvili and S. Rosati, *Phys. Lett.* B **318**, 23 (1993).
14. R. Ya. Kezerashvili, *Phys. Lett.* B **334**, 263 (1994).
15. F. Calogero, *Variable phase approach to potential scattering* (New York, Academic Press,1967).
16. B. V. Martemyanov, *Yad. Fiz.* **30**, 1364 (1979).

A GENERIC WAY TO LOOK AT MANY-BODY THEORY

E. KROTSCHECK
Institut für Theoretische Physik, Johannes Kepler Universität, A4040 Linz, Austria
E-mail: Eckhard.Krotscheck@jk.uni-linz.ac.at

Based on ideas of density functional theory, we formulate a "generic" many-body theory. It is shown that the ground state configuration of a many-body system can be derived from a variational principle for the energy as a functional of the *pair distribution function,* and that such a variational principle is consistent with the Feynman-Hellman theorem. The functional is then constructed by postulating that the pair distribution function satisfies the two fundamental equations of many-body theory, namely the Bethe-Goldstone equation and the RPA equation, in a self-consistent manner. The generating energy functional of the theory is identical to the energy functional derived within the optimized HNC theory.

1 Introduction

The past three decades have seen a tremendous improvement of our *understanding* of strongly interacting many-particle systems, partly due to an equally impressive improvement of the power of the *techniques* available to describe them. These include the variational Jastrow-Feenberg method (most successfully applied in conjunction with the optimization of correlations), coupled-cluster theory, parquet theory, and simulation methods. *Part* of the present high quality of many-body theory is due to the variety of methods that have been developed with an uncompromising commitment to high technical standards. However, the sometimes complicated formulations have the potential of hiding the essentials, it is therefore worthwhile to occasionally step back and try to take the "view from the top". This is the purpose of this contribution which will, starting from the Jastrow-Feenberg variational theory, try to forget the origin of the equations and to examine the questions how general these relationships between the physical variables are.[1] Although many of the statements to be made are independent of the statistics, we will focus on bosons because, in this case, the relevant manipulations can be carried out in closed form.

2 Jastrow-Feenberg Theory

One of the most successful methods for studying strongly interacting quantum many-body systems is the Jastrow-Feenberg theory.[2] Input to the theory is the microscopic Hamiltonian

$$H = \sum_i \left[-\frac{\hbar^2}{2m} \nabla_i^2 + U_{\rm ext}(\mathbf{r}_i) \right] + \sum_{i<j} V(|\mathbf{r}_i - \mathbf{r}_j|) \ , \tag{1}$$

where $V(|\mathbf{r}_i - \mathbf{r}_j|)$ is a local interaction, and $U_{\rm ext}(\mathbf{r})$ is a possible potential describing the interaction with the environment. The ground state wave function is written in the Jastrow-Feenberg form

$$\Psi_0(\mathbf{r}_1, \ldots, \mathbf{r}_N) = \exp \frac{1}{2} \left[\sum_i u_1(\mathbf{r}_i) + \sum_{i<j} u_2(\mathbf{r}_i, \mathbf{r}_j) + \sum_{i<j<k} u_3(\mathbf{r}_i, \mathbf{r}_j, \mathbf{r}_k) + \ldots \right] . \tag{2}$$

Approximations are defined by truncating the sequence of n-body *correlation functions* $u_n(\mathbf{r}_1, \ldots, \mathbf{r}_n)$. These are, in turn, determined by functional minimization of the energy expectation value.

Technically, the calculation is carried out by first establishing relationships between the correlation functions $u_n(\mathbf{r}_1, \ldots, \mathbf{r}_n)$ and the one- and two-body densities or distribution functions $\rho_1(\mathbf{r})$ and $g(\mathbf{r}, \mathbf{r}')$. These are provided by the Born-Green-Yvon and the hypernetted chain equations;[3] at the end one obtains an energy functional that is expressed in terms of physical observables, but contains no reference any more to the correlation functions.

$$E = E[U_{\text{ext}}(\mathbf{r}), v(r), \rho(\mathbf{r}), g(\mathbf{r}, \mathbf{r}')] \tag{3}$$

The densities are then determined by the two variational principles

$$\frac{\delta E}{\delta \rho(\mathbf{r})} = 0\,, \qquad \frac{\delta E}{\delta g(\mathbf{r}, \mathbf{r}')} = 0\,. \tag{4}$$

The procedure is reminiscent of density functional theory, which determines the one-body density by variation of an energy functional, which is a functional of the density alone. It is therefore logical to ask two questions:

- Is there a generalization of the Hohenberg-Kohn theorem to pair-correlation functions, and, if so,
- Is there a practical way to construct a "pair-density functional"?

One might argue that density-functional theory is complete in the sense that there is no need to introduce another variable in the theory. This is in principle true, and, in fact, the introduction of the second variational function $g(\mathbf{r}, \mathbf{r}')$ does not compromise this statement: One can in principle think of solving, for *fixed* one-body density, the second equation (4)

$$\frac{\delta E}{\delta g(\mathbf{r}, \mathbf{r}')} = 0 \qquad \Longrightarrow \qquad g(\mathbf{r}, \mathbf{r}') = g[\rho(\mathbf{r})](\mathbf{r}, \mathbf{r}')\,. \tag{5}$$

Then, the pair distribution function is uniquely determined by the one-body density, and, upon inserting the solution into the energy functional (3) we obtain a functional of the one-body density alone. The *advantage* of introducing a second variational function is that much is known about the pair distribution function, and this knowledge can be built into the systematic construction of an energy functional. Restricting the attention to the one-body density alone sacrifices this knowledge, among others, no statement about the *homogeneous* limit can be made.

3 Pair-density functional theory and the Feynman-Hellman theorem

The first of the two above questions is almost trivial to answer: yes, there is a two-body version of the Hohenberg-Kohn[4,5] theorem. For simplicity, we formulate the theorem for the homogeneous system. Proceeding by analogy with the line of argument that led to the Hohenberg-Kohn theorem for the one-body density, two assertions can be made about the *exact* ground state:

1. The total energy of the system can be written as

$$E = T + V\,, \tag{6}$$

where the potential energy is

$$V = \frac{\rho}{2}\int d^3r v(r) g(r)\,. \tag{7}$$

2. The kinetic energy T, whose form is yet unspecified, depends only on $g(r)$ and not on $v(r)$, and

3. The total energy has a minimum equal to the ground-state energy at the physical ground-state distribution function, which is equivalent to saying that the ground-state distribution function can be obtained from the variational principles (4).

The proof parallels exactly the proof of the original Hohenberg-Kohn theorem and does not need to be repeated here. Thus, the *existence* of a variational principle for the pair distribution function is much more general than the Jastrow-Feenberg theory would naively suggest.

The existence theorem gives no information on the mathematical structure of the pair-density functional (3); let us first discuss how such a functional may be constructed *in principle.* For that purpose, assume that we have an algorithm for calculating $g(r)$ from a microscopic two-body interaction $\lambda v(r)$ at any value of the parameter λ, $0 < \lambda < 1$. Let the resulting pair distribution function be

$$g_\lambda(r) = g[\lambda v(r)](r), \qquad g(r) \equiv g_{\lambda=1}(r)\,. \tag{8}$$

The Feynman-Hellman theorem states that the ground-state energy can be evaluated by coupling-constant integration of the potential energy alone, *i.e.* from

$$\frac{E}{N} = \frac{E_0}{N} + \frac{\rho}{2}\int d^3r v(r)\int_0^1 d\lambda g_\lambda(r)\,, \tag{9}$$

where E_0 is the energy of the non-interacting system. Since the two-body version of the Hohenberg-Kohn theorem tells us that the relationship between $g(r)$ and $v(r)$ is unique, we can also think of $g_\lambda(r)$ as a functional of $g(r)$ and λ. That way, we can assert the first statement of the above theorem.

Replacing now $v(r)$ by $\lambda v(r)$ in Eq. (6) and differentiating with respect to λ, we obtain

$$\frac{d(E/N)}{d\lambda} = \frac{\rho}{2}\int d^3r v(r) g_\lambda(r) + \int d^3r \frac{\delta(E/N)}{\delta g_\lambda(r)}\frac{dg_\lambda(r)}{d\lambda}\,. \tag{10}$$

The second term on the right-hand side is zero due to the minimum principle (4). Integration with respect to λ establishes the desired property (9). The above derivation also shows that Eq. (9) holds not only for the exact ground state, but also for *any approximate* energy functional, as long as the pair distribution function is obtained by minimizing this approximate energy functional. It also proves the reverse statement: *The result of the coupling constant integration is the desired energy functional.*

4 The self-consistency loop

A number of basic definitions and theorems allow us to establish a self-consistency loop for the energy of a many-body system, in which the importance of properties of the pair-correlation function becomes obvious. Let us start by assuming that we have some kind of energy functional (3). Adding a small, harmonic time-dependent external field $\delta U_{\rm ext}(\mathbf{r};\omega)$ to the ground state Hamiltonian and observing the density-response of $\delta\rho(\mathbf{r};\omega)$ the system defines the density-density response function

$$\delta\rho(\mathbf{r};\omega) = \int d^3r' \chi(\mathbf{r},\mathbf{r}';\omega)\delta U_{\rm ext}(\mathbf{r}';\omega)\,. \tag{11}$$

In the translationally invariant and isotropic system, it is convenient to express the above relationship in momentum space,

$$\delta\rho(k;\omega) = \chi(k;\omega)\delta U_{\rm ext}(k;\omega)\,. \tag{12}$$

The density-density response function gives information on the infinitesimal excitations of the system. Note that a physically unstable system (like an unphysical equation of state with negative compressibility) does not preclude the calculation of $\chi(k;\omega)$, but by studying the poles of $\chi(k;\omega)$ we can obtain the energetics and strength of excitations as well as determine whether any mode is unstable, which would lead us to conclude that something is wrong with our energy functional.

From the density-density response function, we can calculate the static structure function

$$S(k) = 1 + \rho\int d^3r e^{i\mathbf{k}\cdot\mathbf{r}}\left[g(r) - 1\right] \tag{13}$$

by frequency integration

$$S(k) = -\int_0^\infty \frac{d\omega}{\pi}\Im m\chi(k,\omega)\,. \tag{14}$$

Carrying out this step requires that all excitations are real, in other words by looking at $S(k)$ we can determine whether the original energy functional is physically meaningful.

In the next step, we have to distinguish between soft-core and hard-core interactions, where we mean with "hard-core" an interaction that has no Fourier transform. For a "soft-core" interaction, we can directly apply the Feynman-Hellman theorem in momentum space

$$E = E_0 + \frac{1}{2}\tilde{v}(0) + \frac{1}{2\rho}\int d^3k\tilde{v}(k)\int_0^1 d\lambda\left[S_\lambda(k) - 1\right] \tag{15}$$

and arrive at the ground state energy. While the step looks innocuous, it is not without traps because $\tilde{v}(k)\left[S(k) - 1\right]$ must fall off sufficiently rapidly such that the momentum integration converges. This is *not* the case for two famous approximations: The Gross-Pitaevskii approximation for weakly interacting bosons, and the "time-dependent local density approximation" used in the density functional theory of excitations. Strictly speaking, these approximations are self-contradictory in the worst sense: they lead to an *infinite* energy upon completion of the loop.

In Eq. (15), we have introduced, as usual, the "tilde" notation for the dimensionless Fourier-transform

$$\tilde{f}(k) = \rho \int d^3 r e^{i\mathbf{k}\cdot\mathbf{r}} f(r) \,. \tag{16}$$

If the bare interaction has a hard core, one must first calculate the pair distribution function $g(r)$ by inverting Eq. (13). Again, the step looks innocuous, except when one looks at popular approximations: More often than not, the pair distribution function does *not* vanish within the repulsive core of the potential, making the integral either very large, or infinite.

It is known that the self-consistency loop described above will *never* return to exactly the same energy as the one we started out with, *except* for an exact evaluation of all ingredients.[6] This exact evaluation is generally not possible, hence one must live with an inconsistency. One can turn this into an advantage by using the value if this inconsistency as a measure for the achieved accuracy of the theory.

The concept of "pair density functionals" bypasses a part of the loop in the sense that it determines the pair distribution function and the structure function directly by variation of the energy functional, and avoids the second-order variation needed for calculating the density-density response function. That way, the full self-consistency of the theory can be achieved.

A schematic flowchart of the self-consistency loop described here is sown in Fig. 1. The break in the "conventional" loop indicates the fact that full self-consistency can never be achieved; it could be placed anywhere on the loop.

5 Construction of a functional

We have above simply outlined a sequence of arguments based on theorems and concepts that are in the average 50 years old. It is therefore natural to ask whether the theory can be made to work in practice. Let us now restrict our attention to Bose systems because, in this case, most of the manipulations can be carried out analytically.

For the response-function, we postulate the simple random-phase approximation

$$\chi(k,\omega) = \frac{\chi_0(k,\omega)}{1 - 2\tilde{V}_{\mathrm{p-h}}(k)\chi_0(k,\omega)}\,, \qquad \chi_0(k,\omega) = \frac{2t(k)}{\omega^2 - t^2(k)}\,, \tag{17}$$

where $t(k) = \hbar^2 k^2/2m$ is the kinetic energy of a free particle. The energy integration (14) leads to the familiar Bogoliubov equation

$$S(k) = \frac{1}{\sqrt{1 + \dfrac{4m\tilde{V}_{\mathrm{p-h}}(k)}{\hbar^2 k^2}}}\,, \tag{18}$$

except that the bare interaction is replaced by a static, effective interaction. Note that we have not made any statements on the origin of the "particle-hole" interaction $\tilde{V}_{\mathrm{p-h}}(k)$; hence Eq. (18) may equally well be taken as a definition for $\tilde{V}_{\mathrm{p-h}}(k)$.

A second piece of information comes from the fact that the *very short ranged* structure of the pair distribution function should be determined by a two-body

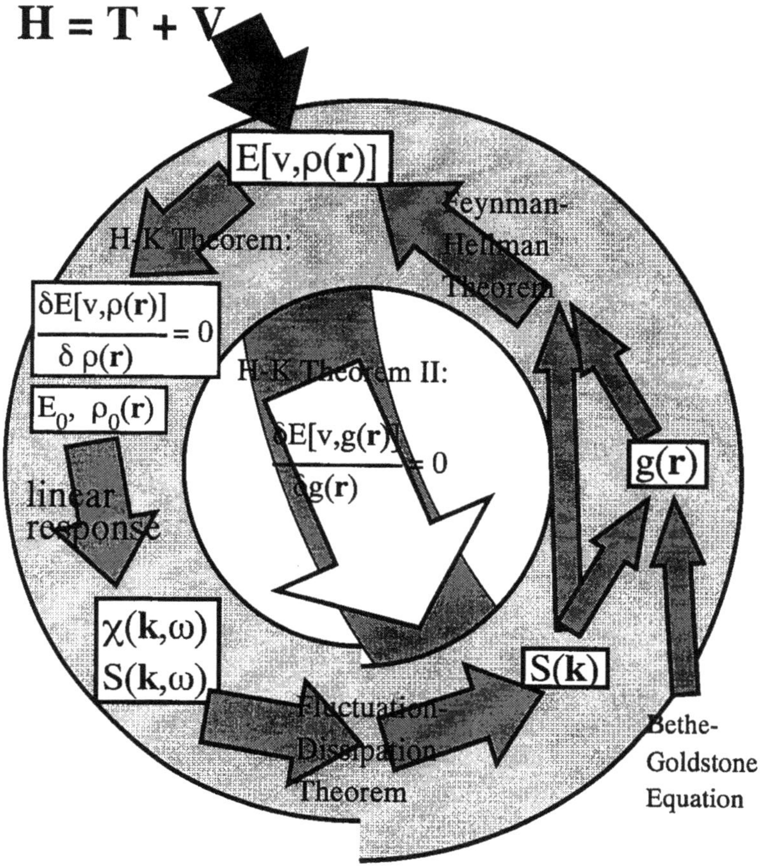

Figure 1. The figure shows a schematic flowchart of the self-consistency loop leading through the first Hohenberg-Kohn theorem, linear response theory, the fluctuation-dissipation theorem, and the Feynman-Hellman theorem. This conventional path is shaded in light-gray, the break in the loop at the bottom indicates that full self-consistency can never be achieved. The dark-gray path through the middle indicates the shortcut offered by the two-body version of the Hohenberg-Kohn theorem.

Schrödinger equation or, to be more precise, by a Bethe-Goldstone equation,

$$-\frac{\hbar^2}{m}\nabla^2\Psi(r) + v(r)\Psi(r) = \lambda\Psi(r) \qquad \text{as} \qquad (r \to 0+) \tag{19}$$

for a loosely defined "pair wave function" $\Psi(r)$. At short distances, the pair distribution function $g(r)$ is proportional to the square of this pair wave function,

$$g(r) \sim |\Psi(r)|^2 \qquad \text{as} \qquad r \to 0+ \,. \tag{20}$$

Thus, one would naturally argue that the form (19) of the Euler equation is the *expected* form of any equation determining the short-range behavior of the pair distribution functions. This equation can be written as

$$\frac{\hbar^2}{m}\nabla^2\sqrt{g(r)} = V_{\rm p-p}(r)\sqrt{g(r)}\,, \tag{21}$$

where $V_{\rm p-p}$ may be taken to define another effective interaction, interpreted as a "particle-particle interaction." All we know about $V_{\rm p-p}$ at this point is that it should, for strongly repulsive potentials, be equal to the bare interaction at short distances. It is immediately clear that the correlation corrections must have a non-negligible effect in Eq. (21): The *correct* pair distribution function goes,[2] for $r \to \infty$, as $g(r) \sim 1+\mathcal{O}(r^{-4})$. Normally (for example if one leaves out the correlation corrections), the solution of Eq. (21) will go as $g(r) \sim 1 + a/r$, where a is related to the *scattering length* of the potential. This means that the correlation corrections to the particle-particle interaction must be "just right" to guarantee that $V_{\rm p-p}(r)$ has zero scattering length.

We now have two equations, (18) and (21), for the pair distribution function $g(r)$ or its Fourier transform $S(k)$. Hence the two effective interactions, $V_{\rm p-p}(r)$ and $V_{\rm p-h}(r)$ are not independent and the question arises what the relationship between these two interactions is. To derive a relationship, multiply Eq. (21) with $\sqrt{g(r)}$ and rewrite it in the form

$$\frac{\hbar^2}{m}\sqrt{g(r)}\nabla^2\sqrt{g(r)} = \frac{\hbar^2}{2m}\left[\nabla^2 g(r) - 2\left|\nabla\sqrt{g(r)}\right|^2\right] = V_{\rm p-p}(r)g(r)\,, \tag{22}$$

or, in momentum space

$$-\frac{\hbar^2 k^2}{2m}\left[S(k)-1\right] = \left[V_{\rm p-p}(r)g(r) + \frac{\hbar^2}{m}\left|\nabla\sqrt{g(r)}\right|^2\right]^{\mathcal{F}}(k)\,. \tag{23}$$

Adding Eq. (13) and rearranging gives

$$\begin{aligned}\tilde{V}_{\rm p-h}(k) &= \left[V_{\rm p-p}(r)g(r) + \frac{\hbar^2}{m}\left|\nabla\sqrt{g(r)}\right|^2\right]^{\mathcal{F}}(k)\\ &\quad + \frac{\hbar^2 k^2}{4m}\left[\frac{1}{S^2(k)}-1\right] + \frac{\hbar^2 k^2}{2m}\left[S(k)-1\right]\\ &\equiv \left[V_{\rm p-p}(r)g(r) + \frac{\hbar^2}{m}\left|\nabla\sqrt{g(r)}\right|^2\right]^{\mathcal{F}}(k) - \tilde{w}_I(k)\,.\end{aligned} \tag{24}$$

Defining now a new, "irreducible" effective potential $V_{\rm I}(r)$ through

$$V_{\rm p-p}(r) \equiv v(r) + V_{\rm I}(r) + w_I(r)\,, \tag{25}$$

we obtain the desired relation

$$V_{\rm p-h}(r) = g(r)\left[v(r) + V_{\rm I}(r)\right] + \frac{\hbar^2}{m}\left|\nabla\sqrt{g(r)}\right|^2 + \left[g(r)-1\right]w(r)\,. \tag{26}$$

With Eqs. (25) and (26), both effective interactions, $V_{\rm p-p}(r)$ and $V_{\rm p-h}(r)$, have been expressed in terms of one unknown interaction $V_{\rm I}(r)$. Eqs. (18) and (21) are equivalent as long as the latter interaction is sufficiently well behaved that all required Fourier-transforms exist. The decomposition $[v(r) + V_{\rm I}(r)]$ is, of course, totally artificial and is made only to emphasize that the bare potential $v(r)$ should be the dominant term at short distances.

So far our considerations of the character of generic Euler equations derived from Eq. (4) has involved no more than manipulation of definitions. Nevertheless, we have been led to exactly the same forms of Euler equations as were derived for Jastrow-Feenberg theory.[2] We have in effect demonstrated that these equations are very general and can be "derived" simply from the consistency of plausible equations that focus on the long-range and on the short-range structure of the pair correlation function. The special character introduced by any particular theory lies in the attendant interpretation of the "irreducible" interaction $V_{\rm I}(r)$ defined by Eq. (26). Diagrammatic many-body theories start, in their simplest versions, with $V_{\rm I}(r) = 0$ and improve upon this approximation by means of diagram expansions. Non-trivial corrections can are due to "elementary diagrams" as well as triplets and higher order correlations.[7]

6 Energy Functional

To complete the formulation of a "pair density functional theory", we must derive the energy functional leading to equations (18), (24), and (26). We can build on the known HNC or parquet energy functional, in particular for $V_{\rm I}(r) = 0$, and assert that the *full* functional is given by

$$E = K + V + E_{\rm Q} + E_{\rm I}\,, \tag{27}$$

where

$$K = \frac{\hbar^2 \rho}{2m} \int d^3r \left|\nabla \sqrt{g(r)}\right|^2 \tag{28}$$

is the kinetic energy due to the curvature of the wave function at short distances,

$$E_{\rm Q} = -\frac{\hbar^2}{8m} \int \frac{d^3k}{(2\pi)^3 \rho} k^2 \frac{(S(k)-1)^3}{S(k)} \tag{29}$$

generates the term $w(r)$, and the $E_{\rm I}$ is a functional of the pair distribution function $g(r)$ which generates the irreducible interaction through

$$V_{\rm I}(r) = \frac{2}{\rho} \frac{\delta E_{\rm I}}{\delta g(r)}. \tag{30}$$

7 Conclusion

What has been gained by this analysis? From the point of view of microscopic many-body theory, we have shown that the HNC-EL equations are effectively a *generic* set of equations for the pair distribution function that can be obtained without lengthy diagram argumentation, and which is valid far beyond the Jastrow-Feenberg theory. It can be derived by merely insisting on consistency between coordinate-space and a momentum-space features of the pair distribution function, but without any reference to the specific form of the wave function. We have also shown how phenomenological components can be built naturally into a theory that has originally been designed to be manifestly microscopic. From the point of view of phenomenological theories, among others density functional methods, we have shown how information on the pair-density can be used to ***supplement*** popular methods in a quite natural manner by including information on two-body quantities. There is some hope that the combination of the efficiency of density functional theory with microscopic information can lead to a truly "next generation" type of theories.

Acknowledgements

The author wishes to congratulate Profs. John W. Clark, Alpo Kallio, Manfred L. Ristig, and Sergio Rosati on the occasion of their 65th birthdays. This work was supported, in part, by the Austrian Science Fund under grant No. P11098-PHY.

References

1. E. Krotscheck, *Phys. Lett.* A **190**, 201 (1994).
2. E. Feenberg, *Theory of Quantum Fluids* (Academic, New York, 1969).
3. J. P. Hansen and I. R. McDonald, *Theory of Simple Liquids* (Academic Press, New York, 1976).
4. P. Hohenberg and W. Kohn, *Phys. Rev.* B **136**, 864 (1964).
5. M. Levy, *Proc. Natl. Acad. Sci. USA* **76**, 6062 (1979).
6. A. D. Jackson and R. A. Smith, *Phys. Rev.* A **36**, 2517 (1987).
7. E. Krotscheck, *Phys. Rev.* B **33**, 3158 (1986).

A VARIATIONAL COUPLED-CLUSTER THEORY

Y. XIAN
Department of Physics, UMIST, P.O. Box 88, Manchester M60 1QD, UK
E-mail: Yang.Xian@umist.ac.uk

We develop a general quantum many-body theory in configuration space by extending the traditional coupled-cluster method to a variational formalism. A set of distribution functions are introduced to evaluate the expectation value of the Hamiltonian. An algebraic technique for calculations of these distribution functions via a set of self-consistent set of equations is given. As a demonstration, we apply this method to a quantum antiferromagnetic spin model. It is shown that one of the lowest-order approximations within this new technique agrees with the traditional coupled-cluster method. Comparison with the method of correlated basis functions is also made.

1 Introduction

The main task of a microscopic quantum many-body theory is to study correlations between the constituent particles of a quantum system in a systematic way. The treatment of these many-body correlations is either in real space or in configuration space. A real-space theory usually focuses on the potential part of many-body Hamiltonians; a configuration space theory often starts from the kinetic part of Hamiltonians. One of the most successful real space quantum many-body theories is the method of correlated basis functions (CBF) in which real-space correlation functions of the ground state are determined variationally.[1] Perhaps, the closest counterpart of configuration space theories to the real space CBF is the coupled-cluster method (CCM)[2–4] in which correlation operators are employed to construct the ground state. One key feature of the CCM is that the bra and ket states are not manifestly hermitian to one another.[5]

In this paper we propose a general variational theory in configuration space by extending the traditional CCM to a variational formalism in which ket and bra states are hermitian to one another. The difficult task of evaluating the Hamiltonian expectation can be done by introducing distribution functions which can then be determined either by a diagrammatic technique or by an algebraic one. The diagrammatic approach developed in this context is quite similar to that of the CBF. In the algebraic approach, one derives a set of self-consistent equations for the distribution functions; these equations can then be tackled by various methods, e.g. iterative methods. Easy comparison can be made with the traditional CCM in this approach. We will mainly discuss the algebraic approach in this article; the diagrammatic approach will be discussed elsewhere.[6] For pedagogical reason, we apply this variational method to a well-known spin model.

2 The Ansatz

We shall take the spin-1/2 antiferromagnetic XXZ model on a bipartite lattice as an example. The model Hamiltonian is given by

$$H = \frac{1}{2}\sum_{l,\rho} H_{l,l+\rho} = \frac{1}{2}\sum_{l,\rho}\left(\Delta s_l^z s_{l+\rho}^z + \frac{1}{2}s_l^+ s_{l+\rho}^- + \frac{1}{2}s_l^- s_{l+\rho}^+\right), \tag{1}$$

where Δ is the anisotropy, the index l runs over all lattice sites, ρ runs over all nearest-neighbor sites, and $s^\pm$ are the usual spin raising (+) and lowering (−) operators. The Hamiltonian at $\Delta = 1$ corresponds to the isotropic Heisenberg model which has been a focus of theoretical study in recent years due to its relevance to high-temperature superconductivity.

In the limit $\Delta \to \infty$, the ground state of Eq. (1) is clearly given by the classical Néel state with alternating spin-up and spin-down sublattices. We shall exclusively use index i for the spin-up sublattice and the index j for the spin-down sublattice. For a finite value of Δ, such as the isotropic point $\Delta = 1$, the many-spin correlations in its ground state can then be included by considering the excited states with respect to the uncorrelated Néel model state. These excited states are constructed by applying the so-called configuration creation operators $C_I^\dagger$ to the Néel model state with the nominal index I labelling these operators. In our spin model, the operators $C_I^\dagger$ are given by any combination of the spin-flip operators to the Néel state, namely s_i^- and s_j^+ and the index I in this case corresponds to the collection of the lattice indices (i's and j's). The hermitian conjugate operators of $C_I^\dagger$ are the configuration destruction operator C_I, given by any combination of s_i^+ and s_j^-. For example, the two-spin flip creation operator is given by $C_{ij}^\dagger = s_i^- s_j^+$.

The traditional CCM is based on the Hubbard, Hugenholtz and Coester ansatz[2] (HHC) for the ground ket state, where the correlations are parametrised by an exponentiated operator,

$$|\Psi_g\rangle = e^S|\Phi\rangle, \quad S = \sum_I F_I C_I^\dagger. \tag{2}$$

For the current model, $|\Phi\rangle$ is the Néel state and F_I are the correlation coefficients (which become correlation functions in configuration space). The configuration creation operator $C_I^\dagger$ in this case is given by a product of any number of pairs of the spin-flip operators

$$\sum_I F_I C_I^\dagger = \sum_{n=1}^{N/2} \sum_{i_1\ldots,j_1\ldots} f_{i_1\ldots,j_1\ldots} \frac{s_{i_1}^-\ldots s_{i_n}^- s_{j_1}^+\ldots s_{j_n}^+}{(2s)^n}, \tag{3}$$

where s is the spin quantum number. Although we are mainly interested in $s = 1/2$, we keep the factor of $1/2s$ for the purpose of comparison with the large-s expansion. Notice also that in Eq. (3) the spin-flip operators of the i-sublattice always pair with that of the j-sublattice to ensure the total z-component $s_{total}^z = 0$. For the bra state, however, the CCM proposes a different, practical form as[3–5]

$$\langle\tilde{\Psi}| = \langle\Phi|\tilde{S}e^{-S} \tag{4}$$

where S is as given in the ket state and the *linear* bra state operator $\tilde{S}$ is constructed by the configuration destruction operators only, namely

$$\tilde{S} = 1 + \sum_I \tilde{F}_I\, C_I = 1 + \sum_{n=1}^{N/2} \sum_{i_1\ldots,j_1\ldots} \tilde{f}_{i_1\ldots,j_1\ldots} \frac{s_{i_1}^+\ldots s_{i_n}^+ s_{j_1}^-\ldots s_{j_n}^-}{(2s)^n}. \tag{5}$$

The coefficients $\{F_I, \tilde{F}_I\} = \{f_{i_1\ldots j_1\ldots}, \tilde{f}_{i_1\ldots j_1\ldots}\}$ are determined variationally through the Hamiltonian expectation $\langle H\rangle$, requiring the normalization condition $\langle\tilde{\Psi}|\Psi\rangle = 1$,

$$\langle\tilde{\Psi}|H|\Psi\rangle = \langle\Phi|\bar{H}|\Phi\rangle, \tag{6}$$

where the similarity-transformed $\bar{H}$ given as

$$\bar{H} = e^{-S}He^{S} = H + \frac{1}{1!}[H,S] + \frac{1}{2!}[[H,S],S] + \cdots . \tag{7}$$

In most cases, H contains a finite order of destruction operators. The above nested commutation series then terminates at a finite order as S contains only the creation operators. Hence, the Hamiltonian expectation value in the CCM is a finite order polynomial function of the coefficients $\{F_I, \tilde{F}_I\}$. More specifically, $\langle H\rangle$ in the CCM is linear in the bra-state coefficients $\tilde{F}_I$ and finite-order polynomial in the ket-state coefficients F_I. Thus, any calculation in the CCM in general is quite straightforward; when an approximation scheme is chosen (i.e., a truncation scheme in $\{F_I, \tilde{F}_I\}$), no further approximation is necessary in most calculations. However, this CCM parametrization of the ground state is problematic in dealing with long range correlations as discussed in the context of our spin model calculations in Ref. 7.

An obvious extension of the CCM is to apply the HHC ansatz to both the ket and bra states. Hence we write

$$|\Psi\rangle = e^{S}|\Phi\rangle, \quad S = \sum_I F_I C_I^\dagger, \tag{8}$$

$$\langle\tilde{\Psi}| = \langle\Phi|e^{\tilde{S}}, \quad \tilde{S} = \sum_I \tilde{F}_I C_I, \tag{9}$$

where the ket-state and bra-state correlation coefficients F_I and $\tilde{F}_I$ are hermitian and independent to one another. For our spin model, the model state $|\Phi\rangle$ is the Néel state, and the correlation operators $\sum_I F_I C_I^\dagger$ and $\sum_I \tilde{F}_I C_I$ are given as in Eqs. (3) and (5). The coefficients $\{F_I, \tilde{F}_I\}$ are then determined by the usual variational equations as

$$\frac{\delta\langle H\rangle}{\delta\tilde{F}_I} = \frac{\delta\langle H\rangle}{\delta F_I} = 0 \tag{10}$$

where energy expectation is defined in the usual way as

$$\langle H\rangle = \frac{\langle\tilde{\Psi}|H|\Psi\rangle}{\langle\tilde{\Psi}|\Psi\rangle} = \frac{\langle\Phi|e^{\tilde{S}}He^{S}|\Phi\rangle}{\langle\Phi|e^{\tilde{S}}e^{S}|\Phi\rangle} . \tag{11}$$

Clearly, the normalization factor $\langle\tilde{\Psi}|\Psi\rangle$ and Hamiltonian expectation $\langle H\rangle$ are highly nontrivial functions of the coefficients F_I and $\tilde{F}_I$. Their calculation in the standard variational approach is in general difficult, contrast to the CCM where the expectation value of the Hamiltonian is a finite-order polynomial of the coefficients, as described earlier.

Hence, the key to the extension of the CCM to the standard variational method as described in Eqs. (8)-(11) is to develop a practical and consistent technique to evaluate the normalization factor and the Hamiltonian expectation. It is known that these evaluations can be done more efficiently by employing distribution functions. One then needs to develop a systematic and consistent scheme to calculate these distribution functions. We have considered two such schemes. One is similar to the traditional technique in statistical mechanics employed by the CBF. In this method, one introduces a generating functional whose functional derivatives are the distribution functions. A diagrammatic technique has been developed to evaluate these distribution functions.[6]

The other approach we have considered is an algebraic technique. In this approach, one derives a self-consistent set of equations for the distribution functions by taking the advantage of the operator nature in the ground state as given in Eqs. (8)-(9). As we shall see, a most simple approximation to this self-consistent set of equations will yield the full CCM results, but one can easily go beyond that.

3 Self-Consistent Set Of Equations

We first introduce the so-called bare distribution functions as expectation value of the configuration operators, namely

$$g_I = \langle C_I^\dagger \rangle, \quad \tilde{g}_I = \langle C_I \rangle \,. \tag{12}$$

In general these bare distribution functions are nontrivial functions of $\{F_J, \tilde{F}_J\}$. (Multiplying by the corresponding coefficients, we obtain $F_I g_I$ and $\tilde{F}_I \tilde{g}_I$ which are the the usual distribution functions useful in the diagrammatic approach.[6])

Direct calculation of these functions is certainly not an easy task. Fortunately, by taking the advantage of the properties of the operators, one can derive a self-consistent set of equations which can then be tackled by various methods. In particular, as $C_I^\dagger$ commutes with $S = \sum_I F_I C_I^\dagger$ and C_I with $\tilde{S} = \sum_I \tilde{F}_I C_I$, one can write

$$g_I = \frac{1}{I}\langle \Phi | e^{\tilde{S}} C_I^\dagger e^S | \Phi \rangle = \frac{1}{I}\langle \Phi | e^{\tilde{S}} e^S C_I^\dagger | \Phi \rangle \tag{13}$$

$$\tilde{g}_I = \frac{1}{I}\langle \Phi | e^{\tilde{S}} C_I e^S | \Phi \rangle = \frac{1}{I}\langle \Phi | C_I e^{\tilde{S}} e^S | \Phi \rangle \,. \tag{14}$$

In order to find another expression, one inserts the identity $e^{-\tilde{S}} e^{\tilde{S}}$ in the expression of g_I as

$$g_I = \frac{1}{I}\langle \Phi | e^{\tilde{S}} C_I^\dagger e^{-\tilde{S}} e^{\tilde{S}} e^S | \Phi \rangle = \frac{1}{I}\langle \Phi | \underline{C}_I^\dagger e^{\tilde{S}} e^S | \Phi \rangle \tag{15}$$

where the similarity-transformed operator $\underline{C}_I^\dagger$ can be expanded in the nested commutator series as

$$\underline{C}_I^\dagger = e^{\tilde{S}} C_I^\dagger e^{-\tilde{S}} = C_I^\dagger + \frac{1}{1!}[\tilde{S}, C_I^\dagger] + \frac{1}{2!}[\tilde{S}, [\tilde{S}, C_I^\dagger] + \cdots \tag{16}$$

and this series is finite as $C_I^\dagger$ is finite and $\tilde{S}$ contains only the destruction operators. By definition, $\langle \Phi | C_I^\dagger = 0$, hence $\langle \Phi | \underline{C}_I^\dagger$ can be expressed in a form linear in the destruction operators C_J and finite order polynomial in the coefficients $\tilde{F}_J$. The expectation values of these finite order terms is therefore linear in $\tilde{g}_J$ and finite order polynomial in $\tilde{F}_J$. This yields a linear relation between g_I and $\{\tilde{g}_J\}$. Hence we write

$$g_I = G(\{\tilde{g}_J\}, \{\tilde{F}_J\}) \tag{17}$$

where G is a function linear in $\tilde{g}_J$ and finite order polynomial in $\tilde{F}_J$.

In a similar fashion we write, by inserting identity $e^S e^{-S}$ in the expression for $\tilde{g}_I$,

$$\tilde{g}_I = \frac{1}{I}\langle\Phi|e^{\tilde{S}}e^S\bar{C}_J|\Phi\rangle \tag{18}$$

with the usual commutation series

$$\bar{C}_I = e^{-S}C_I e^S = C_I + \frac{1}{1!}[C_I, S] + \frac{1}{2!}[C_I, S], S] + \cdots , \tag{19}$$

and we obtain

$$\tilde{g}_I = G(\{g_J\}, \{F_J\}) , \tag{20}$$

where G is the same function as in Eq. (17) but now linear in $\{g_J\}$ and finite order polynomial in $\{F_J\}$. As function G is the same in Eqs. (17) and (20), only one calculation is necessary.

Eqs. (17) and (20) provide self-consistent set of equations for g_I and $\tilde{g}_I$ in terms of the correlation coefficients $\{F_J, \tilde{F}_J\}$. We note that for a particular g_I its equation in general contains a higher-order set $\{\tilde{g}_J\}$, and vice versa, even for a truncated coefficient set $\{F_I, \tilde{F}_I\}$. Therefore, in order to make any practical calculation, one has to make two approximations, a truncation on the number of coefficients $\{F_I, \tilde{F}_I\}$, and a truncation on the number of bare correlation functions $\{g_I, \tilde{g}_I\}$. After these two truncations, one should be able to solve the self-consistent set of equations to obtain g_I and $\tilde{g}_I$ in terms of F_I and $\tilde{F}_I$. This is contrary to the CCM where one needs only one truncation (in $\{F_I, \tilde{F}_I\}$). As we shall see, for a similar truncation in the coefficients $\{F_I, \tilde{F}_I\}$, one of our lowest order truncations in the bare correlation function $\{g_I, \tilde{g}_I\}$ will yield the full CCM results. However, it is a simple step to go beyond this approximation by including some higher-order distribution functions g_I which has proved to be essential to obtain the consistent long-range behaviors of the spin correlation functions (and the low-lying excitation energies) as we shall see in our spin model calculation.

Since Hamiltonian usually contains terms involving both creation operators $C_I^\dagger$ and destruction operators C_I, its expectation can be expressed in general as a function linear in g_I and $\tilde{g}_I$ and a finite-order polynomial in F_I and $\tilde{F}_I$,

$$\langle H\rangle = \mathcal{H}(\{g_I, \tilde{g}_I\}; \{F_I, \tilde{F}_I\}) . \tag{21}$$

Solutions of Eqs. (17) and (20) can be substituted in the expectation value of the Hamiltonian and we obtain $\langle H\rangle$ as a function of $\{F_I, \tilde{F}_I\}$. Variational calculation in Eq. (10) can then be carried out. In the following, we consider a simple application to the spin model as a demonstration.

4 Two-Spin Flip Approximation in Spin Model

For simplicity, we consider a truncation approximation in which the correlation operators S and $\tilde{S}$ retain only the two-spin flip correlations as

$$|\Psi_2\rangle = e^{S_2}|\Phi\rangle, \quad \langle\Psi_2| = \langle\Phi|e^{\tilde{S}_2} , \tag{22}$$

where

$$S_2 = \sum_{ij} f_{ij}\frac{s_i^- s_j^+}{2s}, \quad \tilde{S}_2 = \sum_{ij} \tilde{f}_{ij}\frac{s_i^+ s_j^-}{2s} . \tag{23}$$

Using the usual angular momentum commutations

$$[s_l^z, s_{l'}^{\pm}] = \pm s_l^{\pm}\delta_{ll'}, \quad [s_l^+, s_{l'}^-] = 2s_l^z\delta_{ll'} , \tag{24}$$

and the Néel state eigenequations

$$s_i^z|\Phi\rangle = s|\Phi\rangle, \quad s_j^z|\Phi\rangle = -s|\Phi\rangle , \tag{25}$$

it is a straightforward calculation to derive expectation value of various operators with respect to the states of Eqs. (22). In this approximation, for example, the order parameter is derived as

$$M^z = \langle s_i^z\rangle = s - \sum_r n_r , \tag{26}$$

where n_r is the one-body distribution function given by

$$n_{ij} = f_{ij}g_{ij} = f_{ij}\frac{\langle s_i^- s_j^+\rangle}{2s} , \tag{27}$$

and we have taken the advantage of translational invariance by writing $n_{ij} = n_r$ with $j = i + r$; the two-spin correlation function is given by

$$\langle s_i^z s_j^z\rangle = -s^2 + s(\sum_{i'} n_{i'j} + \sum_{j'} n_{ij'}) - (\sum_{i'j'} G_{ij',i'j} + n_{ij}) , \tag{28}$$

where $G_{ij,i'j'}$ is the two-body distribution function

$$G_{ij,i'j'} = f_{ij}f_{i'j'}g_{ij,i'j'} = f_{ij}f_{i'j'}\frac{\langle s_i^- s_j^+ s_{i'}^- s_{j'}^+\rangle}{(2s)^2} ; \tag{29}$$

and finally, the expectation value of Eq. (1) is then given by

$$\langle H_{ij}\rangle = -\Delta s^2 + s\Big(\tilde{g}_{ij} + g_{ij} + \Delta\sum_{i'} n_{i'j} + \Delta\sum_{j'} n_{ij'}\Big) - \Delta\Big(\sum_{i'j'} G_{ij',i'j} + n_{ij}\Big) . \tag{30}$$

As we can see, these physical quantities involve up to two-body distribution functions.

The self-consistent set of equations for the bare distribution functions are derived as described in Section 3. In particular, the equation for the one-body function $\tilde{g}_{ij}$ is

$$\begin{aligned}\tilde{g}_{i_1j_1} &= f_{i_1j_1} + \sum_{ij} f_{ij_1} f_{i_1j} g_{ij} \\ &- \frac{2}{2s} f_{i_1j_1} \Big(\sum_{i} f_{ij_1} g_{ij_1} + \sum_{j} f_{i_1j} g_{i_1j}\Big) \\ &+ \frac{1}{2s} \sum_{ijj'} f_{ij_1} f_{i_1j} f_{i_1j'} g_{i_1j,ij'} + \frac{1}{2s} \sum_{ii'j} f_{ij_1} f_{i'j_1} f_{i_1j} g_{ij_1,i'j} \\ &+ \frac{2}{(2s)^2} f_{i_1j_1}^2 g_{i_1j_1} + \frac{4}{(2s)^2} f_{i_1j_1} \sum_{ij} f_{ij_1} f_{i_1j} g_{i_1j_1,ij} \\ &+ \frac{1}{(2s)^2} \sum_{ii'jj'} f_{ij_1} f_{i'j_1} f_{i_1j} f_{i_1j'} g_{i_1j_1,ij,i'j'} \ . \end{aligned} \tag{31}$$

The hermitian conjugate of these equations are the self-consistent set of equations for $g_{ij}, g_{ij,i'j'}$ etc. The above equation contains up to the three-body function $g_{ij,i'j',i''j''}$. The equation for the two-body function $\tilde{g}_{ij,i'j'}$ will contain up to twelve-body functions, etc. Clearly, we need to make further truncations for any practical calculation.

Consider a simple truncation in which we retain only the first two terms in Eq. (31),

$$\tilde{g}_{i_1j_1} = f_{i_1j_1} + \sum_{ij} f_{ij_1} f_{i_1j} g_{ij} \ , \tag{32}$$

and similar equation for g_{ij}. Using the Fourier transformation technique and translational symmetry, it is easy to solve the two equations to obtain

$$g_k = \frac{f_k}{1 - f_k \tilde{f}_k}, \quad \tilde{g}_k = \frac{\tilde{f}_k}{1 - f_k \tilde{f}_k} \ , \tag{33}$$

where g_k and f_k are Fourier transforms of g_{ij} and f_{ij}, etc. To the same order in $2s$, the ground-state energy, Eq. (30), is

$$e = \frac{2}{zN} \sum_{i,\rho} \langle H_{i,i+\rho} \rangle = -\Delta s^2 + s(g_1 + \tilde{g}_1) + 2\Delta s \sum_{r} n_r \ , \tag{34}$$

where z is the number of nearest-neighbor sites and

$$g_1 = \sum_{k} \frac{\gamma_k f_k}{1 - \tilde{f}_k f_k}, \quad \tilde{g}_1 = \sum_{k} \frac{\gamma_k \tilde{f}_k}{1 - \tilde{f}_k f_k}, \quad \sum_{r} n_r = \sum \frac{f_k \tilde{f}_k}{1 - \tilde{f}_k f_k} \ , \tag{35}$$

with

$$\gamma_k = \frac{1}{z} \sum_{\rho} e^{i\mathbf{k}\cdot\mathbf{r}_\rho} \ . \tag{36}$$

The variational equations $\frac{\partial e}{\partial f_k} = \frac{\partial e}{\partial \tilde{f}_k} = 0$ reduce to a quadratic equation for f_k,

$$\gamma_k f_k^2 + 2\Delta f_k + \gamma_k = 0 \,, \tag{37}$$

and a similar equation for $\tilde{f}_k$. The physical solution is given by

$$f_k = \tilde{f}_k = \frac{\Delta}{\gamma_k}(-1 + \sqrt{1 - \gamma_k^2/\Delta^2}) \,. \tag{38}$$

The ground-state energy and order parameter are then obtained as

$$e = -\Delta s^2 + s\Delta \sum_k (-1 + \sqrt{1 - \gamma_k^2/\Delta^2}) \tag{39}$$

and

$$M^z = s - \frac{1}{2}\sum_k (\frac{1}{\sqrt{1 - \gamma_k^2/\Delta^2}} - 1) \,. \tag{40}$$

It is not difficult to include the contribution of higher-order many-body distribution functions. Consider the two-spin correlation function of Eq. (28), which contain the important two-body distribution function. The two-body distribution functions $g_{ij,i'j'}$ and $\tilde{g}_{ij,i'j'}$ can be calculated through their self-consistent set of equations by keeping the same order terms in the $(2s)$ expansion as we have done for the one-body distribution function in Eqs. (31)-(32). A simpler way to obtain the same approximation for the two-body functions is to employ the following sequential equation

$$\frac{\partial}{\partial f_{i'j'}} g_{ij} = g_{ij,i'j'} - g_{ij} g_{i'j'} \,. \tag{41}$$

Using Eq. (32) and the sequential equation, we obtain an approximation for the two-body distribution,

$$g_{ij,i'j'} = g_{ij} g_{i'j'} + g_{ij'} g_{i'j} \,. \tag{42}$$

Hence, the normalised two-spin correlation function becomes

$$c_r = \langle s_i^z s_{i+r}^z \rangle - \langle s_i^z \rangle \langle s_{i+r}^z \rangle = -g_r \tilde{g}_r = -g_r^2 \,. \tag{43}$$

In fact, our above results of the ground-state energy, order parameter and the correlation function are the same as that of the spin-wave theory.[8,9] In particular, the long-range behavior of the correlation function $c_r \propto 1/r^2$ for a square lattice system at $\Delta = 1$ can not be obtained without the contribution of the two-body distribution function.

The similar CCM SUB2 approximation in this case is to ignore all two-body and higher-order many-body distribution functions in Eq. (31). Therefore, the two-spin correlation function in the CCM SUB2 approximation has unphysical behavior as discussed in Ref. 7.

5 Conclusion

The new coupled-cluster technique proposed here represents our attempt to formulate a variational many-body theory in configuration space. In many ways, this new formalism resembles the real space CBF; but as they are in configuration space, our correlation functions $F_I, \tilde{F}_I$ are clearly state-dependent, contrary to the real space CBF in which correlation functions are state-independent.

The extension of the CCM has also been considered by Arponen.[5] In Arponen's approach, the bra state is chosen as

$$\langle\tilde{\Psi}| = \langle\Phi|e^{\tilde{S}}e^{-S} , \tag{44}$$

where S is the same as in ket state Eq. (2) and $\tilde{S}$ is chosen as the same as Eq. (9). Like CCM, this formalism is also not manifestly hermitian, and the normalization factor is imposed as $\langle\tilde{\Psi}|\Psi\rangle = 1$. This extension to the CCM represent an improvement since the bra-state correlation operator $\tilde{S}$ is now nonlinear. It will be interesting to compare Arponen's extension with our variational formalism as described in this article. This will be reported elsewhere.

Acknowledgments

Many useful discussions with J. Arponen, R.F. Bishop, and H. Kümmel are acknowledged.

References

1. J. W. Clark and E. Feenberg, *Phys. Rev.* **113**, 388 (1959); E. Feenberg, *Theory of quantum fluids* (Academic Press, New York, 1969).
2. J. Hubbard, *Proc. Roy. Soc.* A **240**, 539 (1957); A. Hugenholtz, *Physica* **23**, 481 and 533 (1957); F. Coester, *Nucl. Phys.* **7**, 421 (1958).
3. F. Coester and K. Kümmel, *Nucl. Phys.* **17**, 477 (1960).
4. J. Čižek, *J. Chem. Phys.* **45**, 4256 (1966); *Adv. Chem. Phys.* **14**, 35 (1969).
5. J. Arponent, *Ann. Phys. (NY)* **151**, 311 (1983).
6. Y. Xian, to be published.
7. R. F. Bishop, J. Parkinson and Y. Xian, *Phys. Rev.* B **43**, 13782 (1991); *ibid.* **44**, 9425 (1991).
8. P. W. Anderson, *Phys. Rev.* **86**, 694 (1952).
9. M. Takahashi, *Phys. Rev.* B **40**, 2494 (1989).

Nuclear and Subnuclear Physics

THE NUCLEAR EQUATION OF STATE AND NEUTRON STAR STRUCTURE

M. BALDO
Istituto Nazionale di Fisica Nucleare, Corso Italia 57, 95129 Catania, ITALY
E-mail: marcello.baldo@ct.infn.it

The microscopic many-body theory of the Nuclear Equation of State is developed in the framework of the Bethe-Brueckner-Goldstone expansion. Both for symmetric and for pure neutron matter strong evidence of convergence in the expansion is found. Once three-body forces are introduced, the phenomenological saturation point is reproduced and the theory is applied to the study of neutron stars static properties. In the interior of a neutron star, where the baryon density increases above two-three times the saturation density, the onset of hyperons occurs. Extending the theory to include strangeness, the resulting Equation of State turns out to be strongly softened. The consequences of these results for the mass and radius of neutron stars are discussed.

1 Introduction

Nuclear matter is one of the systems for which several many-body theories and techniques have been developed and applied since few decades. Despite infinite nuclear matter is obviously an idealized physical system, it is believed that macroscopic portions of (asymmetric) nuclear matter form the interior bulk part of a neutron star, commonly associated with pulsars, thus providing a natural system quite close to the ideal one. Unfortunately, neutron stars are elusive astrophysical objects, and only indirect observations of their structure, including their sizes and masses, are possible. However, the astrophysics of neutron stars is rapidly developing, in view of the observations coming from the new generation of artificial satellites, and one can expect that it will be possible in the near future to confront the theoretical predictions with more and more stringent phenomenological data.

Heavy ion reactions is another field of research where the nuclear Equation of State (EOS) is a relevant issue. In this case, the difficulty of extracting the EOS is due to the complexity of the processes, since the interpretation of the data is necessarily linked to the analysis of the reaction mechanism. An enormous amount of work has been done in the last two decades in the field, but clear indications about the main characteristics of the EOS have still to come.

On the theoretical side, the main difficulty is the treatment of the large repulsive core, which dominates the short range behaviour of the nucleon-nucleon (NN) interaction, typical of the nuclear system, but which is common to other systems like liquid helium. Simple perturbation theory cannot of course be applied, since the matrix elements of the interaction are too large. One way of overcoming this difficulty is to introduce the two-body scattering G-matrix, which has a much smoother behaviour even for a large repulsive core. It is possible to rearrange the perturbation expansion in terms of the reaction G-matrix, in place of the original bare NN interaction, and this procedure is systematically exploited in the Bethe-Brueckner-Goldstone (BBG) expansion.[1] In this contribution to the conference 150

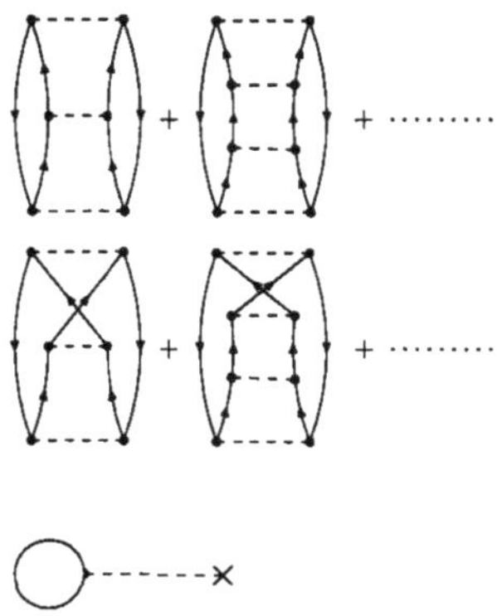

Figure 1. Third and forth order ladder diagrams in the bare interaction (dashed lines) and first order potential insertion (bottom).

Years of QMBT we present the latest results on the nuclear EOS based on the BBG expansion and their application to the physics of neutron stars. In Sec. 2, after a brief historical introduction, we outline the salient features of the BBG theory. In Sec. 2 a report is given on the main results on symmetric nuclear matter and pure neutron matter in the relevant density range. The next two sections are devoted to the physics of neutron stars and the conclusions are drawn at the end of Sec. 4.

2 The BBG expansion and the nuclear EOS

The BBG expansion for the ground state energy at a given density density, i.e. the EOS at zero temperature, can be ordered according to the number of independent hole-lines appearing in the diagrams representing the different terms of the expansion. This grouping of diagrams generates the so-called hole-line expansion.[2] The smallness parameter of the expansion is assumed to be the "wound parameter",[2] roughly determined by the ratio between the core volume and the volume per particle in the system. It gives an estimate of the decreasing factor introduced by an additional hole-line in the diagram series. The parameter turns out to be small enough up to 2-3 times the nuclear matter saturation density. The diagrams with a given number n of hole-lines describe the n-particle correlations in the system. At the two hole-line level of approximation the corresponding summation of diagrams produces the Brueckner-Hartree-Fock (BHF) approximation, which incorporates the two-particle correlations. The BHF approximation includes the self-consistent procedure of determining the single particle auxiliary potential, which is an essential ingredient of the method. Once the auxiliary self-consistent potential is introduced, the expansion is implemented by introducing the set of diagrams which include "potential insertions". The BHF sums the so called "ladder diagrams". Some of them are depicted in Fig. 1.

The self-consistent procedure, first devised by Brueckner,[3] was the real breakthrough towards microscopic calculations of nuclear matter EOS. The summation of the ladder diagrams can be performed by solving the integral equation for the Brueckner G-matrix

$$\langle k_1 k_2|G(\omega)|k_3 k_4\rangle = \langle k_1 k_2|v|k_3 k_4\rangle + \sum_{k_3' k_4'} \langle k_1 k_2|v|k_3' k_4'\rangle \frac{(1-\Theta_F(k_3'))\,(1-\Theta_F(k_4'))}{\omega - e_{k_3'} - e_{k_4'}} \langle k_3' k_4'|G(\omega)|k_3 k_4\rangle, \qquad (1)$$

where $\Theta_F(k) = 1$ for $k < k_F$ and zero otherwise, k_F being the Fermi momentum. The product $Q(k,k') = (1-\Theta_F(k))(1-\Theta_F(k'))$, appearing in the kernel of Eq. (1), enforces the scattered momenta to lie outside the Fermi sphere, and it is commonly referred to as the "Pauli operator". The self-consistent single particle potential $U(k)$ is determined by the equation

$$U(k) = \sum_{k'<k_F} \langle kk'|G(e_{k_1} + e_{k_2})|kk'\rangle_A, \qquad (2)$$

with $|kk'\rangle_A = |kk'\rangle - |kk'\rangle$. The single particle energy $e(k)$ is given by

$$e(k) = \frac{\hbar^2 k^2}{2m} + U(k), \qquad (3)$$

which therefore implies an implicit self-consistent condition for $U(k)$ from Eq. (2). Finally the ground state energy is given by

$$E = \sum_{k<k_F} \left(\frac{\hbar^2 k^2}{2m} + \frac{1}{2} U(k) \right). \qquad (4)$$

The BHF results indicate that already the two hole-line approximation is able to produce reasonable values for the saturation point. The remaining discrepancies can be summarized in the celebrated Coester band,[4] the line along which the results for different "realistic" nucleon-nucleon (NN) forces appear to be approximately concentrated and which misses the phenomenological saturation point. According to the force used, either the saturation density is too high or the binding energy is too small. However, the results depend on the very definition of the single particle potential. The "standard" choice for $U(k)$ assumes that the potential is zero above the Fermi momentum k_F, while in the "continuous choice"[5] the definition of Eq. (2) is extended to momenta k larger than k_F, thus making U a continuous function through the Fermi surface. The final result of a hypothetically exact BBG calculation is independent of the auxiliary potential $U(k)$, but the rate of convergence can of course depend on the particular choice adopted. Therefore, the degree of dependence of the results on the choice of the potential can be considered an indication of the degree of convergence obtained at a given level of the BBG expansion. The results for both choices of $U(k)$ at the BHF level of approximation (Argonne v_{14} potential[6]) are shown in Fig. 2 in the case of symmetric nuclear matter (solid lines).

As one can see, the saturation curves are different for the two prescriptions. It has to be noticed, however, that the apparent discrepancy of 4-5 MeV in the binding energies shown Fig. 2, is about 10% of the calculated potential energy per particle, which is about -40 MeV around saturation. This is the degree of convergence obtained at the Brueckner level according the adopted criterion.

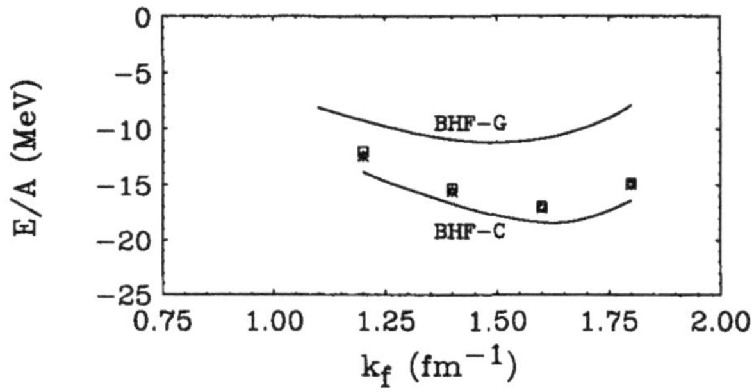

Figure 2. Nuclear matter saturation curve for the Argonne v_{14} NN potential. The solid lines indicate the results at the Brueckner (two hole-lines) level for the standard (BHF-G) and the continuous choices (BHF-C) respectively. The results obtained adding the three hole-line contribution are marked by the squares (standard choice) and the stars (continuous choice).

The results show a substantial improvement if the three hole-line diagrams contribution is added. These diagrams include the correlations among three particles, which cannot be factorized into products of two-body correlations. They can be summed up by introducing the three-body scattering matrix in the medium, in very close analogy with the Brueckner G-matrix, which sums up the two-body correlation diagrams. More details can be found in Ref. 1, where the history of of three hole-line diagrams calculations is summarized. The results both for the standard (square) and the continuous (stars) choices are shown in Fig. 2, for a selected number of densities. The final results are now quite close, indicating that convergence has been practically reached. Notice that the agreement is obtained at all considered densities. These recent results[8] are slightly different from Ref. 7, because a more extended set of two-body and three-body channels has been used, but the discrepancy is small. We have checked that the agreement between the two choices still persists up to nuclear matter density $\rho = 0.75$ fm^{-3}, which is the largest density we considered in the calculations. Similar results have been obtained for pure neutron matter,[9] where the dependence on the two-body forces was also considered.

The phenomenological saturation point for symmetric nuclear matter is, however, not reproduced. The binding energy per particle at the minimum of the saturation curve, displayed in Fig. 2, turns out to be close to the empirical value of about -16 MeV, but the corresponding density comes out about 30-40 % larger than the empirical one. Usually this drawback is corrected by introducing three-body forces in the nuclear hamiltonian. Unfortunately first principle nuclear forces are not available, and one has to rely on a phenomenological approach.

3 Neutron stars within the BBG approach

The nuclear matter equation of state is the fundamental input for building models of neutron stars. These compact objects, among the densest in the universe, are indeed characterized by values of the baryon density which span from the iron density at the surface up to eight-ten times the normal nuclear matter density in the core. Therefore a detailed knowledge of the equation of state over a wide range of densities is required.[10] This is a very hard task from the theoretical point of view. As we have seen in the previous section, the EOS of nucleon matter with two-body forces can be established up to densities relevant to neutron star studies. Furthermore, three-body forces are needed to get the correct saturation point. Furthermore, at higher densities an additional complication arises. In fact, whereas at densities close to the saturation value the matter consists mainly of nucleons and leptons, at higher densities several species of particles may appear due to the fast rise of the nucleon chemical potentials. In our work we perform microscopic calculations of the nuclear matter EOS containing fractions of Λ and Σ^- hyperons in the framework of the Brueckner-Hartree-Fock scheme. The justification of such an approach is based on the results reported in the previous section, which indicate that the three hole-line contribution is quite small within the continuous choice for the single particle potential. Therefore, in first approximation we can adopt the BHF level in the BBG expansion. In this approach, the first basic input is the two-body interaction. We chose the Paris[11] and the Argonne v_{18} potential[12] for the nucleon-nucleon part, whereas the Nijmegen soft-core model[13] has been adopted for the nucleon-hyperon (NY) potential. No hyperon-hyperon interaction is taken into account, since no robust experimental data are available yet. For more details, the reader is referred to Ref. 14 and references therein. As for the three-body forces (TBF's) among nucleons, in our approach we have included a contribution containing a long-range two-pion exchange attractive part and an intermediate range repulsive part.[15] This allows the correct reproduction of the saturation point. In Fig. 3 we show the chemical composition of β-stable and asymmetric nuclear matter containing hyperons (panel (a)) and the corresponding equation of state (panel (b)). The shown calculations have been performed using the Paris potential. We observe that hyperon formation starts at densities $\rho \simeq 2-3$ times normal nuclear matter density. The Σ^- baryon appears earlier than the Λ, in spite of its larger mass, because of the negative charge. The appearance of strange particles has two main consequences, i) an almost equal percentage of nucleons and hyperons are present in the stellar core at high densities and ii) a strong deleptonization of matter, since it is energetically convenient to maintain charge neutrality through hyperon formation than β-decay. The equation of state is displayed in panel (b). The dotted line represents the case when only nucleons and leptons are present in stellar matter, whereas the solid line shows the case when hyperons are included as well. In the latter case the equation of state gets very soft, since the kinetic energy of the already present baryonic species is converted into masses of the new particles, thus lowering the total pressure. This fact has relevant consequences for the structure of the neutron stars.

4 Equilibrium configurations of neutron stars

We assume that a star is a spherically symmetric distribution of mass in hydrostatic equilibrium. The equilibrium configurations are obtained by solving the Tolman-Oppenheimer-Volkoff (TOV) equations[10] for the pressure P and the enclosed mass

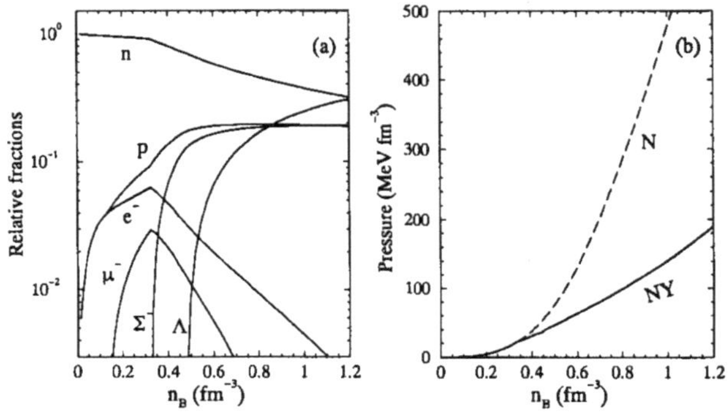

Figure 3. In panel (a) we display the equilibrium composition of asymmetric and β-stable nuclear matter containing Σ^- and Λ hyperons. In panel (b) the solid(dotted) line represents the EOS obtained in the case when nucleons plus hyperons (only nucleons) are present.

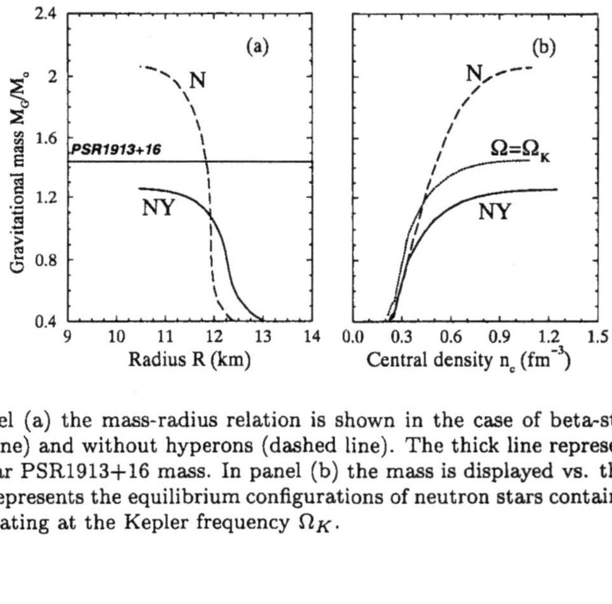

Figure 4. In panel (a) the mass-radius relation is shown in the case of beta-stable matter with hyperons (solid line) and without hyperons (dashed line). The thick line represents the measured value of the pulsar PSR1913+16 mass. In panel (b) the mass is displayed vs. the central density. The dotted line represents the equilibrium configurations of neutron stars containing nucleons plus hyperons and rotating at the Kepler frequency Ω_K.

m,

$$\frac{dP(r)}{dr} = -\frac{Gm(r)\rho(r)}{r^2} \frac{\left[1 + \frac{P(r)}{\rho(r)}\right]\left[1 + \frac{4\pi r^3 P(r)}{m(r)}\right]}{1 - \frac{2Gm(r)}{r}} , \tag{5}$$

$$\frac{dm(r)}{dr} = 4\pi r^2 \rho(r) , \tag{6}$$

where G is the gravitational constant (we assume $c = 1$). Starting with a central mass density $\rho(r = 0) \equiv \rho_c$, we integrate out until the pressure on the surface equals the one corresponding to the density of iron. This gives the stellar radius R

and the gravitational mass is then

$$M_G \equiv m(R) = 4\pi \int_0^R dr\, r^2 \rho(r) \,. \tag{7}$$

For the outer part of the neutron star we have used the equations of state by Feynman-Metropolis-Teller[16] and Baym-Pethick-Sutherland,[17] and for the medium-density regime we use the results of Negele and Vautherin.[18] For density $\rho > 0.08\,\mathrm{fm}^{-3}$ we use the microscopic equations of state obtained in the BHF approximation described above. For comparison, we also perform calculations of neutron star structure for the case of asymmetric and β-stable nucleonic matter. The results are plotted in Fig. 4. We display the gravitational mass M_G (in units of the solar mass M_o) as a function of the radius R (panel (a)) and central baryon density n_c (panel (b)). We note that the inclusion of hyperons lowers the value of the maximum mass from about 2.1 M_o down to 1.26 M_o. This value lies below the value of the best observed pulsar mass, PSR1916+13, which amounts to 1.44 solar masses. However the observational data can be fitted if rotations are included, see the dotted line in panel (b). In this case only equilibrium configurations rotating at the Kepler frequency Ω_K are shown.

In conclusion, the main finding of our work is the surprisingly low value of the maximum mass of a neutron star, which hardly comprises the observational data. This fact indicates how sensitive the properties of the neutron stars are to the details of the interaction. In particular our result calls for the need of including realistic hyperon-hyperon interactions. However, the use of the available hyperon-hyperon interactions seem to introduce only minor changes in the results.[19] Despite the uncertainty on the NY and YY interactions, it is unlikely that one can obtain a neutron star mass substantially larger. The possible occurrence of a quark core is expected to further soften the EOS and lower the maximum mass. Therefore, the observation of a neutron star with a mass much larger than 1.4-1.5 solar mass would indicate that indeed some basic ingredient is missing in our understanding of neutron star structure.

Acknowledgments

The material presented in this contribution is the result of a fruitful collaboration, lasting for several years, with a number of people. I express my acknowledgments and gratitude to Dr. I. Bombaci, Dr. F. Burgio, Prof. L. S. Ferreira, Dr. G. Giansiracusa, Prof. U. Lombardo, Dr. H.-J. Schulze and Prof. H.Q. Song.

References

1. For a pedagogical introduction, see *Nuclear Methods and the Nuclear Equation of State*, International Review of Nuclear Physics, Vol. 9, ed. M. Baldo, (World Scientific, Singapore, 1999)
2. B. D. Day, *Brueckner-Bethe Calculations of Nuclear Matter*, Proceedings of the School E. Fermi, Varenna 1981, Course LXXIX, ed. A. Molinari (Editrice Compositori, Bologna, 1983), p. 1-72; *Rev. Mod. Phys.* **39**, 719 (1967); *Phys.*

Rev. C **24**, 1203 (1981); *Phys. Rev. Lett.* **47**, 226 (1981); R. Rajaraman and H. Bethe, *Rev. Mod. Phys.* **39**, 745 (1967).
3. K. A. Brueckner and J. L. Gammel, *Phys. Rev.* **109**, 1023 (1958).
4. F. Coester, S. Cohen, B. D. Day, and C. M. Vincent, *Phys. Rev.* C **1**, 769 (1970); B. D. Day, *Comm. Nucl. Part. Phys.* **11**, 115 (1983).
5. J. P. Jeukenne, A. Lejeunne and C. Mahaux, *Phys. Rep.* **25**, 83 (1976); M. Baldo, I. Bombaci, L. S. Ferreira, G. Giansiracusa and U. Lombardo, *Phys. Rev.* C **43**, 2605 (1991); M. Baldo, I. Bombaci, G. Giansiracusa and U. Lombardo, *Phys. Rev.* C **16**, L263 (1990).
6. R. B. Wiringa, R. A. Smith and T. L. Ainsworth, *Phys. Rev.* C **29**, 1207 (1984).
7. H. Q. Song, M. Baldo, G. Giansiracusa and U. Lombardo, *Phys. Rev. Lett.* **81**, 1584 (1998).
8. H. Q. Song, M. Baldo, G. Giansiracusa and U. Lombardo, to be published.
9. M. Baldo, G. Giansiracusa, U. Lombardo and H. Q. Song, *Phys. Lett.* B **473**, 1 (2000).
10. S. L. Shapiro and S. A. Teukolsky, *Black Holes, White Dwarfs and Neutron Stars* (John Wiley & Sons, New York, 1983).
11. M. Lacombe, B. Loiseau, J. M. Richard, R. Vinh Mau, J. Côté, P. Pirès, and R. de Tourreil, *Phys. Rev.* C **21**, 861 (1980).
12. R. B. Wiringa, V. G. J. Stocks and R. Schiavilla, *Phys. Rev.* C **51**, 38 (1995).
13. P. Maessen, Th. Rijken, and J. de Swart, *Phys. Rev.* C **40**, 2226 (1989).
14. M. Baldo, G. F. Burgio, and H.-J. Schulze, *Phys. Rev.* C **61**, 055801-1 (2000).
15. M. Baldo, I. Bombaci, and G. F. Burgio, Å **328**, 274 (1997).
16. R. Feynman, F. Metropolis, and E. Teller, *Phys. Rev.* **75**, 1561 (1949);
17. G. Baym, C. Pethick, and D. Sutherland, *Ap. J.* **170**, 299 (1971).
18. J. W. Negele and D. Vautherin, *Nucl. Phys.* A **207**, 298 (1973).
19. I. Vidaña, A. Polls, A. Ramos, L. Engvik, and M. Hjorth-Jensen, preprint University of Barcellona, 1999.

TO DRESS OR NOT TO DRESS ...

W. H. DICKHOFF
Department of Physics, Washington University, St. Louis MO 63130, USA
E-mail: wimd@wuphys.wustl.edu

Evidence from electron scattering experiments is reviewed that unequivocally demonstrates that nucleons behave as correlated particles that can not be described successfully within the confines of the mean field. Theoretical results for nucleon spectroscopic factors are discussed which demonstrate that all qualitative and many quantitative features of the experimental strength distributions are understood. It is argued that a key quantity characterizing nuclear saturation, represented by the central charge density of ^{208}Pb, is dominated solely by the effects of short-range correlations. Based on these considerations it is proposed that nuclear saturation must be studied in nuclear matter without the inclusion of long-range correlations with emphasis on the proper inclusion of short-range correlations. A self-consistent scheme, including full off-shell propagation in the determination of the effective interaction obtained by summing ladderds diagrams, fulfills this requirement. Some consequences of propagating dressing nucleons for pairing correlations and the effective interaction are discussed. A possible implication for the saturation properties of nuclear matter is pointed out.

1 Experimental evidence for dressed nucleons

Before starting with the contents of this contribution it is appropriate to be mindful of the contributions of all many-body theorists who have developed our present understanding of nuclear matter[a]. One of the critical experimental ingredients in clarifying the nature of nuclear correlations has only become available over the last decade and a half. It is therefore not surprising that all schemes that have been developed to calculate nuclear matter saturation properties are not based on the insights that these experiments provide. Exclusive experiments, involving the removal of a proton from the nucleus which is induced by a high-energy electron that is detected in coincidence with the removed proton, have given access to absolute spectroscopic factors associated with quasihole states for a wide range of nuclei.[2] The experimental results indicate that the removal of single-particle strength for quasihole states near the Fermi energy corresponds to about 65%. The spectroscopic factors obtained in these experiments can be directly related to the single-particle Green's function of the system which is given by

$$
\begin{aligned}
g(\alpha,\beta;\omega) = &\sum_m \frac{\langle \Psi_0^A | a_\alpha | \Psi_m^{A+1}\rangle \langle \Psi_m^{A+1} | a_\beta^\dagger | \Psi_0^A \rangle}{\omega - (E_m^{A+1} - E_0^A) + i\eta} \\
&+ \sum_n \frac{\langle \Psi_0^A | a_\beta^\dagger | \Psi_n^{A-1}\rangle \langle \Psi_n^{A-1} | a_\alpha | \Psi_0^A \rangle}{\omega - (E_0^A - E_n^{A-1}) - i\eta} .
\end{aligned}
\tag{1}
$$

This representation of the Green's function is referred to as the Lehmann-representation and involves the exact eigenstates and corresponding energies of

[a] All honorees qualify for this distinction but I dedicate this paper to John Clark who deserves special recognition for his insights into the nuclear matter crisis in the seventies[1] and for his continuing leadership in the many-body community.

the A- and $A \pm 1$-particle systems. Both the addition and removal amplitude for a particle from (to) the ground state of the system with A particles must be considered in Eq. (1). Only the removal amplitude has direct relevance for the analysis of the $(e, e'p)$ experiments. The spectroscopic factor for the removal of a particle in the single-particle orbit α, while leaving the remaining nucleus in state n, is then given by

$$z_\alpha = \left| \langle \Psi_n^{A-1} | \, a_\alpha \, | \Psi_0^A \rangle \right|^2 , \tag{2}$$

which corresponds to the contribution to the numerator of the second sum in Eq. (1) of state n for the case $\beta = \alpha$. Another important quantity, which also contains this information, is the spectral function associated with single-particle orbit α. The part corresponding to the removal of particles, or hole spectral function, is given by

$$S_h(\alpha, \omega) = \sum_n \left| \langle \Psi_n^{A-1} | \, a_\alpha \, | \Psi_0^A \rangle \right|^2 \delta(\omega - (E_0^A - E_n^{A-1})), \tag{3}$$

which is corresponds to the imaginary part of the diagonal elements of the propagator and characterizes the strength distribution of the single-particle state α as a function of energy in the $A - 1$-particle system. From this quantity one can therefore obtain another key ingredient that gauges the effect of correlations, namely the occupation number which is given by

$$n(\alpha) = \int_{-\infty}^{\epsilon_F} d\omega \; S_h(\alpha, \omega) = \langle \Psi_0^A | \, a_\alpha^\dagger a_\alpha \, | \Psi_0^A \rangle . \tag{4}$$

In the experimental analysis the quantum number α is related to the actual Woods-Saxon potential required to both reproduce the correct energy of the hole state as well as the shape of the corresponding $(e, e'p)$ cross section for this particular transition. The remaining parameter required to fit the actual data then becomes the spectroscopic factor associated with this transistion. In this analysis the reduction of the flux associated with the scattering of the outgoing proton is incorporated by the use of empirical optical potentials describing elastic proton-nucleus scattering data. Experiments on ^{208}Pb result in a spectroscopic factor of 0.65 for the removal of the last $3s_{1/2}$ proton.[3] Additional information about the occupation number of this orbit can be obtained by analyzing elastic electron scattering cross sections of neighboring nuclei.[4] The actual occupation number for the $3s_{1/2}$ proton orbit obtained from this analysis is about 10% larger than the quasihole spectroscopic factor.[5] A recent analysis of the $(e, e'p)$ reaction on ^{208}Pb in a wide range of missing energies and for missing momenta below 270 MeV/c yields information on the occupation numbers of more deeply bound orbitals. The data suggest that all deeply bound orbits are depleted by the same amount corresponding to slightly less than 20%.[6]

2 Theoretical description of experimental data

The global depletion of the single-particle strength in about the same amount for all states as observed for ^{208}Pb is expected on the basis of the experience that has been obtained with calculating occupation numbers in nuclear matter with the inclusion of short-range correlations.[7] Such calculations suggest that about 15% of the single-particle strength in heavy nuclei is removed from the Fermi sea leading to the occupation of high-momentum states. This can be interpreted as a clear signature of the influence of short-range correlations. In turn, all these results reflect on one of the key quantities determining nuclear saturation empirically. Elastic electron scattering from ^{208}Pb[8] clearly pinpoints the value of the central charge density in this nucleus. By multiplying this number by A/Z one obtains the relevant central density of heavy nuclei, corresponding to 0.16 nucleons/fm^3 or $k_F = 1.33$ fm^{-1}. Since the presence of nucleons at the center of a heavy nucleus is confined to s nucleons, and their depletion is dominated by short-range correlations, one may conclude that the actual value of the saturation density of nuclear matter must also be closely linked to the effects of short-range correlations. While this argument is particularly appropriate for the deeply bound $1s_{1/2}$ and $2s_{1/2}$ protons, it continues to hold for the $3s_{1/2}$ protons which are depleted predominantly by short-range effects (up to 15%) and by at most 10% due to long-range correlations.

Calculations for the strength distribution for the removal of protons from ^{48}Ca demonstrate that an excellent qualitative agreement with the experimental results is obtained when the coupling of the single-hole states to low-lying collective states is taken into account.[9] This coupling is taken into account by calculating the microscopic RPA phonons and then constructing the corresponding self-energy. The solution of the Dyson equation then provides the theoretical strength distribution. By adding the additional depletion due to short-range correlations a quantitative agreement is obtained. The corresponding occupation numbers calculated for this nucleus also indicate that the influence of collective low-lying states, associated with long-range correlations, on the occupation numbers is confined to single-particle states in the immediate vicinity of the Fermi level. For a global understanding of the strength distribution it is also necessary to account for the appearance of single-particle strength at high momenta as a direct reflection of the influence of short-range correlations. These high-momentum nucleons make up for the missing strength that has been uncovered in $(e, e'p)$ experiments. Results for ^{16}O[10,11] corroborate the expected occupation of high-momenta but put their presence at high missing energy essentially out of the reach for study with the $(e, e'p)$ reaction.

The binding energy of nuclei or nuclear matter usually includes only mean-field contributions to the kinetic energy when the calculations are based on perturbative schemes like the hole-line expansion. With the presence of high-momentum components in the ground state it becomes relevant to ask what the real kinetic and potential energy of the system look like in terms of the single-particle strength distributions. This theoretical result[12] has the general form

$$E_0^A = \langle \Psi_0^A | \hat{H} | \Psi_0^A \rangle = \frac{1}{2} \sum_{\alpha\beta} \langle \alpha | T | \beta \rangle n_{\alpha\beta} + \frac{1}{2} \sum_{\alpha} \int_{-\infty}^{\epsilon_F} d\omega\, \omega S_h(\alpha, \omega) \tag{5}$$

in the case when only two-body interactions are involved. In this equation, $n_{\alpha\beta}$ is the one-body density matrix element which can be directly obtained from the single-particle propagator. A delicate balance exists between the repulsive kinetic energy term and the attractive contribution of the second term in Eq. (5) which samples the single-particle strength weighted by the energy parameter ω. When realistic spectral distributions are used to calculate these quantities surprising results emerge.[11] Such calculations for ^{16}O indicate that the contribution of the quasihole states to Eq. (5), corresponding to the $1s_{1/2}$, $1p_{3/2}$, and $1p_{1/2}$ orbitals, comprise only 37% of the total energy leaving 63% for the continuum terms that represent the spectral strength associated with the coupling to low-energy two-hole–one-particle states. These contributions therefore contain the presence of high-momentum components in the nuclear ground state reflecting the effect of short-range correlations. Although these high momenta account for only 10% of the particles in the case of ^{16}O, their contribution to the energy is extremely important. These results give a first indication of the importance of treating the dressing of nucleons in finite nuclei in determining the binding energy per particle. It is therefore reasonable to conclude that a careful study of short-range correlations including the full fragmentation of the single-particle strength is necessary for the calculation of the energy per particle in finite nuclei. This has the additional advantage that agreement with data from the $(e, e'p)$ reaction can be used to gauge the quality of the theoretical description in determining the energy per particle. This argument can be turned inside out by noting that an exact representation of the spectroscopic strength must lead to the correct energy per particle according to Eq. (5) in the case of the dominance of two-body interactions.

3 Nuclear saturation in nuclear matter

Returning to the saturation problem in nuclear matter, it is important to comment on the recent success of the Catania group in determining the nuclear saturation curve including three hole-line contributions.[13] These calculations demonstrate that a good agreement is obtained at the three hole-line level for calculations that start from different prescriptions for the auxiliary potential. Since the contribution of the three hole-line terms are significant but indicate reasonable convergence properties compared to the two hole-line contribution, one may assume that these results provide an accurate representation of the energy per particle as a function of density. The saturation density obtained in this recent work corresponds to $k_F = 1.565\,\text{fm}^{-1}$ with a binding energy of -16.18 MeV. The conclusion appears to be appropriate that additional physics in the form of three-body forces or the inclusion of relativistic effects is necessary to repair this discrepancy with the empirical saturation properties. Before agreeing with this conclusion it is useful to remember that three hole-line contributions include a third-order ring diagram. The agreement of three hole-line calculations with advanced variational calculations[14] further emphasizes the notion that important aspects of long-range correlations are included in these calculations. This conclusion is based on the observation that hypernetted chain calculations effectively include ring-diagram contributions to the energy per particle although averaged over the Fermi sea.[15] The effect of these long-range correlations on nuclear saturation properties is not small and can be illustrated by quoting explicit results for three- and four-body ring diagrams.[16] These results for the Reid potential,[17] including only nucleons, already demonstrate that these ring-diagram terms are dominated by attractive contributions involving pion quantum numbers propagating around the rings. Furthermore these contributions increase in importance with increasing density. Including the possibility of the coupling to Δ-hole states in these ring diagrams leads to an additional large increase in the binding with increasing density.[16] The importance of these long-range contributions to the binding energy is of course related to the possible appearance of pion condensation at higher nuclear density. These long-range pion-exchange dominated contributions to the binding energy appear because of conservation of momentum in nuclear matter. For a given momentum q carried by a pion around a ring diagram one is able to sample coherently the attractive interaction that exists for q above 0.7 fm^{-1}. All ring diagrams contribute coherently when the interaction is attractive and one may therefore obtain huge attractive contributions at densities which reflect the importance of this collective pion-propagation mode.[18]

No such collective pion degrees of freedom are actually observed in finite nuclei. A simple explanation of this fact is provided by the observation that in finite nuclei both the attractive and repulsive parts of the pion-exchange interaction are sampled before a build-up of long-range correlations can be achieved. Since these contributions very nearly cancel each other, one does not see any marked effect on pion-like excited states in nuclei associated with long-range pion degrees of freedom even when Δ-hole states are included.[19] It seems therefore reasonable to call into question the relevance of these coherent long-range pion-exchange contributions to the binding energy per particle. Since the actual saturation properties of nuclei appear to be dominated by short-range correlations, as discussed above, a critical test of this idea may be to calculate nuclear saturation properties focusing solely on the contribution of short-range correlations. The recent experimental results discussed above demand furthermore that the dressing of nucleons in nuclear matter is taken into account in order to be consistent with the extensive collection of experimental data from the $(e,e'p)$ reaction that have become available in recent years. The self-consistent calculation of nucleon spectral functions obtained from the contribution to the nucleon self-energy of ladder diagrams which include the propagation of these dressed particles, fulfills this requirement.

4 Self-consistently dressed nucleons in nuclear matter

It is straightforward to write down the equation that involves the calculation of the effective interaction in nuclear matter obtained from the sum of all ladder diagrams while propagating fully dressed particles. This result is given in a partial wave representation by the following equation

$$\langle k|\,\Gamma_{LL'}^{JST}(K,\Omega)\,|k'\rangle = \langle k|\,V_{LL'}^{JST}(K,\Omega)\,|k'\rangle$$
$$+\sum_{L''}\int_0^\infty dq\, q^2\ \langle k|\,V_{LL''}^{JST}(K,\Omega)\,|q\rangle\, g_f^{II}(q;K,\Omega)\,\langle q|\,\Gamma_{LL''}^{JST}(K,\Omega)\,|k'\rangle\,, \qquad (6)$$

where k, k', and q denote relative and K the total momentum involved in the interaction process. Discrete quantum numbers correspond to total spin, S, orbital angular momentum, L, L', L'', and the conserved total angular momentum and isospin, J and T, respectively. The energy Ω and the total momentum K are conserved and act as parameters that characterize the effective two-body interaction in the medium. The critical ingredient in Eq. (6) is the noninteracting propagator g_f^{II} which describes the propagation of the particles in the medium from interaction to interaction. For fully dressed particles this propagator is given by

$$g_f^{II}(k_1,k_2;\Omega) = \int_{\epsilon_F}^{\infty} d\omega_1 \int_{\epsilon_F}^{\infty} d\omega_2 \, \frac{S_p(k_1,\omega_1)S_p(k_2,\omega_2)}{\Omega-\omega_1-\omega_2+i\eta} - \int_{-\infty}^{\epsilon_F} d\omega_1 \int_{-\infty}^{\epsilon_F} d\omega_2 \, \frac{S_h(k_1,\omega_1)S_h(k_2,\omega_2)}{\Omega-\omega_1-\omega_2-i\eta}, \tag{7}$$

where individual momenta k_1 and k_2 have been used instead of total and relative momenta as in Eq. (6). The dressing of the particles is expressed in the use of particle and hole spectral functions, S_p and S_h, respectively. The particle spectral function, S_p, is defined as a particle addition probability density in a similar way as the hole spectral function in Eq. (3) for removal. These spectral functions take into account that the particles propagate with respect to the correlated ground state incorporating the presence of high-momentum components in the ground state. This treatment therefore provides the correlated version of the Pauli principle and leads to substantial modification with respect to the Pauli principle effects related to the free Fermi gas. The corresponding propagator is obtained from Eq. (7) by replacing the spectral functions by strength distributions characterized by δ-functions as follows

$$S_p(k,\omega) = \theta(k-k_F)\delta(\omega-\epsilon(k)),$$
$$S_h(k,\omega) = \theta(k_F-k)\delta(\omega-\epsilon(k)), \tag{8}$$

which leads to the Galitski-Feynman propagator including hole-hole as well as particle-particle propagation of particles characterized by single-particle energies $\epsilon(k)$. Discarding the hole-hole propagation then yields the Brueckner ladder diagrams with the usual Pauli operator for the free Fermi gas. The effective interaction obtained by solving Eq. (6) using dressed propagators can be used to construct the self-energy of the particle. With this self-energy the Dyson equation can be solved to generate a new incarnation of the dressed propagator. The process can then be continued by constructing anew the dressed but noninteracting two-particle propagator according to Eq. (7). At this stage one can return to the ladder equation and so on, until self-consistency is achieved for the complete Green's function which is then legitimately called a self-consistent Green's function.

While this scheme is easy to present in equations and words, it is quite another matter to implement it. The recent accomplishment of implementing this self-consistency scheme[20] builds upon earlier approximate implementations. The first nuclear-matter spectral functions were obtained for a semirealistic interaction by employing mean-field propagators in the ladder equation.[21] Spectral functions for the Reid interaction were obtained by still employing mean-field propagators in the ladder equation but with the introduction of a self-consistent gap in the single-particle spectrum to take into account the pairing instabilities obtained for a realistic interaction.[22] The first solution of the effective interaction using dressed propagators was obtained by employing a parametrization of the spectral functions.[23] The calculations employing dressed propagators in determining the effective interaction demonstrate that at normal density one no longer runs into pairing instabilities on account of the reduced density of states associated with the reduction of the strength of the quasiparticle pole, z_{k_F}, from 1 in the Fermi gas to 0.7 in the case of dressed propagators. For two-particle propagation this leads to a reduction factor of $z_{k_F}^2$ corresponding to about 0.5 that is strong enough to push even the pairing instability in tl $^3S_1$-$^3D_1$ channel to lower densities.[24] The consequences for the scattering process of interacting particles in nuclear matter characterized by phase shifts and cross sections are also substantial and lead to a reduction of the cross section in a wide range of energies.[24]

The current implementation of the self-consistent scheme for the propagator across the summation of all ladder diagrams includes a parametrization of the imaginary part of the nucleon self-energy. Employing a representation in terms of two gaussians above and two below the Fermi energy, it is possible to accurately represent the nucleon self-energy as generated by the contribution of relative S-waves (and including the tensor coupling to the $^3D_1$ channel).[20] Self-consistency at a density corresponding to $k_F = 1.36$ fm^{-1} is achieved in about ten iteration steps, each involving a considerable amount of computer time.[20] It is important to reiterate that this scheme isolates the contribution of short-range correlations to the energy per particle which is obtained from Eq. (5). If the assertion is correct that long-range pion-exchange contributions to the energy per particle need not be considered in explaining nuclear saturation properties, it is quite feasible that a very different saturation curve is obtained with the present scheme.

5 Conclusions

A review of experimental data that exhibit clear evidence for the notion that nucleons in nuclei are dressed particles is given. Based on these considerations and the success of theoretical calculations to account for the qualitative features of the single-particle strength distributions it is suggested that this dressing must be taken into account in calculations of the energy per particle. By identifying the dominant contribution of short-range correlations to the empirical saturation density, it is argued that these correlations need to be isolated in the study of nuclear matter. Current three hole-line calculations include contributions of long-range correlations. It is argued that inclusion of such correlations, especially those involving pion propagation, leads to an increase in the theoretical saturation density. Since this collectivity in the pion channel is not observed in nuclei, it is proposed that the corresponding correlations in nuclear matter are not relevant for the study of nuclear saturation and should therefore be excluded. A scheme which fulfills this requirement and includes the propagation of dressed particles, as required by experiment, is outlined. Successful implementation of this scheme has recently been demonstrated.[20] It is pointed out that these new calculations may lead to new insight into the long-standing problem of nuclear saturation.

Acknowledgments

This work was supported by the U.S. National Science Foundation under Grant No. PHY-9900713.

References

1. J. W. Clark, *Nucl. Phys.* A **328**, 587 (1979).
2. L. Lapikás, *Nucl. Phys.* A **553**, 297c (1993).
3. I. Sick and P. de Witt Huberts, *Comm. Nucl. Part. Phys.* **20**, 177 (1991).
4. G. J. Wagner, *AIP Conf. Proc.* **142**, 220 (1986).
5. P. Grabmayr *et al.*, *Phys. Lett.* B **164**, 15 (1985).
6. L. Lapikás, private communication July 2000.
7. B. E. Vonderfecht, W. H. Dickhoff, A. Polls, and A. Ramos, *Phys. Rev.* C **44**, R1265 (1991).
8. B. Frois *et al.*, *Phys. Rev. Lett.* **38**, 152 (1977).
9. G. Rijsdijk, K. Allaart, W. H. Dickhoff, *Nucl. Phys.* A **550**, 159 (1992).
10. H. Müther and W. H. Dickhoff, *Phys. Rev.* C **49**, R17 (1994).
11. H. Müther, A. Polls, and W. H. Dickhoff, *Phys. Rev.* C **51**, 3040 (1995).
12. V. M. Galitski and A. B. Migdal, *Sov. Phys. JETP* **34**, 96 (1958).
13. H. Q. Song, M. Baldo, G. Giansiracusa, and U. Lombardo, *Phys. Rev. Lett.* **81**, 1584 (1998).
14. B. D. Day and R. B. Wiringa, *Phys. Rev.* C **32**, 1057 (1985).
15. A. D. Jackson, A. Lande, and R. A. Smith, *Phys. Rep.* **86**, 55 (1982).
16. W. H. Dickhoff, A. Faessler, and H. Müther, *Nucl. Phys.* A **389**, 492 (1982).
17. R. V. Reid, *Ann. Phys. (NY)* **50**, 411 (1968).
18. W. H. Dickhoff, *Prog. Part. Nucl. Phys.* **12**, 529 (1984).
19. P. Czerski, W. H. Dickhoff, A. Faessler, and H. Müther, *Phys. Rev.* C **33**, 1753 (1986).
20. E. P. Roth, Ph. D. Thesis, Washington University, St. Louis, 2000.
21. A. Ramos, Ph. D. Thesis, University of Barcelona, 1988.
22. B. E. Vonderfecht, Ph. D. Thesis, Washington University, St. Louis, 1991.
23. C. C. Gearhart, Ph. D. Thesis, Washington University, St. Louis, 1994.
24. W. H. Dickhoff, C. C. Gearhart, E. P. Roth, A. Polls, and A. Ramos, *Phys. Rev.* C **60**, 064319 (1999).

FERMI HYPERNETTED CHAIN EQUATIONS AND NUCLEAR MANY-BODY PHYSICS

A. FABROCINI
Dipartimento di Fisica, Università di Pisa,
and Istituto Nazionale di Fisica Nucleare, sezione di Pisa,
I-56100 Pisa, Italy
E-mail: Adelchi.Fabrocini@pi.infn.it

We briefly review some applications of the Correlated Basis Functions theory and Fermi hypernetted chain equations to the study of infinite and finite nuclear systems.

1 Introduction

In Anno Domini MCMLXXIV a paper by Sergio Rosati and Stefano Fantoni[1] provided the large (and ever increasing) community of many-body physicists with a new, powerful tool to address the complexity of strongly interacting systems. Correlated basis functions (CBF) theory had already been put on a firm basis by the work of Feenberg.[2] Jastrow correlated wave functions had allowed studies of boson systems, such as liquid ^{4}He and charged bosons, with an unprecedented accuracy, well beyond the reach of standard perturbative approaches. The cluster expansion techniques had been extended to the quantum case exploiting the analogy between the partition function of a classical gas and the two-body distribution function. So, the well known classical hypernetted chain (HNC) equations had been straightforwardly used in Bose fluids studies.

Fermi fluids were, however, a more severe challenge. A ground state correlated wave function, $\Psi_0(1,2...A)$, for A interacting fermions may be built by applying the many-body correlation operator, $F(1,2...A)$, to an independent particle model function, $\Phi_0(1,2...A)$,

$$\Psi_0(1,2...A) = F(1,2...A)\Phi_0(1,2...A) \quad , \tag{1}$$

where the operator F is meant to take care of the *dynamical correlations*, or the modifications induced by the interaction on Φ_0. Φ_0 includes antisymmetrization effects and, possibly, long-range correlations due to collective excitations (as BCS type states or surface vibrations). In infinite, homogeneous matter, Φ_0 is a Slater determinant of plane waves, but, in general, single particle wave functions, $\phi_\alpha(i)$, obtained by some mean field potential are used.

The simplest form of $F(1,2...A)$ is:

$$F(1,2...A) = \prod_{i<j=1,A} f(r_{ij}) \quad , \tag{2}$$

i.e. a product of two-body correlation operators, $f(r_{ij})$, depending on the interparticle distance, r_{ij}, only (Jastrow correlated wave function). The variational principle teaches us that the best choice for the correlation is obtained by minimizing the ground state energy, $E_0 = \langle\Psi_0|H|\Psi_0\rangle/\langle\Psi_0|\Psi_0\rangle$, where H is the many-body hamiltonian.

Cluster expansion can be used in correlated Fermi systems to evaluate matrix elements, bearing in mind that the antisymmetry of the wave function generates *statistical correlations*, which are embedded in $|\Phi_0|^2$. So, the cluster terms must contain both types of correlations. This unfortunate peculiarity had forced to adopt low order cluster expansions for fermions, and to sum only the very first terms, in contrast with bosons, where the HNC equations actually sum diagrams to all orders.

Fantoni and Rosati carefully studied the properties of Jastrow correlated wave functions for infinite Fermi systems. They demonstrated that the cluster expansion of the two-body distribution function is linked and fully irreducible. This implies that reducible diagrams (*i.e.* those factorizing in a product of two, or more subdiagrams having a common reducibility point) exactly cancel, due to the properties of the plane waves Slater determinant. This is a major difference with boson fluids, where irreducibility is satisfied only at the $1/A$ order. Moreover, they derived a set of integral equations to sum all the cluster diagrams contributing to the two-body distribution function, known, since then, as the Fermi hypernetted chain (FHNC) equations.[3]

Since their derivation, the FHNC equations have been used in a variety of many-body systems, belonging to different fields of physics. We just mention the progress achieved in the description of liquid ^{3}He due to the use of CBF theory and FHNC equations. The interaction between Helium atoms is relatively simple, but exhibits a strong repulsion at short interatomic distances, that induces important dynamical correlations. In addition, the Helium systems densities are usually large. These two properties make CBF the leading theory in Helium physics. Nuclear systems are generally lower density ones. However, the interaction, and, as a consequence, the correlations are very complicated, depending on the relative state of the interacting nucleons. FHNC and its extensions have been used to study nuclear and neutron matter with modern two- and three-nucleon potentials, which fit at best the large body of available nucleon-nucleon scattering data. These studies have produced nuclear equations of state (EOS) in good agreement with the empirical information extracted from the mass formulae and neutron stars structures in accordance with the current observations.[4] Other nuclear matter properties have been examined within the CBF/FHNC approach, such as the momentum distribution,[5] inclusive responses[6–9] and one-body Green's functions.[10,11]

Following the successes obtained by CBF in infinite matter, the very same methodology began to be applied to actual nuclei. The FHNC machinery has been upgraded to finite systems, giving birth to its renormalized version (RFHNC).[12,13] At present, the variational approach to medium-heavy doubly closed shell nuclei has attained the same degree of accuracy as in nuclear matter when realistic interactions are adopted.[14]

In this contribution we will shortly review and comment on the microscopic description of infinite and finite nuclear systems obtained by using the FHNC equations and some of the most accurate modern hamiltonians.

2 Correlated nuclear systems

The increasing number of experimental data and the sophisticated theories developed to address the problem of an accurate description of nuclear systems make the knowledge of the nuclear interaction to improve steadily. The most recent nucleon-nucleon (NN) potentials reproduce a huge amount of NN scattering data (~1800 *pp* and ~2500 *np* data) with χ^2 ~1 and break the charge independence and charge symmetry.[15–17] However, well known features of the two-nucleon interaction are that light nuclei are underbound and the nuclear matter saturation density is not correctly reproduced. Relativistic effects, extra degrees of freedom (Δ's) or many-body forces are usually advocated in order to cure these deficiencies. If we decide to follow the many-nucleon potential strategy, then the first step consists in introducing three-nucleon interactions (TNI). Their knowledge, however, is far less accurate than that of the NN potential, since a much lower number of experimental data is available and building theoretical models is considerably more involved. The available TNI models are fitted to light nuclei binding energy and bring the nuclear matter saturation density very close to the empirical value.

Within this approach, a realistic hamiltonian is written as:

$$H = -\frac{\hbar^2}{2m}\sum_i \nabla_i^2 + \sum_{i<j} v_{ij} + \sum_{i<j<k} v_{ijk} \tag{3}$$

where v_{ij} and v_{ijk} are the two- and three-nucleon potentials. v_{ij} is given by the sum of three pieces: an electromagnetic part, v_{ij}^{em}, containing Coulomb, Darwin-Foldy, vacuum polarization and magnetic moments contributions; the One Pion Exchange potential (OPEP), v_{ij}^{π}, providing the long range part; the remaining intermediate and short range potential, v_{ij}^{R}, originates by two or more pions and heavier mesons (ρ and ω) in standard meson exchange theory. One of the most popular NN potentials is the Argonne v_{18} interaction,[15] which contains 43 parameters and it is given by a sum of 18 terms

$$v_{ij} = \sum_{p=1,18} v_p(r_{ij}) O_{ij}^p \quad , \tag{4}$$

where the first 14 terms are isoscalar, with

$$O_{ij}^{p=1,14} = \left[1, \sigma_i\cdot\sigma_j, S_{ij}, \mathbf{L}\cdot\mathbf{S}, \mathbf{L}^2, \mathbf{L}^2\sigma_i\cdot\sigma_j, (\mathbf{L}\cdot\mathbf{S})^2\right] \otimes \left[1, \tau_i\cdot\tau_j\right] \quad , \tag{5}$$

and the $p \geq 15$ ones are of the isovector and isotensor type.

Most of the TNI attraction is given by two-pion exchange,[18] $v_{ijk}^{2\pi}$, and the remainder of the potential, v_{ijk}^{R}, is often taken as phenomenological. Attempts are under way to enlarge its microscopic grounds by the introduction of three-pion two-Delta diagrams. The Urbana TNI models[19] express $v_{ijk}^{2\pi,R}$ in terms of three parameters, $A_{2\pi}, C_{2\pi}$ e U_0, chosen to reproduce the binding of A=3,4 nuclei.

It appears natural that the choice of the correlation operator may depend on the interaction. A form of $F(1,2...A)$, suitable to those nuclear systems described by the hamiltonian (3), is

$$F(1,2...A) = \mathcal{S}\left[\prod_{i<j=1,A} F_{ij}\right] \quad , \tag{6}$$

i.e. a symmetrized product of two-body correlation operators, F_{ij}, chosen consistently with the interaction. F_{ij} is given by

$$F_{ij} = \sum_{p=1,8} f^p(r_{ij})O^p_{ij} \quad , \tag{7}$$

where the sum runs up to the spin-orbit components. The Jastrow correlation corresponds to the $p = 1$ term.

The non-commutativity of the correlations (7) is the reason underlying the introduction of the symmetrizer in (6). It has not been possible to derive the full FHNC equations with this type of correlation. So, clusters built with the Jastrow components only are completely summed by FHNC, while just the leading contributions in the spin-isospin correlations are considered. This task is accomplished via the single-operator-chain approximation[20] (FHNC/SOC), where chains formed by one operatorial correlation per side are summed at all orders. Originally devised for infinite matter, the FHNC/SOC approximation has been lately extended to finite nuclei.[21] A natural check of the SOC truncation is given by the degree of accuracy in fulfilling the normalizations of quantities like the one- and two-body densities, or the spin-isospin saturations, in the cases where they hold. At typical nuclear densities, or in nuclei as ^{16}O and ^{40}Ca, these sum rules are usually satisfied within a few percent.

2.1 Nuclear and neutron matter

We present now some results obtained by CBF theory in infinite matter made up of nucleons, without considering the Coulomb interaction (nuclear matter, NM). Information on nuclear matter may be only empirical and are extracted by the volume term of the mass formulae. The main parameters characterizing NM are its density, $\rho_{\rm NM}$, and asymmetry, $\beta = (N - Z)/N$, N and Z being the number of neutrons and protons, respectively. So, symmetric nuclear matter (SNM, $N = Z$) has β=0 and pure neutron matter (NeM, $Z = 0$) β=1. Actually, neutron stars are a very good approximation to almost pure NeM, and their structure is one of the most studied topics in modern nuclear many-body physics. In fact, its correct description provides a demanding test of the available NN interactions, even at very high densities.

The SNM saturation point is known to lay at $\rho_{\rm NM}$=0.16 fm^{-3}, with an energy per nucleon of $E_{\rm NM}/A$=-16 MeV. Figure 1 shows the density dependence of the energy per nucleon of symmetric nuclear matter and neutron matter for several interactions in FHNC/SOC. The square in the left panel gives the empirical SNM saturation point. The stars are the results obtained with the Urbana v_{14} NN interaction, supplemented by a density dependent modification, meant to mimic many-body forces and whose parameters are fitted to reproduce the SNM saturation.[22] The dashed lines are the energies obtained with the Argonne v_{14} NN

Table 1. Neutron star properties with the A14+UVII neutron matter EOS. The Table gives the central density, ρ_c, mass-density, ϵ_c, and pressure, P_c, the gravitational and the amu masses, M_G and $M_A - M_G$, (in units of the solar mass, $M_\odot$), the radius, R, the momentum of inertia, I, and the surface redshift, z.

property	value
ρ_c	0.66 fm^{-3}
ϵ_c	1.2×10^{15} g cm^{-3}
P_c	2.1×10^{35} dyn cm^{-2}
M_G	1.4 $M_\odot$
$M_A - M_G$	0.18 $M_\odot$
R	10.4 km
I	1.2×10^{45} g cm^2
z	0.29

potential[23] (A14). A14 is an isoscalar potential, having only the first 14 components of Eq. (4). The solid lines have been obtained by adding to A14 the Urbana VII three-nucleon interaction[24] (A14+UVII). The introduction of UVII brings the saturation density of SNM much closer to its empirical value than the pure A14 model. However, SNM is still underbound by ~3 MeV. The more recent Argonne v_{18} potential and the Urbana IX model of TNI[25] provide similar results (circles).

The A14 model has been carefully studied within the Bethe-Brueckner-Goldstone (BBG) expansion theory in SNM. The convergence of BBG has been checked up to the three hole-line level of the expansion in the standard[26] and continuous[27] choices. The two approaches yield similar results at $\rho_{\rm NM} = 0.15$ fm^{-3} (E/A=-13.6 MeV with the standard choice and E/A=-14.0 MeV with the continuous one), whereas, at the same density, FHNC/SOC gives E/A=-11.5 MeV. The difference may be traced back to CBF perturbative corrections to be added to the purely variational FHNC/SOC value. In fact, corrections to the variational ground state (1) due to correlated two-particle two-hole states, $\Psi_{2p2h}(A) = F(A)\Phi_{2p2h}(A)$, give a correction of about 3 MeV to the energy per nucleon,[28] making the CBF estimate much closer to the BBG one.

The neutron star structure is obtained once the neutron matter EOS is given. The properties derived by the numerical integration of the general relativistic Tolman-Oppenheimer-Volkov (TOV) equation[29,30] for a non-rotating, spherically symmetric neutron star, using the A14+UVII neutron matter EOS, are given in Table 1. All the observational data are consistent with a mass of $1.4 \pm 0.2 M_\odot$ and a redshift of $z = 0.25 - 0.35$. In the case of the source MXB 1336-536 a second mass-ratio can be obtained from gravitational and transverse Doppler red-shifted spectral features, leading to a radius estimate of R=10.3 km with an error of ±10%. These results, even within the uncertainties due to their extraction from the actual observations, may serve as a check of the available EOS and of the underlying hamiltonians. All the data are consistent with the A14+UVII model, and, at present, only soft EOS are ruled out.

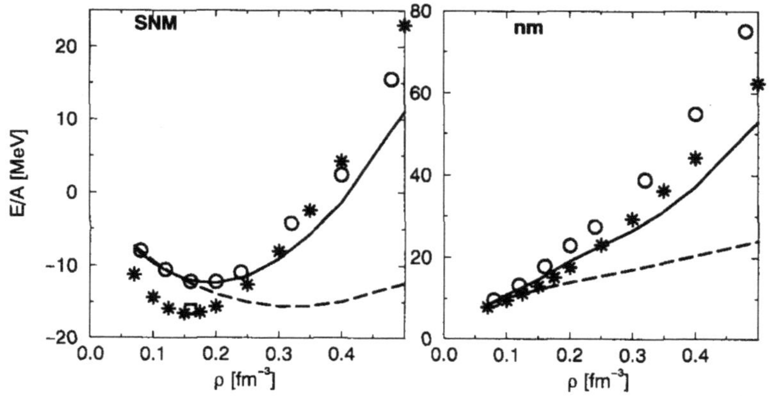

Figure 1. Energy per nucleon vs. density in symmetric nuclear matter (SNM, left panel) and pure neutron matter (NeM, right panel) with different hamiltonians. See text.

Table 2. Ground state energies for doubly closed shell nuclei with the S3 potential and Jastrow correlated wave functions. Energies in MeV.

nucleus	E/A	$(E/A)_{expt}$
^{12}C	-3.84	-7.68
^{16}O	-8.20	-7.98
^{40}Ca	-9.78	-8.55
^{48}Ca	-8.43	-8.67
^{208}Pb	-8.50	-7.87

2.2 Doubly closed shell nuclei

The FHNC formalism has been applied in the last decade to doubly closed shell nuclei, both for Jastrow[13] and spin-isospin dependent[14,21,31] correlations. The cluster expansion for finite systems is no longer irreducible, and reducible diagrams must be explicitly computed as vertex corrections in the renormalized version of the FHNC equations. The ground state of nuclei, ranging from light ^{12}C to heavy ^{208}Pb, has been studied within the RFHNC and RFHNC/SOC approaches.

A collection of the g.s. energies per particle with the semi-realistic, central Afnan and Tang (S3) potential[32,33] is given in Table 2. The RFHNC equations have been solved with correlations depending on τ_z, and distinguishing between the different pairs of nucleon ($f_{nn} \neq f_{pp} \neq f_{np}$). Because of the poor quality of the interaction, the comparison with the experimental values has a limited validity. However, it is interesting to note that a saturation trend of the binding energy along the mass number shows up.

The RFHNC/SOC energies and r.m.s. radii obtained with the A14+UVII and the Argonne v_8' + Urbana IX[25] (A8'+UIX) realistic interactions are shown in Table 3. The A14+UVII results for ^{16}O are compared with the available coupled cluster[34] (CC) and cluster Monte Carlo[35] (CMC) ones, obtained with a similar

Table 3. ^{16}O and ^{40}Ca ground state energies per nucleon and radii for the Argonne v_{14} + Urbana VII and the Argonne v_8' + Urbana IX models with the RFHNC/SOC, coupled cluster (CC) and cluster Monte Carlo (CMC) methods. Energies in MeV and radii in fm.

			E/A	rms
A14+UVII	^{16}O	FHNC	-5.97	2.44
		CC	-6.10	2.86
		CMC	-6.90	2.43
A8'+UIX	^{16}O	FHNC	-5.41	2.67
		expt	-7.98	2.73
	^{40}Ca	FHNC	-6.64	3.39
		expt	-8.55	3.48

wave function and using Monte Carlo sampling to compute the expectation values. The A14+UVII energies are close to the CC estimates, while the differences with CMC come from the approximations in the RFHNC/SOC scheme. FHNC and CMC give similar radii, both of them smaller than CC. Explicit three-body correlations have been found to provide ~0.85 MeV/A extra binding[35] for the A14+UVII model.

Charge densities, $\rho_c(r_1)$, two-body distribution functions, $\rho_2(r_{12})$ and several integrated cross sections have been analyzed within the CBF scheme. In particular, the Coulomb sums, $S_L(q)$, given by

$$S_L(q) = 1 + \frac{1}{Z} \int d^3r_1 \int d^3r_2 \;\; e^{i\mathbf{q}\cdot\mathbf{r}_{12}} \left[\rho_{pp}(\mathbf{r}_1, \mathbf{r}_2) - \rho_c(r_1)\rho_c(r_2)\right] \quad , \tag{8}$$

where $\rho_{pp}(\mathbf{r}_1, \mathbf{r}_2)$ is the proton-proton two-body density, totally agree with those extracted from the world data on inclusive quasi-elastic electron scattering[36] experiments in ^{12}C, ^{40}Ca, and ^{56}Fe.

3 Conclusions

Microscopic many-body theories have reached such a level of sophistication and accuracy that many features of strongly interacting systems are now quantitatively understood. The Correlated Basis Function approach has played a pivotal role in this game. An enormous boost in the application of CBF came from the derivation of the Fermi hypernetted chain equations by Fantoni and Rosati.

The use of FHNC in nuclear physics has led to the derivation of accurate equations of state for infinite nuclear matter and to the description of the ground state of doubly closed shell nuclei in terms of realistic nuclear hamiltonians. Moreover, it has been possible to study other important quantities in nuclear matter, as inclusive or semi-exclusive cross sections. In general, it is now possible to go beyond the simple mean field picture and to investigate the correlation effects induced by the nuclear interaction. The FHNC equations are among those advances in many-body theories that have made realistic this exciting perspective.

References

1. S. Fantoni and S. Rosati, *Nuovo Cim.* A **20**, 179 (1974).
2. E. Feenberg, *Theory of quantum fluids*, (Academic Press, New York, 1969).
3. S. Fantoni and S. Rosati, *Nuovo Cim.* A **25**, 593 (1975).
4. R. B. Wiringa, V. Fiks and A. Fabrocini, *Phys. Rev.* C **38**, 1010 (1988).
5. S. Fantoni and V. R. Pandharipande, *Nucl. Phys.* A **427**, 473 (1984).
6. A. Fabrocini and S. Fantoni, *Nucl. Phys.* A **503**, 375 (1989).
7. O. Benhar *et al.*, *Phys. Rev.* C **44**, 2328 (1991).
8. A. Fabrocini, *Phys. Lett.* B **322**, 171 (1994).
9. A. Fabrocini, *Phys. Rev.* C **55**, 338 (1997).
10. O. Benhar, A. Fabrocini and S. Fantoni, *Nucl. Phys.* A **505**, 267 (1989).
11. O. Benhar, A. Fabrocini and S. Fantoni, *Nucl. Phys.* A **552**, 201 (1992).
12. E. Krotscheck, *Nucl. Phys.* A **465**, 461 (1987).
13. G. Co', A. Fabrocini, S. Fantoni and I. E. Lagaris, *Nucl. Phys.* A **549**, 439 (1992).
14. A. Fabrocini, F. Arias de Saavedra and G. Co', *Phys. Rev.* C **61**, 044302 (2000).
15. R. B. Wiringa, V. G. J. Stoks and R. Schiavilla, *Phys. Rev.* C **51**, 38 (1995).
16. V. G. J. Stoks, R. A. M. Klomp, C. P. F. Terheggen and J. J. DeSwart, *Phys. Rev.* C **49**, 2950 (1994).
17. R. Machleidt, F. Sammarruca and Y. Song, *Phys. Rev.* C **53**, R1483 (1996).
18. J. Fujita and H. Miyazawa, *Prog. Theor. Phys.* **17**, 360 (1957).
19. J. Carlson, V. R. Pandharipande and R. B. Wiringa, *Nucl. Phys.* A **401**, 59 (1983).
20. V. R. Pandharipande and R. B. Wiringa, *Rev. Mod. Phys.* **51**, 821 (1979).
21. A. Fabrocini, F. Arias de Saavedra, G. Co' and P. Folgarait, *Phys. Rev.* C **57**, 1668 (1998).
22. I. E. Lagaris and V. R. Pandharipande, *Nucl. Phys.* A **359**, 349 (1981).
23. R. B. Wiringa, R. A. Smith and T. L. Ainsworth, *Phys. Rev.* C **29**, 1207 (1984).
24. R. Schiavilla, V. R. Pandharipande and R. B. Wiringa, *Nucl. Phys.* A **449**, 219 (1986).
25. B. S. Pudliner, *et al.*, *Phys. Rev.* C **56**, 1720 (1997).
26. B. D. Day, *Phys. Rev.* C **24**, 1203 (1981).
27. H. Q. Song, M. Baldo, G. Giansiracusa and U. Lombardo, *Phys. Rev. Lett.* **81**, 1584 (1998).
28. A. Fabrocini and S. Fantoni, *Phys. Lett.* B **298**, 263 (1993).
29. R. C. Tolman, *Proc. Natl. Acad. Sci. U. S. A.* **20**, 3 (1934).
30. J. Oppenheimer and G. Volkoff, *Phys. Rev.* **55**, 374 (1939).
31. F. Arias de Saavedra, G. Co', A. Fabrocini and S. Fantoni, *Nucl. Phys.* A **605**, 359 (1996).
32. I. R. Afnan and Y. C. Tang, *Phys. Rev.* **175**, 1337 (1968).
33. R. Guardiola *et al.*, *Nucl. Phys.* A **371**, 79 (1981).
34. J. H. Heisenberg and B. Mihaila, *Phys. Rev.* C **59**, 1440 (1999).
35. S. C. Pieper, R. B. Wiringa and V. R. Pandharipande, *Phys. Rev.* C **46**, 1741 (1991).
36. J. Jourdan, *Nucl. Phys.* A **603**, 117 (1996); private communication.

NUCLEAR MATTER WITH THE AUXILIARY FIELD DIFFUSION MONTE CARLO METHOD

S. FANTONI, A. SARSA
International School for Advanced Studies, SISSA
Via Beirut 2/4, I-34014 Trieste, Italy

K. E. SCHMIDT
Department of Physics and Astronomy
Arizona State University
Tempe, AZ, USA

The auxiliary-field diffusion Monte Carlo method is applied to nuclear matter and the alpha particle. Simulations have been carried out with up to 66 neutrons for the pure neutron matter case and 76 nucleons for symmetrical nuclear matter. Results are given for spin-isospin dependent central interactions, which show a lowering of the energy per particle of about 1 MeV with respect to the best variational estimates for infinite matter. An analysis of finite size effects, carried out with the FHNC method for nuclear systems in a periodic box, is also presented. The effect of the constrained path approximation on the energy has been analyzed by a simulation for the alpha particle.

1 Introduction

Several important processes occurring in matter with very large densities, particularly in stellar and primordial environments of our universe are essentially not reproducible in our laboratories and rely heavily on theoretical simulations.

Properties of this matter, such as the equation of state and neutrino and photon opacities, seem to govern the evolution of supernovae and the neutron stars physics.[1]

The nuclear astrophysics phenomenology, requires more quantitative studies of the above processes, than those made in the past. It is believed that the strong correlations present amongst the nucleons in dense matter changes their dynamics.

In recent years, important developments have been made in many-body theories for strongly interacting particles, and some of these are discussed in this book.

However, the strong spin-isospin dependence of the $N - N$ interaction is still posing non trivial difficulties to practically all the existing many-body methods, if we want to reach the accuracy required today. This is particularly true for the case of medium to heavy nuclei and nuclear matter. The non-perturbative methods, such as Green's function Monte Carlo (GFMC), Faddeev theory, hyperspherical theory are limited in the number of nucleons they can treat. The largest systems they will able to deal with in the near future is $A \sim 14$, which is already a tremendous achievement if one consider the complexity of the problem. The approaches based on perturbation theory, like correlated basis function, Brueckner-Hartree-Fock or coupled cluster theory, suffer from the same spin problem or from uncertain perturbative convergence. It is extremely hard to imagine next order corrections in these theories.

Recently, we have developed a new diffusion Monte Carlo method, based on auxiliary field variables (AFDMC), which can handle nuclear Hamiltonians and a relatively large number of nucleons. In this approach the scalar parts of the hamiltonian are propagated as in standard diffusion Monte Carlo (DMC). Auxiliary fields are introduced to replace the spin-isospin dependent interactions between pairs of particles with interactions between particles and auxiliary fields. Integrating over the auxiliary fields reproduces the original spin-isospin dependent interaction. The method consists of a Monte Carlo sampling of the auxiliary fields and then propagating the spin-isospin variables at the sampled values of the auxiliary fields. This propagation results in a rotation of each particle's spin-isospin spinor. In addition, a constraint analogous to the fixed-node approximation for standard GFMC is introduced, which requires that the real part of the overlap with a trial function remains positive. The method can deal with a number of neutrons different from that of protons, as well as for non-cubic boxes or non-spherical mean field wave functions. A more detailed description of the method is given in Ref. 2.

We have already demonstrated that we can calculate energies with a low variance (< 0.1 MeV per nucleon) for a neutron drop with $A = 7, 8$ and for neutron matter with fairly realistic interactions that include tensor, spin-orbit and three-body terms. The neutron matter calculations have already been done with up to 66 neutrons in a periodic box, and they scale in particle number roughly like fermion Monte Carlo calculations with central forces.[3]

In this contribution, we report progress made for (N, Z)-matter. We will present results for v_4 model interactions. Such interactions do not include tensor, spin-orbit, three-body interaction. Therefore, our results will not be directly comparable with experimental data. The extension to more realistic interactions, although more difficult than for the neutron matter case, does not present overwhelming difficulties.

We will also present new results obtained with the recently developed periodic-box Fermi hypernetted chain (PB-FHNC) method. In view of (i) the fact that the FHNC cluster expansion is valid for any finite value of the number of particles A (and not only in the thermodynamic limit),[4] and (ii) the translational invariance of the correlation function, the PB-FHNC equations have exactly the same structure as the FHNC ones, with the convolutions done over the box volume and the Slater function changed to the equivalent one in the box.

The AFDMC method, along with more traditional GFMC methods[5] typically uses fewer than 100 particles in a box with periodic boundary conditions to simulate nuclear and neutron matter. To better estimate the finite size effects missing from this type of calculations, we have used PB-FHNC theory to compute the expectation values for the same systems as that used for Monte Carlo simulations - a fixed number of particles A in a periodic box. The finite size effects of PB-FHNC are expected to be close to those in AFDMC. The PB-FHNC method is described in Ref. 6.

2 Results

We have made calculations for two v_4 model interactions. The first one is given by the first four components of the v_8' potential,[7] hereafter denoted by v_4'. Cutting out the tensor and the spin-orbit components results in a potential that has no direct connection with the N-N data. We consider this case, because, from a methodological point of view, it is useful to study v_4 and v_6 partitions of the v_8' interaction.

The second interaction considered is the S3 potential by Afnan and Tang.[8] This is a semi-realistic interaction of the Serber type, and, therefore, defined in the even channels only. It has been built to reproduce the s-wave scattering data up to about 60 MeV, and it provides a reasonable description of both light nuclei and the binding energy of nuclear matter at saturation density. It has been used in a number of Monte Carlo and FHNC calculations on nuclei and nuclear matter. As in these calculations we have added to the original S3 potential an interaction for the odd channels, given by the repulsive term of the even channels. This modified potential is generally denoted by MS3.

2.1 AFDMC calculations

The guiding function in our AFDMC calculation is a simple trial function given by a Slater determinant of one-body space-spin orbitals multiplied by a central Jastrow correlation,

$$|\Psi_T\rangle = \left(\prod_{i<j} f_c(r_{ij})\right) \mathcal{A}\left(\prod_i |\phi_i, s_i, \tau_i\rangle\right) . \tag{1}$$

The overlap of a walker with this wave function is the determinant of the space-spin orbitals, evaluated at the walker position and spinor for each particle, and multiplied by the scalar Jastrow product.

For nuclear matter in a box of side L, the orbitals are plane waves that fit in the box times four component spinors, corresponding to neutron-up, neutron-down, proton-up, proton-down states. For the alpha particle, and, in general for a nucleus the orbitals are mean field single particle wave functions.

The guiding function is used for the importance sampling and for the path constraint. The overlap of our walkers with the trial function is complex, so the usual fermion sign problem becomes a phase problem. We constrain the path of the walkers to a region where the real part of the overlap with our trial function is positive. For spin-independent potentials this reduces to the fixed-node approximation. It is straightforward to show that if the sign of the real part is that of the correct ground state, we get the correct answer, and small deviations give second order corrections to the energy. We have not been able to prove that this constraint always gives an upper bound to the ground state energy, although it appears to do so for the calculations we have done so far. Note however that an upper bound can always be produced by calculating the expectation value of the Hamiltonian without constrained wave function. This should be possible using forward walking methods.

The use of a realistic Jastrow wave function will lower the variance in the calculations, but we can get a reasonable estimate even with a very poor Jastrow correlation. To improve the description of the nodal surface we can use combinations of $|\Psi_T>$. The number of components is only limited by the computer time needed for the calculation.

For the infinite matter cases at $\rho = 0.16\ \mathrm{fm}^{-3}$ we find little sensitivity to the time step below $\Delta\tau \sim 5 \times 10^{-5}\ \mathrm{MeV}^{-1}$. At this time step the mixed and the growth energies agree within the statistical error. We have not attempted in our preliminary calculations a systematic time step extrapolation.

The calculations are performed using the full volume of the box and not within the sphere of radius $L/2$. Moreover, summations over the potential are done over the neighboring boxes. The inclusion of the first 26 neighboring boxes were always sufficient for all the cases considered. Therefore there are no tail corrections to add to the AFDMC results reported in this paper.

2.2 FHNC calculations

The PB-FHNC calculations are performed using a simple Jastrow correlation. We have used the scalar component f_c of the correlation operator $\hat{F}(1,2)$ resulting from an FHNC/SOC calculation for infinite matter. We will denote such trial function by F_1.

We have taken the lattice version of the correlation factor $f_c(r)$ and of the $v_c(r)$, $v_\sigma(r)$, $v_\tau(r)$ and $v_{\sigma\tau}(r)$ potentials, namely

$$
\begin{aligned}
f(x,y,z) &= \prod_{mno} f_c(|(x+mL_x)\hat{x} + (y+nL_y)\hat{y} + (z+oL_z)\hat{z}|), \\
V_\alpha(x,y,z) &= \sum_{mno} v_\alpha(|(x+mL_x)\hat{x} + (y+nL_y)\hat{y} + (z+oL_z)\hat{z}|). \qquad (2)
\end{aligned}
$$

For the calculation shown we found it adequate to include only the 26 additional neighboring cells corresponding to m, n and o taking the values -1,0 and 1, exactly as in the AFDMC calculations.

The elementary diagrams have been neglected in the PB-FHNC equations, although they may give a sizable contribution in nuclear and neutron matter, as discussed below.

The FHNC/SOC results have been obtained using a standard FHNC/SOC code.[9] We have minimized the energy per particle with respect to the two main variational parameters: the healing distance of the correlations and the quenching factor of the spin-isospin correlations. In this case the trial function has scalar, spin, isospin and spin-isospin components and we will denote it by F_4.

Recent FHNC/SOC calculations on $N = Z$ nuclei[10] show that the four-body elementary diagram, occurring at the lowest order of the FHNC expansion, is not negligible and gives a repulsive contribution of 0.6–0.8 MeV to the energy per nucleon. We have computed the leading part of this diagram, namely

$$
\begin{aligned}
E_{ee}^{\mathrm{exch}}(r_{12}) &= -\frac{\rho^2}{d^3}\int d\vec{r}_{13}d\vec{r}_{14} < \hat{F}(12)\hat{H}(12)\hat{F}(12) > \\
&\times (f_c^2(r_{34})-1)l(r_{13})l(r_{32})l(r_{24})l(r_{41}), \qquad (3)
\end{aligned}
$$

where $l(r)$ is the slater function given by

$$
l(r) = j_0(k_f r) + j_2(k_f r), \qquad (4)
$$

Table 1. Results for the v_4' model of symmetrical nuclear matter at $\rho = 0.16$.The AFDMC column reports the mixed energy at a time step of 5×10^{-5} MeV^{-1}. The PB-FHNC and the FHNC/SOC results refer to the Jackson-Feenberg energy. PB-FHNC is calculated with the F_1 trial function and FHNC/SOC with the F_4. The corrections (in parentheses) for PB-FHNC and FHNC-SOC correspond to the contribution from the elementary diagram of Eq. (3). For AFDMC the number in parentheses gives the statistical error. The energies per particle are in MeV.

A	E_{free}	PB-FHNC	FHNC/SOC	AFDMC
28	22.427	1.30	-	0.34(3)
2060	22.136	1.95	-	↓
∞	22.108	1.92(+.56)	1.45(+.60)	0.96

$k_f = (6\pi^2\rho/d)^{1/3}$ and d is the degeneracy of the system, which is equal to 2 for neutron matter and 4 for nuclear matter.

It turns out that the contribution of this elementary diagram is of the same order as in finite nuclei. A more complete treatment of the FHNC elementary diagrams may reduce this value. However, a check performed for neutron matter at $\rho = 0.16$ fm^{-3} with a corresponding variational Monte Carlo calculation with 14 neutrons[11] seems to indicate that E_{ee}^{exch} of Eq. (3) gives at least 80% of the elementary diagrams contributions.

Table 1 reports the results for the v_4' potential. The AFDMC calculation has been performed for 28 nucleons only. The estimated value given for $A = \infty$ is obtained by adding to 0.34 MeV the energy difference between the $A = 28$ and the $A = \infty$ cases in the PB-FHNC calculation.

The results obtained with the MS3 potential for symmetrical nuclear matter, asymmetrical matter and pure neutron matter are given in Tables 2, 3, and 4 respectively. The AFDMC results for 28 and 76 nucleons, apart from a roughly constant energy shift, closely follow the variational PB-FHNC estimates, suggesting that PB-FHNC theory can be very effective to estimate the finite size effects. To get the extrapolated values we have first added the PB-FHNC finite size corrections to each AFDMC mixed energies reported in the Tables and then taken an average.

We have no PB-FHNC calculations for asymmetric matter. We give in Table 3 the difference of the free energy per particle between the infinite case and the box case. The two cases considered miss the Fermi energy by roughly the same amount, so one may expect that finite size effects are similar for the two cases.

The dependence of the energy per particle on the asymmetry parameter $\alpha = (N - Z)/(N + Z)$ are shown in Fig.1.

The lower plot of Fig. 1 shows the function $\delta(\alpha)$ given by

$$\delta(\alpha) = E_{\text{AFDMC}}(\alpha) - E_{\text{FHNC/SOC}}(\alpha), \tag{5}$$

where

$$E_{\text{FHNC/SOC}}(\alpha) = 40.59\alpha^2 - 16.10 \tag{6}$$

corresponds to the quadratic fit of $E_{\text{FHNC/SOC}}(\alpha)$ for nuclear matter ($\alpha = 0$) and pure neutron matter ($\alpha = 1$). This gives us a symmetry energy of 40.59 MeV. If we include the elementary diagram of Eq. (3) the corresponding symmetry energy

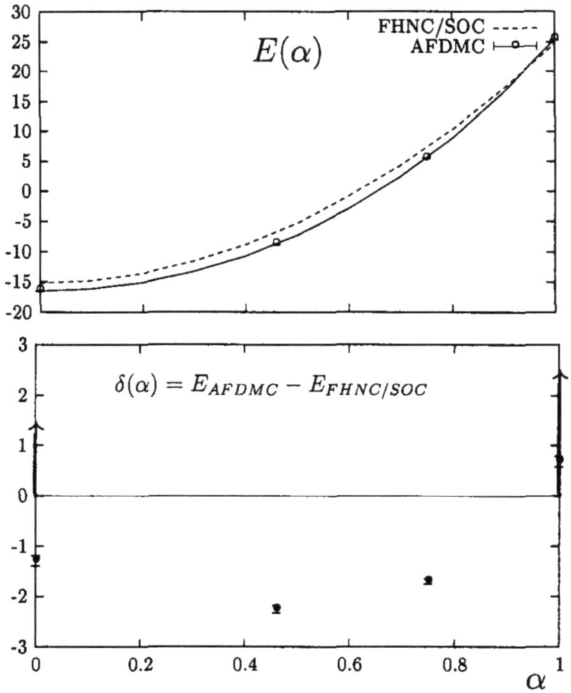

Figure 1. Energy per particle of nuclear matter as a function of the asymmetry parameter α. The FHNC/SOC results do not include the correction from the elementary diagram of Eq. (3). Such contributions are indicated by the arrows in the lower plot.

Table 2. Results for the MS3 model of symmetrical nuclear matter at $\rho = 0.16$. See caption of Table 1.

A	PB-FHNC	FHNC/SOC	AFDMC
28	-14.79	-	-16.17(6)
76	-16.83	-	-18.08(3)
2060	-15.15	-	↓
∞	-15.20(+1.2)	-16.10(+1.2)	-16.5

becomes 41.59 MeV. The AFDMC results seem to indicate that $E(\alpha)$ is not fully quadratic with a symmetry energy of 36.4 MeV. However, one should consider the possibility that the nodal surface adopted for the neutron matter is better than that of nuclear matter as well as the fact that the AFDMC energies for $N \neq Z$ are not finite size corrected.

In order to estimate the quality of the guiding function used in AFDMC calculations, we have also made simulations for the alpha particle, for which there are extremely good upper bounds in the literature. We compare in Table 5 our results with two existing estimates. One is the variational Monte Carlo calculation of Ref. 12 performed for a trial function resulting from the so-called J-TICI2 coupled cluster truncation[13] and denoted by JLO. The second is from correlated

Table 3. Time step dependence of AFDMC calculations for the MS3 model of asymmetrical nuclear matter at $\rho = 0.16$. The energies per particle are in MeV.

N	Z	α	$\Delta\tau(10^{-5}\ \mathrm{MeV}^{-1})$	$E(N,Z)$	$E_{\mathrm{free}}(\infty) - E_{\mathrm{free}}(N,Z)$
14	2	0.75	10	5.52(4)	0.69
14	2	0.75	5	5.55(5)	0.69
38	14	0.46	10	-8.69(6)	0.77
38	14	0.46	5	-8.79(5)	0.77

Table 4. Results for the MS3 model of pure neutron matter at $\rho = 0.16$. See captions of Table 1.

N	PB-FHNC	FHNC/SOC	AFDMC
14	24.42	-	25.46(2)
38	22.49	-	23.15(1)
66	24.30	-	24.80(1)
1030	24.79	-	↓
∞	24.72(+2.2)	24.49(+2.2)	25.4

Table 5. AFDMC results for the MS3 model of the alpha particle, compared with the variational Monte Carlo and the hyperspherical harmonics results. The statistical errors are given in parentheses.

Method	Ref.	$\Delta\tau(10^{-5}(\mathrm{MeV}^{-1})$	E (MeV)
JLO	12	-	-30.41(2)
HM	14	-	-30.299
AFDMC		1	-29.95(7)
AFDMC		2	-29.42(6)
AFDMC		3	-29.28(6)

hyperspherical-harmonic theory,[14] and is denoted by HM.

The single particle wave functions used in our calculation are the eigenfunctions of the one-body Hamiltonian given by

$$H = -\frac{\hbar^2}{2m}\nabla^2 + V(r),$$
$$V(r) = \frac{V_0}{1+\exp\left(\frac{r-R}{a}\right)}, \tag{7}$$

with $V_0 = -56.2$ MeV, $R = 1.8$ fm and $a = 0.22$ fm.

The correlation factor in the Jastrow product of Eq. (1) is taken from a FHNC/SOC nuclear matter calculation at $\rho = 0.16\ \mathrm{fm}^{-3}$. We have tested other functional forms, in particular the J-model of Ref. 12 and the results remain unchanged. We report results at three different time steps. The linear extrapolated value, -30.22 ± 0.07 MeV, is consistent with the variational results.

3 Conclusions

We have calculated the properties of nuclear matter and the alpha particle using the auxiliary field diffusion Monte Carlo method with the spin-isospin dependent v_4 model potentials. While these potentials do not contain a tensor interaction, the full set of auxiliary fields needed to include the tensor interactions is already included at the v_4 level so that extension to the semi-realistic v_6 interaction is straightforward. Similarly, the dominant part of the Urbana 3-body potentials can also be included in this way.

We have found that the auxiliary field method gives energies somewhat lower than FHNC methods, and we are able to estimate finite size effects using the periodic box FHNC method.

The calculation of the asymmetry energy of nuclear matter is straightforward using the auxiliary field method. To demonstrate this we calculated the asymmetry coefficient for a v_4 model of nuclear matter, and observe apparent deviations from a simple quadratic form.

We expect to complete studies with v_6 models for nuclei and nuclear matter in the near future.

Acknowledgements

We wish to thank E. Buendía, J. Carlson, A. Fabrocini and V. R. Pandharipande for helpful conversations. Portions of this work were supported by MURST-National Research Projects, and the CINECA computing center.

References

1. G. G. Raffelt, *Stars as Laboratories for Fundamental Physics*, (University of Chicago Press, Chicago&London, 1996).
2. K. E. Schmidt and S. Fantoni, *Phys. Lett.* B **446**, 99 (1999).
3. S. Fantoni, A. Sarsa and K. E. Schmidt, *Prog. Part. Nucl. Phys.* **44**, 63 (2000).
4. S. Fantoni and S. Rosati, *Nuovo Cim.* A **20**, 179 (1974).
5. K. E. Schmidt and M. H. Kalos, in *Monte Carlo Methods in Statistical Physics II, Topics in Current Physics*, Ed. K. Binder (Springer-Verlag, Berlin, Heidelberg, New York, 1984) p. 125.
6. S. Fantoni and K. E. Schmidt, submitted to *Nucl. Phys.* A.
7. A. Smerzi, D. G. Ravenhall, V. R. Pandharipande, *Phys. Rev.* C**56**, 2549 (1997).
8. I. R. Afnan and Y. C. Tang, *Phys. Rev.* **175**, 1337 (1968).
9. R. B. Wiringa, V. Ficks and A. Fabrocini, *Phys. Rev.* C**38**,1010 (1988).
10. A. Fabrocini, F. Arias de Saavedra, G. Có and P. Folgarait, *Phys. Rev.* C**57**, 1668 (1998).
11. J. Carlson, private communication;
12. E. Buendía, F. J. Gálvez, J. Praena and A. Sarsa, *J. Phys. G (Nucl. Part. Phys.)*, in press.
13. R. F. Bishop, R. Guardiola, I. Moliner, J. Navarro, M. Portesi, A. Puente and

N. R. Walet, *Nucl. Phys.* A**643**, 243 (1998).
14. S. Rosati and M. Viviani, in *Advances in Quantum Many-Body Theory* **2**, Eds. R. F. Bishop and N. R. Walet (World Scientific, Singapore, in press).

THREE-BODY FORCE EFFECTS IN FEW-NUCLEON SYSTEMS

A. KIEVSKY
Istituto Nazionale di Fisica Nucleare, Via Buonarroti 2, 56100 Pisa, Italy
E-mail: kievsky@pi.infn.it

The effects of three-nucleon interactions are analyzed in the three-nucleon system. The binding energy, asymptotic constants as well as scattering lengths and polarization observables at low energy are compared to experimental data. An LS three-body force is introduced in order to improve the description of the vector analyzing powers.

1 Introduction

The new generation of NN potentials describes the two-nucleon (2N) observables with a χ^2 per datum ≈ 1.[1-3] This high accuracy obtained in the description of the 2N system does not imply that a similar accuracy will be achieved in the description of larger nuclear systems, in particular the three-nucleon (3N) data. In fact, the simplest observable in the 3N system, the binding energy, is underpredicted by each of the new NN potentials. The energy deficit ranges from 0.5 to 0.9 MeV, depending on the off-shell and short range parametrization of the NN interaction. This underbinding problem has not yet been solved, and a number of effects beyond the static NN interaction have been considered (a review is given in Ref. 4). For example, considerable efforts have been put into calculating relativistic corrections and three-nucleon force (3NF) contributions to the 3N binding energy.

It is common practice to look at the 3N bound state problem as the solution of the non-relativistic Schrödinger equation using phenomenological NN interactions and then to introduce a 3NF to provide supplementary binding. The models for the 3NF are usually based on two-pion exchange with intermediate Δ-isobar excitation, and the strength of the interaction is adjusted to reproduce the ^{3}H binding energy.

Once the 3N binding energy is well reproduced, the description of several other observables improves as well. For example, the $A = 3$ r.m.s radii,[5] the asymptotic normalization constants η,[6] and the doublet $n - d$ scattering lengths[7] are now in much better agreement with the experimental values. These observables have the property to scale with 3N binding energy (the so-called Phillips lines).[8]

With respect to the 3N continuum, a complete quantitative analysis in terms of χ^2 of the 3N data versus theory has not yet been made for any of the new NN potentials. Therefore, there is a need to evaluate in detail the ability of those interactions to describe the 3N scattering data. In Ref. 9 a detailed analysis has been performed for the total $n - d$ cross section in which calculations solving the Faddeev equations have been compared to the data. This analysis has been recently repeated[10] by taking into account new high-precision measurements.[11] The analysis could not be extended to the differential cross section, due to lack of an adequate data set. In Ref. 12 a new set of precise measurements of $d - p$ elastic observables at $E_d = 270$ MeV has been presented. The differential cross section as well as some polarization observables have been analyzed with Faddeev calculations using modern NN potentials including 3NF contributions. The χ^2 per datum has been studied in a limited angular range ($\theta_{c.m.} = 50° - 180°$) in order to avoid the effects of the Coulomb interaction, which has been neglected in these calculations. At this very high energy a definite sensibility to three-body forces has been observed.

Recently a rigorous solution of the $p - d$ scattering problem has been obtained by the Pisa group[7,13] allowing for a detailed study of this reaction, for which an extensive and high precision data set exists. In Refs. 14 and 15 phase shift analyses have been performed in order to reproduce the $p - d$ differential cross section and vector and tensor analyzing powers. From these analyses it was possible to make comparisons to the theoretical phase-shift and mixing parameters and quantitatively relate the found differences in the P-wave parameters to the so-called "A_y puzzle".[15] Moreover, the usual terms present in 3NF's coming from two-pion-exchange and included in the Tucson-Melbourne, Brazil and Urbana models can not explain the discrepancy, even if the parametrization is changed by a large amount. However, other terms that have been recently introduced, as for example a 3NF depending on the LS operator,[16] are shown to give a large contribution to the vector analyzing powers and, eventually remove the discrepancy.

2 Bound states and scattering lengths in the three-nucleon system

The description of bound states in the three-nucleon system can be performed using an expansion of the wave function in terms of the Correlated Hyperspherical Harmonic basis (CHH) as described in Ref. 13. The energy of the system is obtained using the Raleigh-Ritz variational principle. The same technique can be used to describe continuum states in the framework of the Kohn variational principle in its complex form.[17] The accuracy of this method has been tested several times (see for example Ref. 18). The results for bound states are accurate up to 1 keV, whereas for scattering states the S-matrix is obtained within 0.1%.

In Table 1 the results for the binding energy of ^{3}H and ^{3}He are given, obtained from calculations using the Argonne v_{18} (AV18) interaction. In order to fit the experimental value, the 3NF of Urbana (UR)[19] has been considered as well. It is interesting to note that the mass difference $\Delta B = B(^3\text{H}) - B(^3\text{He})$ is close to the experimental value of 764 keV. The calculated value for the AV18+UR potential is 730 keV. An additional 14 keV results from the $n-p$ mass difference. The predicted value is then off by 20 keV that should be obtained from other mechanisms not included in the charge dependence of the AV18 potential. Let us recall that if only the Coulomb repulsion is taken into account, and all the other electromagnetic terms are neglected, the predicted mass difference is 640 keV.

Another quantity that is much better described when 3NF's are considered is the asymptotic constant η. In Table 1 the theoretical results are compared to the experimental data, showing a nice agreement. The relative large error of the data do not allow to make a distinction between the agreement obtained with charge independent or charge dependent potentials (see Ref. 6). For the $n - d$ scattering lengths the situation is slightly different, since accurate data exist. In Table 1 we see that the AV18+UR potential reproduces the experimental values for the doublet and quartet scattering lengths. When a charge independent potential is used the doublet scattering length is not well reproduced. The motivation is that the $n - p$

Table 1. The ^{3}H and ^{3}He binding energy B, kinetic energy T and occupation probabilities $P_{S'}$, P_D and P_P, as well as the asymptotic constant η and the doublet and quartet scattering lengths ^2a and ^4a calculated using the AV18 and AV18+UR potentials. The experimental data are given for the sake of comparison.

^{3}H	AV18	AV18+UR	exp.
B(MeV)	7.623	8.479	8.48
T(MeV)	46.72	51.27	
$P_{S'}$(%)	1.293	1.055	
P_D(%)	8.510	9.300	
P_P(%)	0.066	0.135	
η		0.0430	0.0411(13)(12)
			0.0431(25)
^2a(fm)		0.63	0.65(4)
^4a(fm)		6.33	6.35(2)

^{3}He	AV18	AV18+UR	exp.
B(MeV)	7.623	7.749	7.72
T(MeV)	46.72	50.21	
$P_{S'}$(%)	1.293	1.243	
P_D(%)	8.510	9.248	
P_P(%)	0.066	0.132	
η		0.0400	0.0386(45)(12)
^2a(fm)		-0.02	-0.13(4)
^4a(fm)			14.7(23)

force is stronger than the $n-n$ one and this fact prevents a simultaneous description of the ^{3}H binding energy and the doublet $n-d$ scattering length without including charge dependence. For the $p-d$ reaction the doublet and quartet scattering lengths have been recently extracted from an extrapolation of the low energy data.[20]

3 $n-d$ scattering at low energies

The study of $n-d$ provides a further test of our understanding of the NN interaction and 3NF's. In the low energy regime experimental data exist for cross section and vector and tensor polarization observables. The differential cross section is in general well described reflecting the fact that the nuclear interaction is basically understood. In Fig.1 the theoretical $n-d$ and $p-d$ cross section calculated with the AV18+UR are compared to the experimental data.[21] There is a nice agreement between both calculations and the measurements. A quantitative analysis has shown that it is possible to describe the low energy differential cross section with a χ^2 per datum of one.[22]

The polarization observables are sensitive to specific parts of the interaction, for example to spin-orbit terms as the tensor or LS interactions. When the calcu-

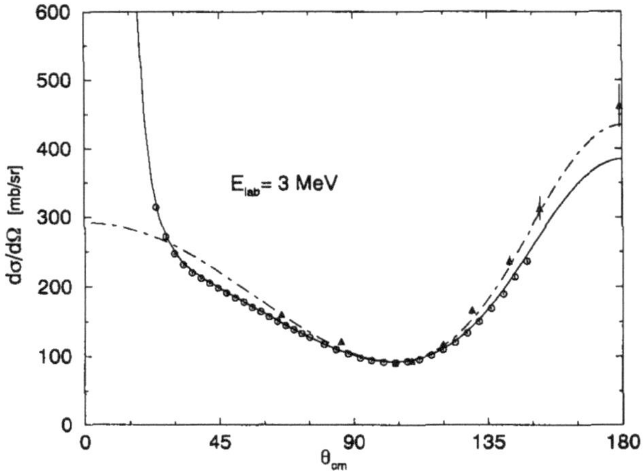

Figure 1. Comparison between the calculated $p-d$ (solid line) and $n-d$ (dashed line) differential cross section and the experimental data.

lations are applied to describe these observables some disagreements are evident. An emblematic example is the vector analyzing powers for which a systematic underprediction of about 30% has been observed in $n-d$ scattering below 30 MeV.[9] In order to improve the description of the vector observables a 3NF with LS dependence has been recently introduced.[16] The LS interaction in the NN potential in channels with $S_{ij}=1$ and $T_{ij}=1$,

$$V_{3N}^{ls}=\sum_{i<j} v_{11}^{ls}(r_{ij})\mathbf{L}_{ij}\cdot\mathbf{S}_{ij}P_{11}(ij) \;, \tag{1}$$

has been modified due to the presence of the third particle. The following form has been proposed

$$V_{3N}^{ls}=\sum_{i<j}\frac{1}{2}[w_{11}^{ls}(r_{ijk})\mathbf{L}_{ij}\cdot\mathbf{S}_{ij}+\mathbf{L}_{ij}\cdot\mathbf{S}_{ij}w_{11}^{ls}(r_{ijk})]P_{11}(ij) \;, \tag{2}$$

where r_{ijk} is a scalar function of the three interparticle distances r_{ij}, r_{jk}, r_{ki}. The symmetric form has been introduced since, in general, the LS operator does not commute with an operator depending on r_{ijk}. Different forms are possible for the three-body interaction $w_{11}^{ls}(r_{ijk})$ provided that $w_{11}^{ls}(r_{ijk})\to v_{11}^{ls}(r_{ij})$ when $r_{ik}, r_{jk}\to\infty$. A simple two-parameter form is

$$w_{11}^{ls}(r_{ijk})=v_{11}^{ls}(r_{ij})+W_0\mathrm{e}^{-\alpha\rho} \;, \tag{3}$$

where the hyperradius ρ is

$$\rho^2=\frac{2}{3}(r_{12}^2+r_{23}^2+r_{31}^2) \tag{4}$$

and W_0 and α are parameters characterizing the strength and range of the three-body term. When the dependence in the scalar function r_{ijk} is limited to r_{ij} and

ρ, the operators $w_{11}^{ls}(r_{ij},\rho)$ and $\mathbf{L}_{ij}\cdot\mathbf{S}_{ij}$ commute. Accordingly, the spin-orbit force becomes

$$V_{3N}^{ls}=\sum_{i<j}v_{11}^{ls}(r_{ij})\mathbf{L}_{ij}\cdot\mathbf{S}_{ij}P_{11}(ij)+W_0\mathrm{e}^{-\alpha\rho}\sum_{i<j}\mathbf{L}_{ij}\cdot\mathbf{S}_{ij}P_{11}(ij)\ . \qquad (5)$$

Three different choices of the exponent α in the hyperradial spin-orbit interaction defined in Eq. (3) have been selected with the intention of constructing forces with different ranges. The strength W_0 has been adjusted in each case in an attempt to improve the description of the vector observables. The analysis has been performed at $E_{lab}=3.0$ MeV. The selected ranges are $\alpha=0.7,1.2,1.5$ fm^{-1}, so as to simulate a long, medium and short range force. The corresponding values for the depth are $W_0=-1,-10,-20$ MeV. The calculations have been performed using the nuclear part of AV18 plus the Coulomb interaction.

The results for the proton and neutron analyzing powers A_y and the deuteron analyzing power iT_{11} are given in Fig.2 together with the experimental data of Ref. 23. The four curves correspond to the AV18 potential and the three different choices for the parameters (α, W_0). The dotted line is the AV18 prediction and shows the expected discrepancy. The solid line corresponds to the AV18 plus the long range force (AV18+LS1), the long-dashed line to the AV18 plus the medium range force (AV18+LS2) and the dotted-dashed line to the AV18 plus the short range force (AV18+LS3). The inclusion of the spin-orbit force improves the description of the vector observables. Moreover, in the bottom panel of Fig.2 the $n-d$ analyzing power has been calculated using the same potential models as before. Again, there is an improvement in the description of A_y equivalent to that one obtained in the $p-d$ case.

In Fig. 3 the tensor analyzing powers T_{20}, T_{21}, T_{22} are shown at the same energy and compared to the data of Ref. 23. The inclusion of the spin-orbit 3NF has no appreciable effect and the four curves are practically on top of each other. These observables are not very sensitive to the splitting in ^4P_J-waves. They are sensitive to scattering in D-waves and higher partial waves, which are only weakly distorted by the LS operator in the 3NF.

4 Conclusions

The three-nucleon bound state and elastic $n-d$ scattering in the low energy region have been studied with a realistic interaction including charge dependence and a 3NF. This interaction provides a reasonable description of the binding energy, the asymptotic constants and scattering lengths. Moreover the $n-d$ differential cross section and polarization observables are well described with the exception of the vector analyzing powers. An improvement in the predictions for A_y and iT_{11} have been obtained including a new term in the three-nucleon potential with LS dependence.

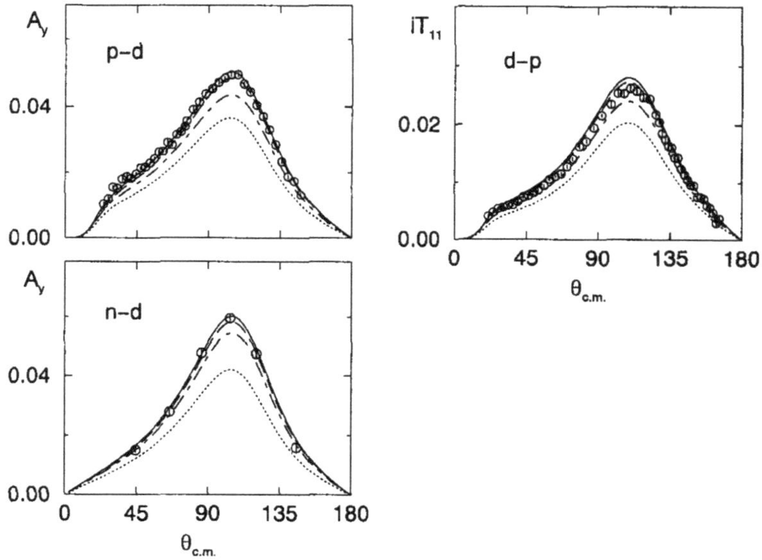

Figure 2. Vector analyzing powers in $p-d$ and $n-d$ scattering at $E_{lab} = 3$ MeV. For the different curves see text.

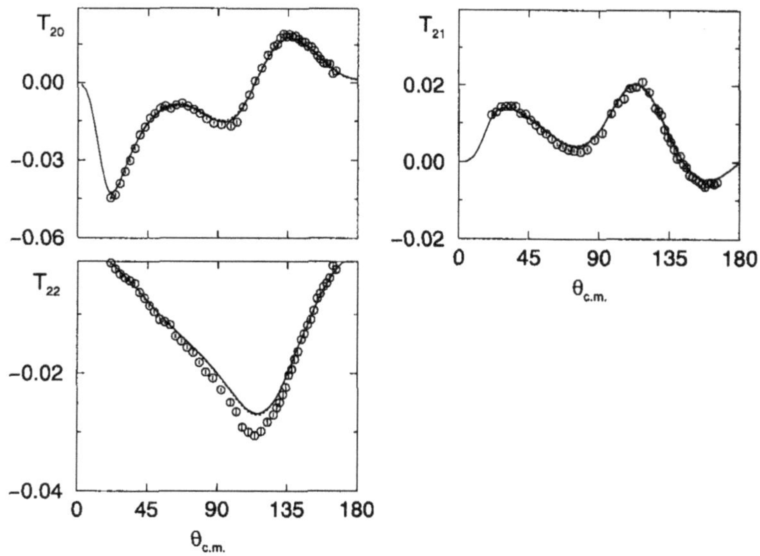

Figure 3. Tensor analyzing powers in $d-p$ scattering at $E_{lab} = 3$ MeV. For the different curves see text.

Acknowledgments

I would like to thank S. Rosati and M. Viviani for useful discussions and collaboration during the realization of this work.

References

1. V. G. J. Stoks, R. A. M. Klomp, C. P. F. Terheggen and J. J. de Swart, *Phys. Rev.* C **49**, 2950 (1994).
2. R. B. Wiringa, V. G. J. Stoks and R. Schiavilla, *Phys. Rev.* C **51**, 38 (1995).
3. R. Machleidt, F. Sammarruca and Y. Song, *Phys. Rev.* C **53**, R1483 (1996).
4. J. Carlson and R. Schiavilla, *Rev. Mod. Phys.* **70**, 743 (1998).
5. J. L. Friar, B. F. Gibson, C. R. Chen and G. L. Payne, *Phys. Lett.* B **161**, 241 (1985).
6. A. Kievsky *et al.*, *Phys. Lett.* B **406**, 292 (1997).
7. A. Kievsky, M. Viviani and S. Rosati, *Phys. Rev.* C **52**, R15 (1995).
8. R. C. Phillips, *Rep. Prog. Phys.* **40**, 905 (1977)
9. W. Glöckle *et al.*, *Phys. Rep.* **274**, 107 (1996).
10. H. Witała *et al.*, *Phys. Rev.* C **59**, 3035 (1999).
11. W. P. Abfalterer *et al.*, *Phys. Rev. Lett.* **81**, 57 (1998).
12. H. Sakai *et al.*, *Phys. Rev. Lett.* **84**, 5288 (2000).
13. A. Kievsky, M. Viviani and S. Rosati, *Nucl. Phys.* A **577**, 511 (1994).
14. L. D. Knutson, L. O. Lamm and J. E. McAninch, *Phys. Rev. Lett.* **71**, 3762 (1993).
15. A. Kievsky, S. Rosati, W. Tornow and M. Viviani, *Nucl. Phys.* A **607**, 402 (1996).
16. A. Kievsky, *Phys. Rev.* C **60**, 034001 (1999).
17. A. Kievsky, *Nucl. Phys.* A **624**, 125 (1997).
18. A. Kievsky *et al.*, *Phys. Rev.* C **58**, 3085 (1998).
19. B. S. Pudliner, V. R. Pandharipande, J. Carlson and R. B. Wiringa, *Phys. Rev. Lett.* **74**, 4396 (1995).
20. ~~T.~~ C. Black *et al.*, *Phys. Lett.* B **471**, 103 (1999).
21. K. Sagara *et al.*, *Phys. Rev.* C **50**, 576 (1994); P. Schwarz *et al.*, *Nucl. Phys.* A **398**, 1 (1983).
22. A. Kievsky *et al.*, to be published.
23. S. Shimizu *et al.*, *Phys. Rev.* C **52**, 1193 (1995); J. E. McAninch, L. O. Lamm and W. Haeberli, *Phys. Rev.* C **50**, 589 (1994).

CORRELATIONS IN NUCLEAR MATTER WITH TWO-TIME GREEN'S FUNCTIONS

H. S. KÖHLER
Physics Department, University of Arizona, Tucson, Arizona 85721, USA
E-mail: kohler@physics.arizona.edu

K. MORAWETZ
LPC-ISMRA, Bld Marechal Juin, 14050 Caen and GANIL, Bld Becquerel, 14076 Caen Cedex 5, France

The Kadanoff-Baym (KB) equations are solved numerically for infinite nuclear matter. In particular we calculate correlation energies and correlation times. Approximating the Green's functions in the KB collision kernel by the free Green's functions the Levinson equation is obtained. This approximation is valid for weak interactions and/or low densities. It relates to the extended quasi-particle approximation for the spectral function. The Levinson correlation energy reduces for large times to a second order Born approximation for the energy. Comparing the Levinson, Born and KB calculations allows for an estimate of higher order spectral corrections to the correlations.

1 Introduction

The quantum Kadanoff-Baym equations (KB)[1] describe the time-evolution of the two-time (one-particle) Green's functions $G(\mathbf{p}, t, t')$. Imposing various approximations they have played an important role in the past developing corrections to the classical Boltzmann equation such as memory-effect and damping. With some restrictions it is now however feasible to solve these equations numerically without approximations.

Numerical results of the quantum KB-equations already have been compared in the past with the classical Markovian dynamics as well as with other frequently used approximations.[2,3] Since the first numerical applications of the KB-equations by Danielewicz[2] several contributions to this evolving new field have been published with applications to nuclear matter,[3–6] to one- and two-band semiconductors,[7,8] to phonon-production in $e-e$ collisions in plasmas[9] as well as to electron plasmas in general.[10,11] A paper on the details of the computational methods is published in Computer Physics Communications.[12]

The KB-equations are designed to study time-dependent non-equilibrium phenomena but they can also be used to study the system in its final equilibrium state. The Green's functions contain a wealth of information such as correlation energy and particle distribution. Spectral functions are also easily derived. Although the collision term basically implies a second order calculation with respect to the potential the propagators are by the process of time-iteration dressed with second order insertions (with their proper energy-dependence) up to all orders, in this respect superseding conventional perturbation expansions.

I shall in this presentation focus on the correlation energies obtained with the KB-equations and compare with some approximations, in particular the second order Born. As will be shown later in this talk the importance of the higher order dressing of the propagator lines will be exemplified.

2 The KB-equations

We show some of the equations regarding the KB-formalism needed for our presentation. For further details see for example Refs. 1, 2 and 13.

In a homogeneous medium neglecting the mean field the KB-equations reduce (with $\hbar = 1$) to:

$$(i\frac{\partial}{\partial t} - \frac{p^2}{2m})G^{\gtrless}(\mathbf{p},t,t') = \int_{t_0}^{t} dt''(\Sigma^{>}(\mathbf{p},t,t'') - \Sigma^{<}(\mathbf{p},t,t''))G^{\gtrless}(\mathbf{p},t'',t') - \int_{t_0}^{t'} dt''\Sigma^{\gtrless}(\mathbf{p},t,t'')(G^{>}(\mathbf{p},t'',t') - G^{<}(\mathbf{p},t'',t')) \tag{1}$$

$$(-i\frac{\partial}{\partial t'} - \frac{p^2}{2m})G^{\gtrless}(\mathbf{p},t,t') = \int_{t_0}^{t} dt''(G^{>}(\mathbf{p},t,t'') - G^{<}(\mathbf{p},t,t''))\Sigma^{\gtrless}(\mathbf{p},t'',t') - \int_{t_0}^{t'} dt''G^{\gtrless}(\mathbf{p},t,t'')(\Sigma^{>}(\mathbf{p},t'',t') - \Sigma^{<}(\mathbf{p},t'',t')). \tag{2}$$

The notations are the conventional ones. $G^{>}$ and $G^{<}$ are essentially the occupation-numbers for holes and particles respectively. The particle distribution function $\rho(\mathbf{p},t)$ is given by

$$\rho(\mathbf{p},t) = -iG^{<}(\mathbf{p},t,t). \tag{3}$$

The Green's functions $G^{>}$ and $G^{<}$ are related on the diagonal in the t,t' plane by

$$G^{>}(\mathbf{p},t,t) = -i + G^{<}(\mathbf{p},t,t). \tag{4}$$

Between these two Green's functions exists a useful relation

$$G^{\gtrless}(\mathbf{p},t,t') = [G^{\gtrless}(\mathbf{p},t',t)]^{*}. \tag{5}$$

The scattering rates Σ are given by

$$\Sigma^{\gtrless}(\mathbf{p},t,t') = -i\int \frac{d^3\mathbf{p}_1}{(2\pi)^3}\left\langle\frac{1}{2}(\mathbf{p}-\mathbf{p}_1)\mid T^{\gtrless}(\mathbf{p}+\mathbf{p}_1,t,t')\mid\frac{1}{2}(\mathbf{p}-\mathbf{p}_1)\right\rangle\times G^{\lessgtr}(\mathbf{p}_1,t',t). \tag{6}$$

Here $T^{\gtrless}$ is defined by

$$\left\langle\mathbf{p}\mid T^{\gtrless}(\mathbf{P},t,t')\mid\mathbf{p}\right\rangle = \int dt''dt'''d\mathbf{p}''d\mathbf{p}'''\left\langle\mathbf{p}\mid T^{+}(\mathbf{P},t,t'')\mid\frac{1}{2}(\mathbf{p}''-\mathbf{p}''')\right\rangle\times G^{\gtrless}(\mathbf{p}'',t'',t''')G^{\gtrless}(\mathbf{p}''',t'',t''')\left\langle\frac{1}{2}(\mathbf{p}''-\mathbf{p}''')\mid T^{-}(\mathbf{P},t,''' t')\mid\mathbf{p}\right\rangle. \tag{7}$$

The effective interaction $T^{\pm}$ is usually defined in a binary collision (ladder) approximation by an integral equation formally written as

$$T_{12}^{\pm} = V + VG_1^{\pm}G_2^{\pm}T_{12}^{\pm}, \tag{8}$$

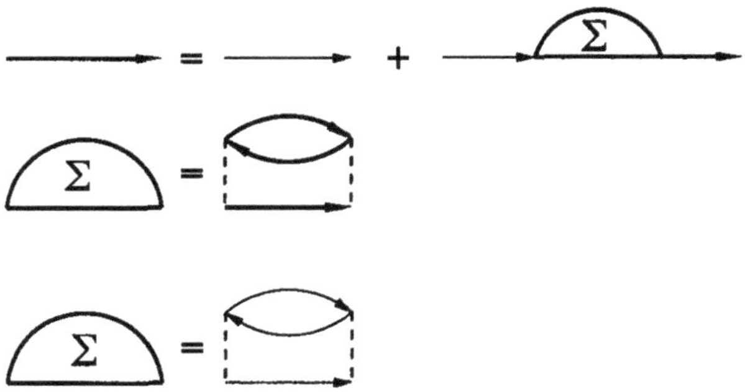

Figure 1. The integral form of Kadanoff-Baym equation (above) and the two discussed approximations for the selfenergy (below). The first line describes the selfconsistent Born approximation (thick lines) and the second line the non-selfconsistent Born approximation (thin lines). The latter one leads to the Levinson equation.

where V is the 'free' interaction potential.

In the following $T^{\pm}$ will be approximated by a local time-independent effective interaction

$$V(\mathbf{p}) = \pi^{3/2}\eta^3 V_0 e^{-\frac{1}{4}\eta^2 p^2} \tag{9}$$

with $\eta = 0.57$ fm and $V_0 = -453$ MeV. Considering the full dynamical T-matrix approximation one obtains more involved time integrals which give rise to nonlocal effects.[14–16]

The exchange term is not included. Equation (6) for the scattering rates then simplifies to

$$\Sigma^{\gtrless}(\mathbf{p},t,t') = -i\int \frac{d^3\mathbf{p}_1 d^3\mathbf{q}}{(2\pi)^6} V^2(\mathbf{q}) G^{\lessgtr}(\mathbf{p}_1,t',t) G^{\gtrless}(\mathbf{q}+\mathbf{p}_1,t,t') G^{\gtrless}(\mathbf{p}-\mathbf{q},t,t'). \tag{10}$$

The momentum integrations are conveniently evaluated using the convolution theorem for Fourier transforms. The diagrammatic representation can be seen in Fig. 1. We will compare in the paper the selfconsistent with the non-selfconsistent approximation which are given in Fig. 1 by thick and thin lines respectively.

The total energy is given by[1]

$$E_{\rm tot}(t) = \frac{1}{2}K_\rho(t) + \frac{1}{4}\int \frac{d^3\mathbf{p}}{(2\pi)^3}\left(\frac{\partial}{\partial t_1} - \frac{\partial}{\partial t_2}\right)G^{<}(\mathbf{p},t_1,t_2)). \tag{11}$$

In the results shown in this paper the total energy is calculated from this latter expression. The kinetic energy K_ρ for the correlated medium is

$$K_\rho(t) = \int \frac{d^3\mathbf{p}}{(2\pi)^3}\frac{p^2}{2m}\rho(\mathbf{p},t) \tag{12}$$

and the correlation energy is defined by

$$E_{\rm corr}(t) = E_{\rm tot}(t) - K_\rho(t). \tag{13}$$

Note that the mean Hartree-Fock field is not included in our work and the total energy therefore contains only the correlated energy. We also define the *uncorrelated* kinetic energy by

$$K_f(t) = \int \frac{d^3\mathbf{p}}{(2\pi)^3} \frac{p^2}{2m} f(\mathbf{p}, t). \tag{14}$$

The relation between the reduced density matrix ρ and the quasi-particle distribution f will be discussed below (Sec. 4) Several numerical applications of this formalism are published. For some references see the discussion in the introduction. A detailed description of numerical details has been published in Ref. 12.

In this paper we concentrate on correlation energies and correlation times. All (with one exception) calculations are performed with an initially uncorrelated nuclear matter system and a momentum-distribution $f(\mathbf{p}, t)$ specified by a density ρ and temperature T. The system is then time-evolved beyond equilibrium. The selfenergies are conserving[1,17] so that the total energy is conserved. Therefore we have during the time evolution

$$K_f^i = E_{\text{tot}}(t) = E_{\text{tot}}^{\text{eq}} \tag{15}$$

where K_f^i is the kinetic (and total) energy of the initial unperturbed and uncorrelated system and $E_{\text{tot}}^{\text{eq}}$ is the total energy after equilibration ($t \to \infty$).

3 The Levinson equation

By replacing the correlated (damped) Green's functions in the collision term of the KB-equations by free Green's functions one obtains the Levinson equation.

The correlation energy in this case resembles the second order estimate of the potential energy but with two important differences. The densities ρ_{eq} are the final densities after the system has equilibrated and there is a factor of one-half missing. The first item is not of major concern, because for weak interactions for which the Levinson equation and the Born approximation are valid the difference between initial uncorrelated and final correlated densities is of a higher order. We shall however address this question below showing numerical results for Levinson, KB as well as initial densities. The second item may at first sight be puzzling but has been clarified.[18] The proof relies on the fact that the correlated and uncorrelated kinetic energies at the end of the time-evolution are related by

$$K_\rho^{\text{eq}} = K_f^{\text{eq}} - \frac{1}{2} E_{\text{corr}}^{\text{eq}} \tag{16}$$

This relation is obtained using the Extended Quasi-Particle (EQP) approximation for the spectral function.

4 Numerical results

In this section we show some results of our calculations. The equations (KB and Levinson) were time-evolved starting at time $t = 0$ with an uncorrelated Fermi gas of specified density and temperature. In addition to KB and Levinson correlation energies we have also calculated the second order Born energy.

Table 1. Correlation energies as a function of the density of nuclear matter. At normal density the temperature dependence is also shown. All results are here with $V_0 = 453.0$MeV. The energies E^{eq}_{corr} are (the negative of) the equilibrium correlation energies. The Born energy E^{eq}_{corr} (Born) is calculated with three different distribution-functions as discussed in the text.

ρ	T	E^{eq}_{corr} (KB)	E^{eq}_{corr} (Lev)	E^{eq}_{corr} (Born) MeV			t_c (KB)	$\frac{\hbar}{2E_F}$
fm^{-3}	MeV	MeV	MeV	KB	Lev.	Init.	fm/c	fm/c
0.380	0	53.63	—	-	-	65.16	2.0	1.5
0.183	0	35.95	49.67	-	49.69	43.97	2.4	2.4
0.181	10	36.03	48.60	50.52	48.54	49.16	—	—
0.182	20	35.94	46.74	–	46.65	52.84	—	—
0.182	40	34.31	42.14	–	42.09	50.52	—	—
0.182	60	31.59	37.22	38.00	37.20	44.65	—	—
0.095	0	23.55	31.33	–	31.76	28.83	3.4	3.8
0.047	0	14.40	17.84	18.24	18.26	17.47	5.2	6.2
0.023	0	8.42	9.80	9.96	9.96	9.96	8.5	9.7

A comparison with the approximate Levinson and Born results requires a high precision of the calculations to be meaningful. To minimize the relative errors all calculations are made with essentially the same KB computer-program. To perform the Levinson calculations all that has to be changed is to replace the selfconsistently calculated Green's functions used in the KB-code in the collision-kernel with the free Green's functions. In the Born-approximation calculations the Green's functions on the time-diagonal were replaced by constant (time-independent) distribution functions as described below.

It was already shown above that the Levinson correlation energy for large times approaches a second order Born value. But the Born calculation then has to be made with the Levinson final reduced density but *not* with the initial quasiparticle distribution as would normally be done in a perturbative expansion. In the low-density and/or weak interaction limit, where the Born and Levinson approximations both become valid this difference should be irrelevant. This is verified by the results shown in columns 5,6, and 7 in Table 1 showing Born energies calculated with occupation numbers from the KB, Levinson and initial distributions respectively. As expected the three energies agree exactly at the lowest density but the agreement becomes progressively worse as the density is increased. As is also expected the Born-column indicated “Lev.” (column 6) agrees nearly exactly with the Levinson result (column 4) at all densities.

We again emphasize that the Born(Init) and Born(Lev) calculations only differ because of a difference between the occupation-numbers in the two cases. There are two distinct effects contributing to this difference. One is because of the heating of the system as it correlates in the Levinson case. The uncorrelated quasiparticle distribution $f(\omega)$ is consequently of a higher temperature than the $f(\omega)$ in the Born(Init) case. In order to make a meaningful comparison between the two results we have to correct for this difference in temperature. Fortunately this is straightforward using relation (16) between the two kinetic energies valid for the

Table 2. Comparison of Levinson $E^{\rm eq}_{\rm corr}$ and Born results at equal uncorrelated kinetic energies ($K^{\rm eq}_f$). The initial temperature T_i of the Levinson calculation is increased to T_f as a consequence of the correlations. Born(Init) is the Born correlation energy at this same temperature T_f. The remaining difference (Diff.) between $E^{\rm eq}_{\rm corr}$ and "Born" is due to the correlational spreading of the spectral function and is discussed in the text. All energies are in MeV.

T_i	$E^{\rm eq}_{\rm corr}$	K^i_f	$K^{\rm eq}_f$	T_f	Born(Init)	Diff.
0	49.67	24.18	49.03	27	53.0	3.3
10	48.60	29.51	53.78	31	52.5	4.0
20	46.74	40.91	64.24	38	51.0	4.3
40	42.14	67.93	88.98	54	46.5	4.4
60	37.22	96.40	115.00	—	—	—

Levinson equation. Knowing the uncorrelated kinetic energy $K^{\rm eq}_f$ we can deduce the temperature and compare with the Born(Init) correlation energy at that same final temperature. The result of this comparison is shown in Table 2. With the Born and Levinson calculations compared at the same temperature i.e. with the same distribution $f(\omega)$ there is however a remaining difference, in Table 2 indicated by $Diff$. This is due to the second of the two separate effects referred to above and it is attributed to the difference between the reduced density ρ and the corresponding quasiparticle distribution f, the former containing spectral corrections. Such a spectral correction can be included in e.g. Brueckner type calculations by iteration but is rarely done. Table 2 shows a decrease of 3.3 MeV of the correlation energy at zero temperature. This implies a decrease in binding energy of 1.6 MeV. We point out that in the EQP-approximation a change δ in correlation energy for a given distribution $f(\omega)$ changes the total energy by $\delta/2$. It changes the kinetic energy K_ρ by $-\delta/2$.

Although the KB collision term in our calculations is formally up to second order in the interaction the correlation energy at equilibrium is actually of a much higher (infinite) order. This is because the correlated Green's functions are formed by iterative time-stepping functionals of the interaction.

The effect of these higher order terms can be assessed from comparing the KB and Levinson correlation-energies. The difference between the two stems from the difference between the Green's functions in the collision kernel for the two separate cases. In the KB case they are selfconsistent (correlated) while in the latter they are free Green's functions. In diagrammatic language the presence of the correlated Green's functions in the collision kernel means that hole and particle lines are dressed with the two-point (second order) insertions *to all orders* and *all time orderings* . In the ω-representation the proper dependence on ω of the insertion is maintained. This dressing is for example not included in a perturbative summation of say Brueckner-type. There the only corresponding diagram that is included would be an on-shell second order insertion $\Sigma^+(\mathbf{p}, \omega_0)$ in hole and particle lines. This would however modify the result in a negligible way in our case because of the local interaction that we use. This has been verified by recomputing our Born-result with such an insertion. The conclusion is that the higher order diagrams contained in the KB results are quite important. Our result implies that the binding energy of nuclear matter of normal density should be reduced by about 7 MeV/A compared to standard Brueckner calculations. At half nuclear density the correction would be about 4 MeV/A. Referring to the discussion above in Table 2, these results do actually refer to nuclear matter excited to about 20 MeV temperature. (When comparing the Levinson and KB-results, there should in principle also be applied a correction similar to the one applied in Table 2, when comparing the Born-results above. But in this case this correction is negligible.)

An important difference between the Levinson and KB-collision terms is the damping, which is related to the width of the spectral-function. Above we deduced a 1.6 MeV difference between the Born (Init) and the Born (Levinson) calculations stemming from a spectral correction of the time diagonal density-matrix. The present correction is also a spectral correction but on the off-diagonal elements i.e. the ω dependence of Σ^+.

The total energy given by Eq. (11) includes besides a kinetic energy only a correlation energy. There is of course also a first order term, the Hartree-Fock term, contributing to the energy of the many-body system. This term also contributes to the mean field that can be included in the Green's functions.[12] We point out that besides the usual momentum independent Hartree shift another effect appears usually referred to as the dispersion-effect. This effect stemming from the momentum-dependence of the Fock (or Brueckner-Hartree-Fock) field is well-known and not of interest in our present calculation. It decreases the binding energy in nuclear many-body calculations.

All of our nuclear matter calculations have until now been restricted to using a time-local interaction. The correlations appear however to be similar to those for more realistic interactions. This is illustrated by the spectral functions which we also calculated. They show a width comparable to calculations with more realistic interactions and the expected behavior as a function of momentum.[19] Our interaction does not have a short-ranged repulsion or tensor-part, but we believe the long-ranged part to be a reasonable representation of the true interaction. We do however envision a future extension to more realistic nucleon forces as well as a T-matrix.

It may finally be relevant to point out that an important difference between the present KB-calculations and more conventional Green's function work is that the present are made in t, t' rather than ω -space. This is found to be very practical.

This work was supported in part by the National Science Foundation Grant No. PHY-9722050

References

1. L. P. Kadanoff and G. Baym, *Quantum Statistical Mechanics* (Benjamin, New York, 1962).
2. P. Danielewicz, *Ann. Phys. (NY)* **152**, 305 (1984).
3. H. S. Köhler, *Phys. Rev.* E **53**, 3145 (1996).
4. H. S. Köhler, *Phys. Rev.* C **51**, 3232 (1995).
5. H. S. Köhler, *Nucl. Phys.* A **583**, 339 (1995).
6. H. S. Köhler, in *Proc. of the 7th International Conference on Nuclear Reaction*

Mechanisms, Varenna, 1994 Ed. E. Gadioli.
7. R. Binder, H. S. Köhler, M. Bonitz, N. Kwong, *Phys. Rev.* **B55**, 5110 (1997).
8. N. H. Kwong, M. Bonitz, R. Binder and H. S. Köhler, *Phys. Stat. Sol* **206**, 197 (1998).
9. H. S. Köhler and R. Binder, *Cont. Plasma Phys.* B **37**, 167 (1997).
10. M. Bonitz, D. Kremp, D. C. Scott, R. Binder, W. D. Kraeft and H. S. Köhler, *J. Phys. C (Cond. Matt.)* **8**, (1996).
11. M. Bonitz, R. Binder,and H. S. Köhler, *Cont. Plasma Phys.* **B37**, 101 (1997).
12. H. S. Köhler, N. H. Kwong and Hashim A. Yousif, *Comp. Phys. Com.* **123**, 123 (1999).
13. W. D. Kraeft, D. Kremp, W. Ebeling and G. Röpke, *Quantum Statistics of Charged Particle Systems*, (Akademie-Verlag, Berlin, 1986); D. Kremp, W. D. Kraeft and A. J. D. Lambert, *Physica* A **127**, 72 (1984).
14. P. Lipavský, K. Morawetz, and V. Špička, (1999), book sub. to *Annales de Physique*; K. Morawetz, Habilitation, University of Rostock 1998.
15. V. Špička, P. Lipavský, and K. Morawetz, *Phys. Lett.* A **240**, 160 (1998).
16. K. Morawetz *et al.*, *Phys. Rev. Lett.* **82**, 3767 (1999).
17. Gordon Baym and Leo P. Kadanoff,*Phys. Rev.* **124**, 287 (1961); Gordon Baym,*Phys. Rev. Lett.* **1391**, 1962 (.)
18. H. S. Köhler and K. Morawetz, submitted for publication in *Phys. Rev.* C.
19. H. S. Köhler, *Phys. Rev.* C **46**, 1687 (1992).

WEAK PROTON CAPTURE ON ^{3}He AND THE SOLAR NEUTRINO PROBLEM

LAURA ELISA MARCUCCI

Department of Physics, Old Dominion University, Norfolk, Virginia 23529, USA *

E-mail: marcucci@df.unipi.it

The results of a new study of the proton weak capture on ^{3}He and their implications for the Super-Kamiokande solar neutrino data are here reviewed.

1 Introduction

In the present talk, I will review a recent study of the process $^3\text{He}(p, e^+\nu_e)^4\text{He}$, also known as *hep* reaction.[1] This study finds its main motivation in the Super-Kamiokande (SK) collaboration measurements of the energy spectrum of electrons recoiling from scattering with solar neutrinos.[2] Accordingly to the Standard Solar Model (SSM),[3] the *hep* reaction is the only source of solar neutrinos with energy higher than 14 MeV, and its astrophysical S-factor is 2.3×10^{-20} keV b, as obtained from the calculation of Ref. 4. Since an apparent excess of events has been observed in the highest energy bins of the SK spectrum, this fact has led to questions about the reliability of the SSM *hep* S-factor.[5] In particular, the SK collaboration[2] has shown that a large enhancement, by a factor of about 17, of the *hep* S-factor would essentially fit this observed excess of events.

Since in the past few years significant progresses have been made in the modelling of the nuclear interactions and weak currents, and in the description of the bound and continuum four-nucleon wave functions, the *hep* reaction has been re-examined and the astrophysical S-factor has been recalculated in Ref. 1. Here I will review the main aspects of this calculation, and I will summarize the S-factor results and discuss their implications for the SK solar neutrino data.

2 Review of the Calculation

The astrophysical S-factor at center-of-mass (c.m.) energy E is defined as

$$S(E) = E\,\sigma(E)\,\mathrm{e}^{2\,\pi\,\eta} \ , \tag{1}$$

where $\sigma(E)$ indicates the cross section and η is given by $\eta = 2\alpha/v_{\rm rel}$, α being the fine structure constant and $v_{\rm rel}$ the $p\,^3$He relative velocity. The transition amplitude for the *hep* process is given by

$$\langle f|H_W|i\rangle = \frac{G_V}{\sqrt{2}}\, l^\sigma \langle -\mathbf{q};{}^4\text{He}|j^\dagger_\sigma(\mathbf{q})|\mathbf{p};p\,^3\text{He}\rangle \ . \tag{2}$$

Here G_V is the Fermi constant, $\mathbf{q} = \mathbf{p}_e + \mathbf{p}_\nu$, $|\mathbf{p};p\,^3\text{He}\rangle$ and $|-\mathbf{q};{}^4\text{He}\rangle$ represent, respectively, the $p\,^3$He scattering state with relative momentum $\mathbf{p}$ and ^{4}He bound state recoiling with momentum $-\mathbf{q}$, l_σ is the leptonic weak current, $l_\sigma = \overline{u}_\nu\gamma_\sigma(1-\gamma_5)v_e$ (the lepton spinors are normalized as $v_e^\dagger v_e = u_\nu^\dagger u_\nu = 1$), and $j^\sigma(\mathbf{q})$ is the

Table 1. Binding energies in MeV of ^{4}He calculated with the CHH method using the AV18, the AV18/UIX, and the older AV14/UVIII Hamiltonian models. Also listed are the corresponding "exact" GFMC results[13,14] and the experimental value.

Model	CHH	GFMC
AV18	24.01	24.1(1)
AV18/UIX	27.89	28.3(1)
AV14/UVIII	27.50	28.3(2)
Exp	28.3	

nuclear weak current, $j^\sigma(\mathbf{q}) = (\rho(\mathbf{q}), \mathbf{j}(\mathbf{q}))$. In the low energy regime (the Gamow-peak energy for the *hep* reaction is 10.7 keV), it is convenient to expand the $p\,^3$He scattering state into partial waves, and perform a multipole decomposition of the nuclear weak charge and current operators. Standard manipulations lead to[1]

$$\frac{1}{4}\sum_{s_e s_\nu}\sum_{s_1 s_3} |\langle f \,|\, H_W \,|\, i\rangle|^2 = (2\pi)^2 \rangle G_V^2 \rangle L_{\sigma\tau} \rangle N^{\sigma\tau}, \tag{3}$$

where the lepton tensor $L^{\sigma\tau}$ is written in terms of electron and neutrino four-velocities, while the nuclear tensor $N^{\sigma\tau}$ is given in terms of the reduced matrix elements (RMEs) of the Coulomb ($C_{\ell\ell_z}$), longitudinal ($L_{\ell\ell_z}$), transverse electric ($E_{\ell\ell_z}$), and transverse magnetic ($M_{\ell\ell_z}$) multipole operators between the initial $p\,^3$He state with orbital angular momentum L, channel spin S (S=0,1), and total angular momentum J, and final ^{4}He state. The study of Ref. 1 includes S- and P-wave capture channels, i.e. the ^{1}S$_0$, ^{3}S$_1$, ^{3}P$_0$, ^{1}P$_1$, ^{3}P$_1$, and ^{3}P$_2$ states, and retains all contributing multipoles connecting these states to the J^π=0^+ ground state of ^{4}He.

The bound and scattering state wave functions are obtained with the correlated-hyperspherical-harmonics (CHH) method, developed for the four-body problem in Refs. 6, 7. The CHH method essentially consists in expanding the wave function on a suitable basis, and in determining variationally the expansion coefficients, applying the Rayleigh-Ritz and Kohn variational principles in the bound-state and scattering-state problems, respectively.

The nuclear Hamiltonian H consists of the Argonne v_{18} two-nucleon[8] and Urbana-IX three-nucleon[9] interactions. To make contact with earlier studies,[4,10] however, and to have some estimate of the model dependence of the results, the older Argonne v_{14} two-nucleon[11] and Urbana-VIII three-nucleon[12] interaction models have also been used. Both these Hamiltonians, the AV18/UIX and AV14/UVIII, reproduce the experimental binding energies and charge radii of the trinucleons and ^{4}He in exact Green's function Monte Carlo (GFMC) calculations.[13,14] The results of the ^{4}He binding energy calculated with the CHH method are given in Table 1 and compared with the GFMC values, while the ^{3}He binding energy and the $p\,^3$He S-wave singlet and triplet scattering lengths are listed in Table 2. Generally, there is good agreement between the CHH results and the results obtained using other techniques, as well as the available experimental data.

The nuclear weak current $j^\sigma(\mathbf{q}) = (\rho(\mathbf{q}), \mathbf{j}(\mathbf{q}))$ has vector (V) and axial-vector (A) parts, with corresponding one- and many-body components. All the one-body terms can be obtained in a standard way from a non-relativistic reduction of the

Table 2. Binding energies, B_3, of ^{3}He, and $p\,^3$He singlet and triplet S-wave scattering lengths, a_s and a_t, calculated with the CHH method using the AV18, the AV18/UIX, and the older AV14/UVIII Hamiltonian models. The corresponding experimental values are also listed.[15,16]

Model	B_3(MeV)	a_s(fm)	a_t(fm)
AV18	6.93	12.9	10.0
AV18/UIX	7.74	11.5	9.13
AV14/UVIII	7.73		9.24
Exp	7.72	10.8±2.6	8.1±0.5
			10.2±1.5

covariant single-nucleon vector and axial-vector currents, including terms proportional to $1/m^2$. The two-body components of the weak vector current $\mathbf{j}(\mathbf{q};V)$ are constructed from the isovector two-body electromagnetic currents in accordance with the conserved-vector-current (CVC) hypothesis, and consist of "model-independent" (MI) and "model-dependent" (MD) terms.[1] The MI terms are obtained from the nucleon-nucleon interaction, and by construction satisfy current conservation with it. The leading MI two-body contributions, the "π-like" and "ρ-like" operators, have been included. The MD currents are purely transverse, and therefore cannot be directly linked to the underlying two-nucleon interaction. Among the MD currents, those associated with excitation of Δ isobars have been included. However, these terms are found to give a rather small contribution in weak-vector transitions, as compared to that due to the π-like current.

The many-body weak vector charge operators can also be obtained from their electromagnetic correspondents applying the CVC hypothesis. However, while the main parts of the two-body electromagnetic or weak vector current are linked to the form of the nucleon-nucleon interaction through the continuity equation, the most important two-body electromagnetic or weak vector charge operators are model-dependent, and should be viewed as relativistic corrections. The model commonly used[17] for the electromagnetic many-body charge operators includes the π-, ρ-, and ω-meson exchange terms with both isoscalar and isovector components, as well as the (isoscalar) $\rho\pi\gamma$ and (isovector) $\omega\pi\gamma$ charge transition couplings. At moderate values of momentum transfer ($q < 5$ fm^{-1}), the contribution due to the "π-like" exchange charge operator has been found to be typically an order of magnitude larger than that of any of the remaining two-body mechanisms and one-body relativistic corrections.[20] Therefore, only the "π-like" and "ρ-like" weak vector charge operators have been retained.

The axial charge operator $\rho(\mathbf{q};A)$ includes, in addition to the one-body component, the long-range pion-exchange term,[18] required by low-energy theorems and the partially-conserved-axial-current relation, as well as the (expected) leading short-range terms constructed from the central and spin-orbit components of the nucleon-nucleon interaction, following a prescription due to Kirchbach *et al.*[19] The Δ-excitation terms have also been included, but they have been found to be unimportant.[1]

In contrast to the electromagnetic case, the axial current operator $\mathbf{j}(\mathbf{q};A)$ is not conserved. Thus, its many-body components cannot be linked to the nucleon-nucleon interaction and, in this sense, should be viewed as model-dependent.

Table 3. The *hep* S-factor, in units of 10^{-20} keV b, calculated with CHH wave functions corresponding to the AV18/UIX Hamiltonian model, at $p\,^3$He c.m. energies E=0, 5, and 10 keV. The rows labelled "one-body" and "full" list the contributions obtained by retaining the one-body only and both one- and many-body terms in the nuclear weak current. The contributions due the $^3S_1$ channel only and all S- and P-wave channels are listed separately.

	E=0 keV		E=5 keV		E=10 keV	
	$^3S_1$	S+P	$^3S_1$	S+P	$^3S_1$	S+P
one-body	26.4	29.0	25.9	28.7	26.2	29.3
full	6.38	9.64	6.20	9.70	6.36	10.1

Among the many-body axial current operators, those due to π- and ρ-meson exchanges and to the $\rho\pi$-transition mechanism have been included. However, the leading many-body terms in the axial current are due to Δ-isobar excitation. These contributions have been treated non-perturbatively, using the transition-correlation-operator (TCO) scheme, originally developed in Ref. 4 and further extended in Ref. 20. In the TCO scheme–essentially, a scaled-down approach to a full N+Δ coupled-channel treatment–the Δ degrees of freedom are explicitly included in the nuclear wave functions.

The largest model dependence of the weak transition operator is in the axial current. To minimize it, the poorly known $N\Delta$ transition axial coupling constant g_A^* has been adjusted to reproduce the experimental value of the Gamow-Teller matrix element in tritium β-decay.[1,21] While this procedure is model dependent, its actual model dependence is in fact very weak, as has been shown in Refs. 1,21.

3 Results

I present here the results for the *hep* astrophysical S-factor, and their implications to the SK solar neutrino spectrum.

3.1 Results for the S-factor

The results for the astrophysical S-factor, calculated using CHH wave functions with the AV18/UIX Hamiltonian model, at three different c.m. energies, are given in Table 3. By inspection of the table, it can be noted that: (i) the energy dependence is rather weak: the value at 10 keV is only about 4 % larger than that at 0 keV; (ii) the P-wave capture states are found to be important, contributing about 40 % of the calculated S-factor. However, the contributions from D-wave channels are expected to be very small.[1] (iii) The many-body axial currents play a crucial role in the (dominant) $^3S_1$ capture, where they reduce the S-factor by more than a factor of four.

The different contributions from the S- and P-wave capture channels to the zero energy S-factor are given in Table 4. The results obtained using the two-nucleon AV18 and the older two- and three-nucleon AV14/UVIII interaction models are also listed. Note that the sum of the channel contributions is a few % smaller than the total result reported at the bottom of the table, due to the presence of interference

Table 4. Contributions of the S- and P-wave capture channels to the *hep* S-factor at zero $p\,^3$He c.m. energy in 10^{-20} keV b. The results correspond to the AV18/UIX, AV18 and AV14/UVIII Hamiltonian models.

	AV18/UIX	AV18	AV14/UVIII
$^1S_0$	0.02	0.01	0.01
$^3S_1$	6.38	7.69	6.60
$^3P_0$	0.82	0.89	0.79
$^1P_1$	1.00	1.14	1.05
$^3P_1$	0.30	0.52	0.38
$^3P_2$	0.97	1.78	1.24
TOTAL	9.64	12.1	10.1

terms among multipole operators connecting different capture channels.[1]

The dominant contribution to the S-factor is obtained from the $^3S_1$ capture channel. Among the P-wave capture channels, the $^3P_0$ does not give the largest contribution, as instead expected in previous studies,[5] although this is the only contribution surviving in the limit q=0.

By comparison of the AV18 to the AV18/UIX results, it can be concluded that inclusion of the three-nucleon interaction reduces the total S-factor by about 20%. This decrease is mostly in the $^3S_1$ contribution, and can be traced back to a corresponding reduction in the magnitude of the one-body axial current matrix elements. The latter are sensitive to the triplet scattering length, for which the AV18 and AV18/UIX models predict, respectively, 10.0 fm and 9.13 fm (see Table 2). This 20 % difference in the total S-factor values for AV18 and AV18/UIX emphasizes the need for performing the calculation using a Hamiltonian model that reproduces the binding energies and low-energy scattering parameters for the three- and four-nucleon systems. This is true for the AV18/UIX model, but not for the AV18 model.

The different contributions to the astrophysical S-factor, when the older AV14/UVIII potential model is used, are given in the last column of Table 4. By comparing these results with the ones obtained using the AV18/UIX, it can be observed that both the S- and P-wave contributions are not significantly changed; in particular, the $^3S_1$ capture S-factor values differ by only about 3%. It is important to emphasize that this is due to the procedure of constraining the model dependent many-body axial currents by fitting the Gamow-Teller matrix element of tritium β-decay, as discussed at the end of the previous section. Note that the AV14/UVIII Hamiltonian also reproduces the low-energy properties for the three- and four-nucleon systems.

The chief conclusion of this analysis[1] is that the best estimate for the S-factor at 10 keV, close to the Gamow-peak energy, is 10.1 $\times 10^{-20}$ keV b. This value is $\simeq$ 4.5 times larger than the value adopted in SSM, based on Ref. 4, of 2.3 $\times 10^{-20}$ keV b. It is therefore important to point out the differences between the study reviewed here and the previous calculation of Ref. 4: (i) all P-wave contributions are included; (ii) the CHH method has been used to describe the initial and final state wave functions, corresponding to the latest generation of realistic interactions.

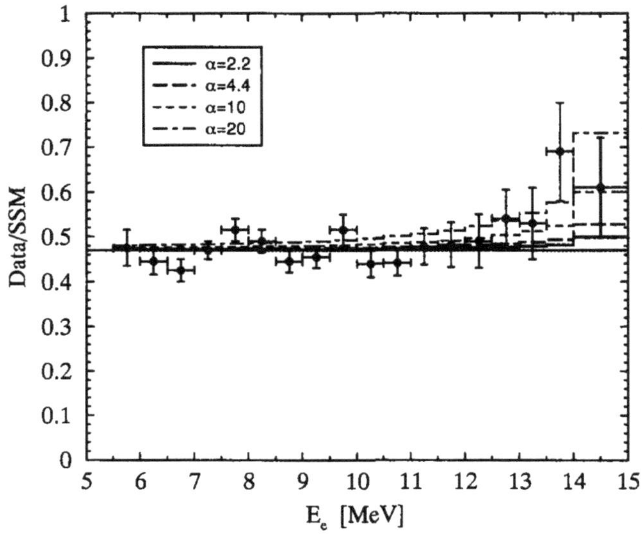

Figure 1. Electron energy spectrum for the ratio between the Super-Kamiokande 825-days data and the expectation based on unoscillated ^{8}B neutrinos.[3] The data were extracted graphically from Fig. 8 of Ref.[2] The 5 curves correspond respectively to no *hep* contribution (dotted line), and an enhancement α of 2.2 (solid line), 4.4 (long-dashed line), 10 (dashed line) and 20 (dot-dashed line).

The CHH method is known to be more accurate than the variational Monte Carlo (VMC) technique used in Ref. 4. (iii) The $1/m^2$ relativistic corrections in the one-body axial current operator are included. In $^3S_1$ capture, for example, these terms increase by 25 % the L_1 and E_1 matrix elements calculated with the one-body axial current operator.

3.2 Implications for the Super-Kamiokande solar neutrino spectrum

The Super-Kamiokande (SK) experiment detects solar neutrinos by neutrino-electron scattering. It is sensitive, according to the SSM,[3] to the very energetic neutrinos from the ^{8}B weak decay ($^8\text{B} \to {}^4\text{He} + {}^4\text{He} + e^+ + \nu_e$) and from the *hep* reaction. The SK results are presented as ratio of the measured electron spectrum to that expected in the SSM, when no neutrino oscillations are included. Over most of the spectrum, this ratio is constant at $\simeq 0.5$.[2] At the highest energies, however, there is an excess of events relative to the 0.5×SSM prediction. This is seen in Fig. 1 where the SK results from 825 days of data acquisition[2] are shown by the points (the error bars denote the combined statistical and systematic error); the dotted line is the 0.5×SSM prediction. In the figure, it is convenient to introduce the ratio α of the *hep* flux to its SSM value, defined as $\alpha \equiv S_{\text{new}}/S_{\text{SSM}} \times P_{\text{osc}}$, where P_{osc} is the observed suppression factor due to neutrino oscillations. Using for S_{new} the value quoted above, $\alpha = (10.1 \times 10^{-20}$ keV b$)/(2.3 \times 10^{-20}$ keV b$) = 4.4$ (long-dashed line), if *hep* neutrino oscillations are ignored, while $\alpha = 2.2$ (solid line) if the *hep* neutrinos are suppressed by $\simeq 0.5$. Two other arbitrary values of α (10 and 20) are shown for comparison.

From these results, it can be concluded that the enhancement of the S-factor reported in Ref. 1, although large, is not enough to completely resolve the discrepancies between the present SK results and the SSM predictions. However, this accurate calculation of the S-factor, and the consequent absolute prediction for the *hep* neutrino flux, will allow much greater discrimination among the proposed solutions to this problem, based on different solar neutrino oscillation scenarios.

Acknowledgments

I wish to thank R. Schiavilla, M. Viviani, A. Kievsky, S. Rosati, and J.F. Beacom for their many important contributions to the work reported here. I also would like to gratefully acknowledge the support of the U.S. Department of Energy under Contract No. DE-AC05-84ER40150.

References

1. L. E. Marcucci, R. Schiavilla, M. Viviani, A. Kievsky, and S. Rosati, *Phys. Rev. Lett.* **84**, 5959 (2000); L. E. Marcucci, R. Schiavilla, M. Viviani, A. Kievsky, S. Rosati, and J. F. Beacom, nucl-th/0006005, *Phys. Rev.* C, in press.
2. Y. Suzuki, contribution to Lepton-Photon Symposium 99 (1999), http://www-sk.icrr.u-tokyo.ac.jp/doc/sk/pub/index.html.
3. J. N. Bahcall, S. Basu, and M. H. Pinsonneault, *Phys. Lett.* B **433**, 1 (1998).
4. R. Schiavilla, R. B. Wiringa, V. R. Pandharipande, and J. Carlson, *Phys. Rev.* C **45**, 2628 (1992).
5. J. N. Bahcall and P. I. Krastev, *Phys. Lett.* B **436**, 243 (1998); G. Fiorentini, V. Berezinsky, S. Degl'Innocenti, and B. Ricci, *Phys. Lett.* B **444**, 387 (1998); R. Escribano, J. M. Frère, A. Gevaert, and P. Monderen, *Phys. Lett.* B **444**, 397 (1998); C. J. Horowitz, *Phys. Rev.* C **60**, 022801 (1999); W. M. Alberico, J. Barnabéu, S. M. Bilenky, and W. Grimus, *Phys. Lett.* B **478**, 208 (2000).
6. M. Viviani, A. Kievsky, and S. Rosati, *Few-Body Sys.* **18**, 25 (1995).
7. M. Viviani, S. Rosati, and A. Kievsky, *Phys. Rev. Lett.* **81**, 1580 (1998).
8. R. B. Wiringa, V. G. J. Stoks, and R. Schiavilla, *Phys. Rev.* C **51**, 38 (1995).
9. B. S. Pudliner, V. R. Pandharipande, J. Carlson, and R. B. Wiringa, *Phys. Rev. Lett.* **74**, 4396 (1995).
10. J. Carlson, D. O. Riska, R. Schiavilla, and R. B. Wiringa, *Phys. Rev.* C **44**, 619 (1991).
11. R. B. Wiringa, R. A. Smith, and T. L. Ainsworth, *Phys. Rev.* C **29**, 1207 (1984).
12. R. B. Wiringa, *Phys. Rev.* C **43**, 1585 (1991).
13. B. S. Pudliner, V. R. Pandharipande, J. Carlson, S. C. Pieper, and R. B. Wiringa, *Phys. Rev.* C **56**, 1720 (1997).
14. J. Carlson, private communication.
15. M. T. Alley and L. D. Knutson, *Phys. Rev.* C **48**, 1901 (1993).
16. P. E. Tegnér and C. Bargholtz, *Ap. J.* **272**, 311 (1983).

17. R. Schiavilla, D. O. Riska, and V. R. Pandharipande, *Phys. Rev.* C **41**, 309 (1990).
18. K. Kubodera, J. Delorme, and M. Rho, *Phys. Rev. Lett.* **40**, 755 (1978).
19. M. Kirchbach, D. O. Riska, and K. Tsushima, *Nucl. Phys.* A **542**, 616 (1992).
20. L. E. Marcucci, D. O. Riska, and R. Schiavilla, *Phys. Rev.* C **58**, 3069 (1998).
21. R. Schiavilla *et al.*, *Phys. Rev.* C **58**, 1263 (1998).

GENERALIZED MOMENTUM DISTRIBUTION OF INFINITE AND FINITE NUCLEON SYSTEMS

E. MAVROMMATIS, M. PETRAKI, AND P. PAPAKONSTANTINOU

Physics Department, Division of Nuclear and Particle Physics, University of Athens, GR-15771 Athens, Greece

T. S. KOSMAS

Physics Department, Theoretical Physics Division, University of Ioannina, GR-45110 Ioannina, Greece

CH. MOUSTAKIDIS

Physics Department, University of Thessaloniki, GR-54006 Thessaloniki, Greece

The generalized momentum distribution $n(\vec{p},\vec{Q})$, a Fourier transform of the half-diagonal two-body density matrix $\rho_{2h}(\vec{r}_1,\vec{r}_2;\vec{r}_1')$ in the variables $\vec{r}_1-\vec{r}_1'$ and $\vec{r}_1'-\vec{r}_2$ is studied for nuclear systems. In the case of infinite nuclear matter, the calculation has been carried out within the Fermi hypernetted-chain procedure assuming Jastrow correlations, and the role of the short-range correlations has been revealed in certain kinematic domains. In the case of finite nuclei, the calculation of $n(\vec{p},\vec{Q})$ for protons has been carried out, first within the independent-particle model, assuming a harmonic oscillator basis. Results for the closed shell nuclei ^{4}He, ^{16}O, ^{40}Ca are presented and are expected to be reliable in certain regions of the momenta $\vec{p}$ and $\vec{Q}$, where the finite size and the Fermi statistics play the dominant role. Next, a calculation of $n(\vec{p},\vec{Q})$ of protons for the nucleus ^{4}He is presented, based on Jastrow correlations and the first two terms of a low-order cluster approximation. Work is in progress to study $n(\vec{p},\vec{Q})$ of other nuclei incorporating the effect of short range correlations. The evaluation of $n(\vec{p},\vec{Q})$ is useful for the study of the propagation mechanism of struck nucleons through the nuclear medium in various scattering processes, and for the understanding of elementary excitations of nuclei.

1 Introduction

This paper deals with the study of the generalized momentum distribution (GMD) $n(\vec{p},\vec{Q})$, a momentum-space transform of the half-diagonal two-body density matrix, in the case of infinite nuclear matter and finite nuclei. It is our contribution to the Festschrift issue in honour of the 65th birthday of our colleagues John Clark, Alpo Kallio, Manfred Ristig, and Sergio Rosati. It is in particular dedicated to John Clark for many years of fruitful collaboration, continuous inspiration, and friendship.

The description of a quantum many-body system in terms of the n-body density matrices has been introduced by Löwdin.[1] Starting from the one-body density matrix $\rho_1(\vec{r}_1,\vec{r}_1')$ and the associated momentum distribution $n(\vec{p})$, substantial effort has been devoted to their determination both theoretically and experimentally for different many-body systems.[2] Clark and Ristig have studied these quantities within the variational theory in a series of ten papers starting with Ref. 3. As a logical next step, the interest has turned to the two-body density matrix $\rho_2(\vec{r}_1,\vec{r}_2;\vec{r}_1',\vec{r}_2')$. In a topical Workshop on Momentum Distributions that took place at Argonne in

1988, Clark and Ristig conclude their review of momentum distributions calculations saying: "...man cannot live by momentum distributions alone. Accordingly, microscopic calculation of ρ_2 should be high on the agenda of the many-body theorist".[4] Shortly after, they published two studies of the two-body density matrices of Bose[5] and normal Fermi fluids[6] within variational theory.[7,8] In a unit-normalized state $|\Psi\rangle$ of a system of A identical particles, the generalized momentum distribution $n(\vec{p},\vec{Q})$ is defined by

$$n(\vec{p},\vec{Q}) = \langle\Psi|\sum_{\hat{k}} a^{+}_{\hat{k}+\vec{Q}}a^{+}_{\hat{p}-\vec{Q}}a_{\hat{p}}a_{\hat{k}}|\Psi\rangle. \tag{1}$$

Here $\hat{k}$ labels single-particle orbitals with wave vector $\vec{k}$ and spin/isospin label σ,τ, while $\hat{k}+\vec{Q} = (\vec{k}+\vec{Q},\sigma,\tau)$. The function $n(\vec{p},\vec{Q})$ is connected to the two-body density matrix in momentum space $n_2(\vec{p}_1,\vec{p}_2;\vec{p}_1^{\,\prime},\vec{p}_2^{\,\prime})$ through the relation

$$n(\vec{p},\vec{Q}) = \int n_2(\vec{p},\vec{k};\vec{p}-\vec{Q},\vec{k}+\vec{Q})\,d\vec{k}. \tag{2}$$

Introducing the half-diagonal two-body density matrix in coordinate space,

$$\rho_{2h}(\vec{r}_1,\vec{r}_2;\vec{r}_1^{\,\prime}) = A(A-1)\int \Psi^*(\vec{r}_1,\vec{r}_2,\ldots,\vec{r}_A)\Psi(\vec{r}_1^{\,\prime},\vec{r}_2,\ldots,\vec{r}_A)d\vec{r}_3\ldots d\vec{r}_A, \tag{3}$$

and performing a Fourier transformation in the variables $\vec{r}_1-\vec{r}_1^{\,\prime}$ and $\vec{r}_1^{\,\prime}-\vec{r}_2$, we obtain for the GMD

$$n(\vec{p},\vec{Q}) = c\int \rho_{2h}(\vec{r}_1,\vec{r}_2;\vec{r}_1^{\,\prime})\,e^{-i\vec{p}\cdot(\vec{r}_1-\vec{r}_1^{\,\prime})}\,e^{-i\vec{Q}\cdot(\vec{r}_1^{\,\prime}-\vec{r}_2)}\,d\vec{r}_1 d\vec{r}_1^{\,\prime} d\vec{r}_2. \tag{4}$$

(In writing Eq. (3), we have suppressed spin/isospin labels and a sum over spin/isospin variables. In the case of nuclear matter, we have adopted for the coefficient c in Eq. (4) the value $\frac{1}{\nu}\frac{\rho}{A}$ (ν is the degeneracy of single-particle states); and in the case of finite nuclei the value $1/(2\pi)^3$). As has been discussed in Ref. 6, the GMD is directly involved in final-state interaction mechanisms, being expressed in terms of the transition matrix element for scattering a particle out of orbital $\hat{p}$ into the orbital $\hat{p}-\vec{Q}$, the process being mediated by a spin-independent density fluctuation.

The generalized momentum distribution $n(\vec{p},\vec{Q})$ has some important formal properties that result from the corresponding properties of the two-body density matrix in coordinate or momentum space.[6,9,10] In particular, we mention the sequential relation

$$n(\vec{p},\vec{Q}=0) = (A-1)\,n(\vec{p}). \tag{5}$$

The GMD in the case of infinitely extended ideal Fermi gas (Eq. (4) with $c=\frac{1}{\nu}\frac{\rho}{A}$) equals

$$n_F(\vec{p},\vec{Q}) = (A-1)\delta(Q)\theta(k_F-p) - (1-\delta(Q))\theta(k_F-p)\theta(k_F-|\vec{p}-\vec{Q}|). \tag{6}$$

The last few years have seen increasing interest in both $n(\vec{p},\vec{Q})$ and $\rho_{2h}(\vec{r}_1,\vec{r}_2;\vec{r}_1^{\,\prime})$ as key descriptors of the nucleon-nucleon correlations in the nuclear medium. This interest arises mainly from the need to properly analyze recent and future experiments focusing on inclusive scattering in (e,e')[11,12] and (p,p')[13] reactions, as well as on exclusive scattering processes such as $(e,e'N)$,[14,15] $(p,2p)$[13,16] and (γ,N), $(\gamma,2N)$,[17,18] and to extract reliable values for the momentum distribution, the transparency and other quantities. To achieve these goals, one must take into account the final-state interaction (FSI) of the struck nucleons as they propagate through the nuclear medium. The quantities $n(\vec{p},\vec{Q})$ and $\rho_{2h}(\vec{r}_1,\vec{r}_2;\vec{r}_1')$ appear in almost all quantitative microscopic "post-mean-field" treatments of the FSI (see, for example, Refs. 19–24 in the case of inclusive (e,e') scattering). Moreover, in analogy with other quantum many-body systems, these quantities are expected to enter in fundamental sum rules that furnish insight into the nature of the elementary excitations of nuclear matter.[25,26]

The half-diagonal two-body density matrix and the generalized momentum distribution were initially investigated within the context of final-state effects in inelastic neutron scattering from quantum fluids. Three approximations have been proposed. In particular, we mention Silver's approximation[27]

$$n(\vec{p},\vec{Q}) \simeq n(\vec{p})(S(\vec{Q})-1) \tag{7}$$

($S(\vec{Q})$ being the static structure factor). With regard to nuclear systems, calculations based on the work of Clark and Ristig[6] have been performed for Jastrow-correlated infinite nuclear matter. This has been done for the GMD using low-order cluster truncations,[28] and later for the GMD and ρ_{2h} using Fermi hypernetted-chain procedures.[9,24,29] Rather recently, calculations of $n(\vec{p},\vec{Q})$ of finite nuclei within the independent-particle model have been reported.[10]

This paper is organized as follows. In Section 2 we briefly present our calculation of the GMD of infinite nuclear matter using the Fermi hypernetted-chain procedure. In Section 3 the independent-particle model with a harmonic-oscillator basis is adopted for the study of the GMD of ℓ-closed nuclei and results are presented for the magic nuclei ^{16}O, ^{40}Ca. In Section 4 an attempt is made to introduce Jastrow correlations using a low-order cluster expansion,[30] and results are presented in the case of the nucleus ^{4}He. A summary and conclusions are given in Section 5.

2 The generalized momentum distribution $n(\vec{p},\vec{Q})$ of model nuclear matter

Our calculation of the generalized momentum distribution $n(\vec{p},\vec{Q})$ of the ground state of uniform, isospin-symmetrical, spin-saturated nuclear matter rests on the work by Ristig and Clark[6] and is reported in detail in Ref. 9. Assuming a Jastrow-Slater wave-function, cluster expansion and suitable resummation processes, and the satisfaction of the sequential relation, the following closed-form expression is derived, which manifestly exhibits the physical content of $n(\vec{p},\vec{Q})$ by collecting into separate addends the contributions from various virtual scattering processes:

$$\begin{aligned} n(\vec{p},\vec{Q}) = {} & (A-1)\,\delta(Q)\,n(p) + (1-\delta(Q))\,\{F_{Qdd}(Q)[n(p)+n(\vec{p}-\vec{Q})] \\ & +F_{Qde}(Q)[n_{Dl}(p)+n_{Dl}(\vec{p}-\vec{Q})] \\ & -n_0[\theta(k_F-p)-F_{Qcc}(p)][\theta(k_F-|\vec{p}-\vec{Q}|)-F_{Qcc}(|\vec{p}-\vec{Q}|)] \\ & +n^{(2)'}(\vec{p},\vec{Q})+n^{(3)'}(\vec{p},\vec{Q})\}. \end{aligned} \tag{8}$$

In this expression, n_0 is the strength factor that occurs in the structural formula of the ordinary momentum distribution $n(\vec{k})$, and $n_{Dl}(\vec{k})$ is a modified momentum distribution. In addition, we have the "two-point" quantities $F_{xy}(k)$ (with $xy = dd, de$ or cc) which serve as form factors and the "three-point" quantities $n^{(2)'}(\vec{p}, \vec{Q})$ and $n^{(3)'}(\vec{p}, \vec{Q})$. (Here, "two-point" and "three-point" refer to the graphical topology of the corresponding configuration space functions entering Fourier transformations.) Employing various results from the Fermi hypernetted-chain (FHNC) analyses of the one-body density matrix and the radial distribution function, $n(\vec{p}, \vec{Q})$ within FHNC is constructed. We have implemented FHNC theory at the level in which elementary diagrams are not included (FHNC/0), since the latter are expected to be important only at higher densities.[7,8] By similar reasoning, we have omitted the three-point quantity $n^{(3)'}(\vec{p}, \vec{Q})$ from our calculations.

Numerical calculations have been carried out for two simple models of nuclear matter near its saturation density (namely $\rho = 0.182$ fm^{-3}). The "Monte Carlo" (MC) model,[31] connected with the v_2 potential, is specified by the correlation function

$$f(r) = \exp\left[-C_1\, e^{-C_2 r}\frac{(1 - e^{r/C_3})}{r}\right], \tag{9}$$

with parameter values $C_1 = 1.7$ fm $C_2 = 1.6$ fm^{-1}, and $C_3 = 0.1$ fm. The "Gaussian" model[32] (G2) should be associated with a potential having a softer core than v_2 and is specified by the correlation function

$$f(r) = 1 - \exp(-\beta^2 r^2) \tag{10}$$

(with $\beta = 1.478$ fm^{-1}). The wound parameter κ_{dir} is given by 0.297 and 0.111 for the MC and G2 models, respectively. Figure 1 provides a view of $n(\vec{p}, Q\vec{p}/p)$ for the MC model as given by the FHNC/0 approximation. Figure 2 displays $n(\vec{p}, Q\vec{p}/p)$ as a function of Q at $p = k_F^-$ and $p = 2k_F$, the results for the MC and G2 models being compared with that of infinitely extended ideal Fermi gas (Eq. (6)).

The effect of dynamical correlations (larger for the MC model than for G2) are measured for $p < k_F$ by the deviations of $-n(\vec{p}, Q\vec{p}/p)$ from unity for $Q < p + k_F$ and from zero for $Q > p + k_F$; and for $p > k_F$ by its deviations from zero. A comparison with the results of the lowest-order cluster prescription of Ref. 28 shows that the higher-order contributions present in the FHNC/0 treatment reduce the correlation correction to the Fermi gas limit and lead to significant improvement towards satisfaction of the sequential relation. From the remaining violations, it seems that the omission of elementary diagrams and of the term $n^{(3)'}$ is rather well justified in our calculation. Our results for $n(\vec{p}, Q\vec{p}/p)$ have been compared with those derived by Silver's formula (Eq. 7) by inserting the FHNC/0 versions of $n(p)$ and $S(Q)$ in this formula. The shortcomings of Silver's estimate are displayed in Fig. 3 for the case of $\vec{Q}//\vec{p}$ and $p = k_F^-$.

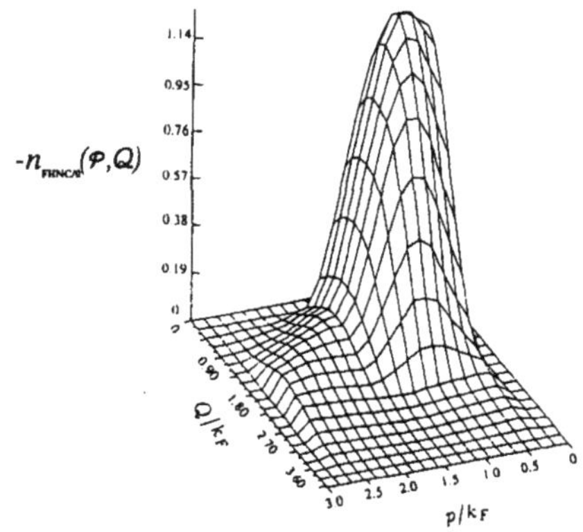

Figure 1. Generalized momentum distribution calculated by the FHNC/0 procedure for $\vec{p}$ parallel to $\vec{Q}$, ($n_{FHNC/0}(p,Q)$) as a function of momentum variables p/k_F and $Q/k_F(>0)$. The calculation is based on MC correlations and nucleon density $\rho = 0.182$ fm^{-3}.

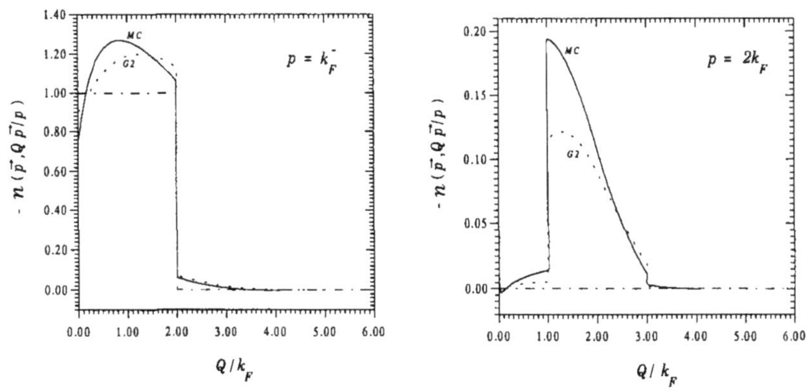

Figure 2. Generalized momentum distribution $n(\vec{p}, Q\vec{p}/p)$ as a function of $Q/k_F(>0)$, for $p = k_F^-$ and $p = 2k_F$, calculated by the FHNC/0 procedure based on the MC and G2 models, and nucleon density $\rho = 0.182$ fm^{-3}. The result for the infinitely extended ideal Fermi gas (Eq. (6)) (dot dashed line) are included for comparison.

3 The generalized momentum distribution $n(\vec{p}, \vec{Q})$ of finite nuclei in the harmonic oscillator model

Our study of the generalized momentum distribution $n(\vec{p}, \vec{Q})$ of the ground state of finite nuclei has started from the consideration of the independent-particle model and is reported in detail in Ref. 10. Our approach is an extension of that applied in Refs. 33,34 to calculate the charge form factor, the charge density, the momentum distribution, and the one-body density matrix and has the advantage of yielding closed analytical expressions. We have used the harmonic oscillator potential, the spin-orbit potential and fully occupied shells,and we have derived closed analytical expressions in the case of j-closed nuclei. Here we focus on the case of protons in

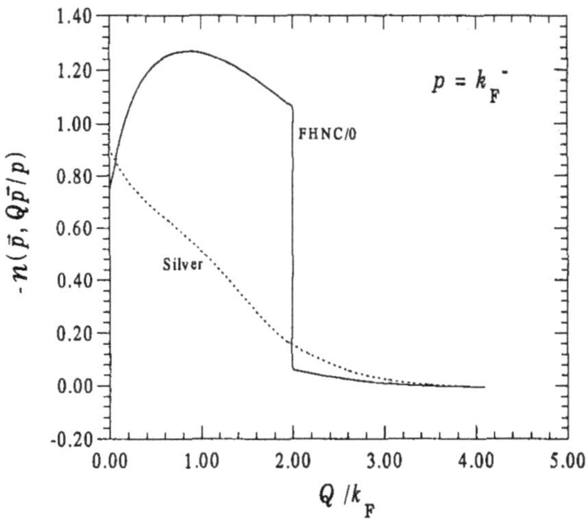

Figure 3. Generalized momentum distribution $n(\vec{p}, Q\vec{p}/p)$ as a function of $Q/k_F(> 0)$ for $p = k_F^-$, calculated from Silver's formula (Eq. 7) and in FHNC/0 approximation, for the MC model at $\rho = 0.182$ fm^{-3}.

ℓ-closed nuclei. In this case, the GMD is given by

$$n(\vec{p}, \vec{Q}) = ZF(\vec{Q})n_1(\vec{p}, \vec{p} - \vec{Q}) - \frac{1}{\nu} \int n_1(\vec{p}, \vec{k} + \vec{Q})n_1(\vec{k}, \vec{p} - \vec{Q})d\vec{k}, \tag{11}$$

where $F(\vec{Q})$ is the form factor, $n_1(\vec{p}, \vec{p}')$ is the one-body density matrix, and ν is the degeneracy due to spin. Eq. (11) has been derived by using the relation of $n(\vec{p}, \vec{Q})$ to the two-body density matrix in the momentum-space $n_2(\vec{p}_1, \vec{p}_2; \vec{p}_1', \vec{p}_2')$ (Eq. (2)) and the corresponding expression of the latter for the case of a system of non-interacting fermions. The first term on the right of Eq. (11) will be denoted by $n^{\rm d}$; the second term, denoted by $n^{\rm st}$, is an exchange term, arising from the statistical correlations among the noninteracting fermions generated by the Pauli exclusion principle. The analytical expression of $n(\vec{p}, \vec{Q})$ in the case of l-closed shell nuclei and $\vec{p}$ parallel to $\vec{Q}$ ($\vec{Q} = Q_p\vec{p}/p$) is derived from Eq. (11) if we insert the corresponding expressions for $F(Q)$ and $n_1(p, Q)$. We have for the form factor[10,33]

$$F(Q) = \frac{1}{Z} e^{Q^2b^2/4} \sum_{\lambda=0}^{N_{\rm max}} \theta_\lambda (Qb)^{2\lambda}, \tag{12}$$

where b is the harmonic oscillator parameter, $N_{\rm max} = (2n + \ell)_{\rm max}$ is the number of energy quanta of the highest occupied proton level, and the coefficients θ_λ are rational numbers varying with Z. Their values are reported in Ref. 10. The one-body density matrix in the case of $\vec{p}$ parallel to $\vec{p}'$ ($\vec{p}' = p_p'\vec{p}/p$) is given by[10,34]

$$n_1(p, p_p') = \frac{b^3}{\pi^{3/2}} e^{-p^2b^2/2} e^{-p_p'^2 b^2/2} \sum_{\mu=0}^{N_{\rm max}} (pb)^\mu \sum_{\mu'=0}^{N_{\rm max}} (p_p'b)^{\mu'} K_{\mu\mu'}. \tag{13}$$

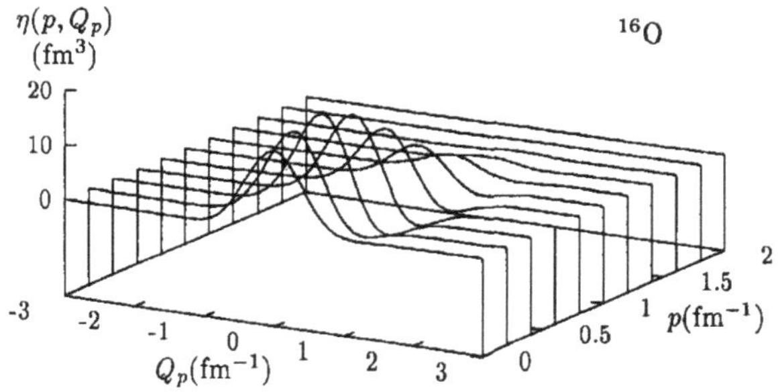

Figure 4. Generalized momentum distribution $n(p, Q_p)$ of ^{16}O for $\vec{p}$ parallel to $\vec{Q}$, calculated in the harmonic oscillator model.

The values of the coefficients $K_{\mu\mu'}$ are discussed in the Appendix of Ref. 10. Inserting the above expression of n_1 and choosing the z-axis along $\vec{p}$ one obtains for the exchange term[10]

$$n^{\rm st}(p, Q_p) = -\frac{b^3}{\pi^{3/2}} e^{-p^2b^2/2} e^{-w_p^2b^2/2} e^{-Q_p^2b^2/4} \times \sum_{\mu=0}^{N_{\rm max}} (pb)^{\mu} \sum_{\mu'=0}^{N_{\rm max}} (w_p b)^{\mu'} \sum_{\rho=0}^{2N_{\rm max}} (Q_p b)^{\rho} C_{\mu\mu'\rho} \tag{14}$$

($w_p \equiv p - Q_p$). The coefficients $C_{\mu\mu'\rho}$ are equal to zero if $\mu + \mu' + \rho =$ odd. The above expressions have been applied to the calculation of the GMD of protons in the magic nuclei ^{4}He, ^{16}O, and ^{40}Ca. Fig 4 illustrates the variation of $n(p, Q_p)$ for ^{16}O in the regions 0 to 2 fm^{-1} and -3 to 3 fm^{-1} in the variables of p and Q_p respectively.

One can see the behaviour in more detail in Fig. 5, which displays $n(p, Q_p)$ (continuous line) and the exchange term $n^{\rm st}(p, Q_p)$ (dotted line) as functions of Q_p at $p = 0.0, 1.2, 2.0$ fm^{-1} in the case of ^{16}O using both ordinary and logarithmic scales. The GMD exhibits a bump centered at $Q_p = 0$ for $p = 0$ and shifted to higher values of Q_p for $p > 0$. There is also a negative part in the GMD arising mainly from the term $n^{\rm st}(p, Q_p)$. It seems that the positive bump and the negative part at positive Q_p are bulk properties of the GMD and are due to Fermi statistics. A comparison with the GMD of the infinitely extended ideal Fermi gas (Eq. (6)) of equal Fermi wave number k_F, which exhibits discontinuities at $Q_p = 0, p = k_F$ and $Q = p + k_F$ or $|p - k_F|$, shows that finite size leads to disappearance of the discontinuities. An indication of the effect of dynamical correlations neglected in our calculation can be drawn from a comparison with the results of Section 2 for the GMD of infinite nuclear matter. In Fig. 6, a comparison is made of the GMD per particle of ^{16}O, ^{40}Ca, and of infinite nuclear matter (see Section 2) at density $\rho^{NM} = 0.182$ fm^{-3} ($k_F^{NM} = 1.3915$ fm^{-1}) for $p = 0, k_F^{NM^+}, 3/2k_F^{NM}$. The deviations of the results of the GMD for ^{16}O and ^{40}Ca from those of nuclear matter in the kinematical domains mentioned also in Section 2 (namely, for $p < k_F$ deviations from -1 or 0 for $Q < p + k_F$ and $Q > p + k_F$ respectively, and for $p > k_F$ deviations from 0) are mainly due to the effect of dynamical correlations.

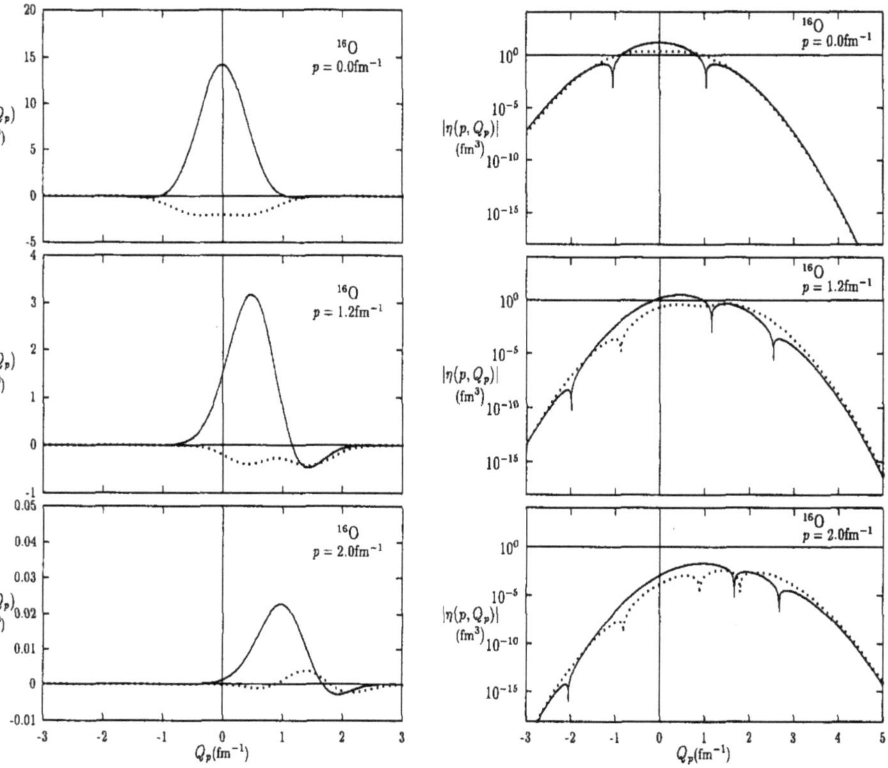

Figure 5. Generalized momentum distribution $n(p, Q_p)$ of ^{16}O for $\vec{p}$ parallel to $\vec{Q}$ as a function of Q_p for $p = 0, 1.2, 2$ fm^{-1} (continuous line), calculated in the harmonic oscillator model. The exchange term n^{st} is plotted separately (dotted line). Both ordinary (left) and logarithmic (right) scales are used.

4 The generalized momentum distribution $n(\vec{p}, \vec{Q})$ of ^{4}He including Jastrow correlations

The above study of the GMD of finite nuclei within the independent-particle model showed that dynamical short-range correlations are rather important in certain kinematical domains. As in the case of infinite nuclear matter, we consider Jastrow correlations and start by using only the first two terms of the so called low-order approximation (LOA) of the two-body density matrix (Eq. (14) of Ref. 30). Taking the corresponding term for ρ_{2h},

$$\rho_{2h}^{c}(\vec{r}_1, \vec{r}_2; \vec{r}_1') = f(|\vec{r}_1 - \vec{r}_2|) f(|\vec{r}_1' - \vec{r}_2|) \rho_{2h}^{\rm unc}(\vec{r}_1, \vec{r}_2; \vec{r}_1') \tag{15}$$

($\rho_{2h}^{\rm unc}$ equals ρ_{2h} calculated in the independent-particle model), we obtain the corresponding generalized momentum distribution by Fourier transformation. The calculation has been carried out for the GMD of protons in the nucleus ^{4}He using the correlation function $f(r) = 1 - e^{-r^2/\beta^2}$, and the following analytical expression

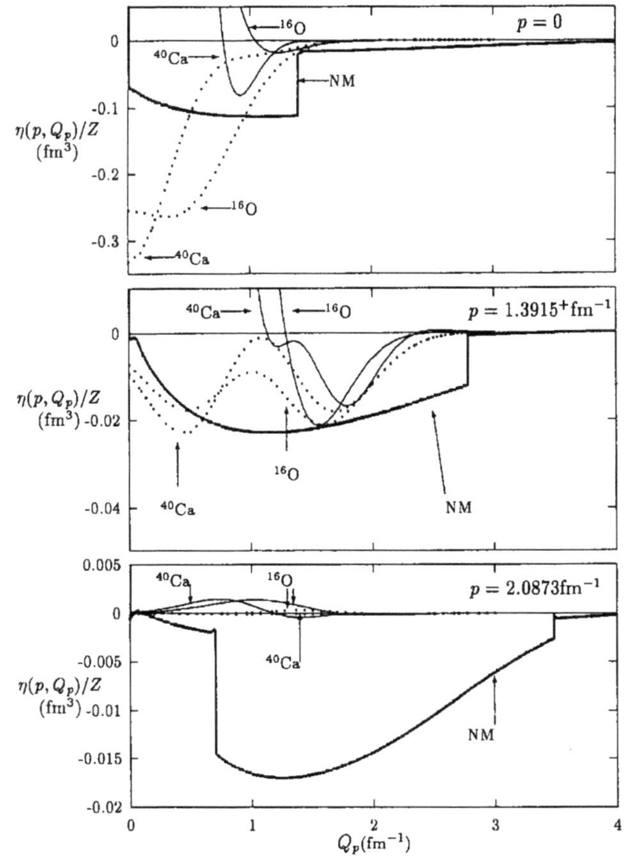

Figure 6. Generalized momentum distribution per particle $n(p, Q_p)/Z$, for $\vec{p}$ parallel to $\vec{Q}$, as a function of Q_p for $p = 0$, k_F^{NM+} and $\frac{3}{2}k_F^{NM}$ ($k_F^{NM} = 1.3915$ fm^{-1}), calculated for ^{16}O and ^{40}Ca in the harmonic oscillator model, and calculated for infinite nuclear matter with Fermi wave number k_F^{NM} in the Fermi hypernetted-chain approximation (FHNC/0). The exchange term n^{st} is plotted separately (dotted line).

has been obtained

$$
\begin{aligned}
n^{c}(\vec{p},\vec{Q}) &= n^{unc}(\vec{p},\vec{Q}) \\
&- \frac{2b^3}{\pi^{3/2}}\Big(\frac{1}{1+3y}\Big)^{3/2} e^{-\frac{p^2b^2}{2}(\frac{1+y}{1+3y})} e^{-\frac{w^2b^2}{2}} e^{-\frac{Q^2b^2}{4}(\frac{1+2y}{1+3y})} e^{\vec{Q}\cdot\vec{p}b^2\frac{y}{1+3y}} \\
&- \frac{2b^3}{\pi^{3/2}}\Big(\frac{1}{1+3y}\Big)^{3/2} e^{-\frac{p^2b^2}{2}} e^{-\frac{w^2b^2}{2}(\frac{1+y}{1+3y})} e^{-\frac{Q^2b^2}{4}(\frac{1+2y}{1+3y})} e^{-\vec{Q}\cdot\vec{w}b^2\frac{y}{1+3y}} + \\
&+ \frac{2b^3}{\pi^{3/2}}\left(\frac{1}{(1+2y)(1+4y)}\right)^{3/2} e^{-\frac{p^2b^2}{2}(\frac{1}{1+2y})} \\
&\qquad \times e^{-\frac{w^2b^2}{2}(\frac{1}{1+2y})} e^{-\frac{Q^2b^2}{4}(\frac{1}{(1+2y)(1+4y)})}, \qquad (16)
\end{aligned}
$$

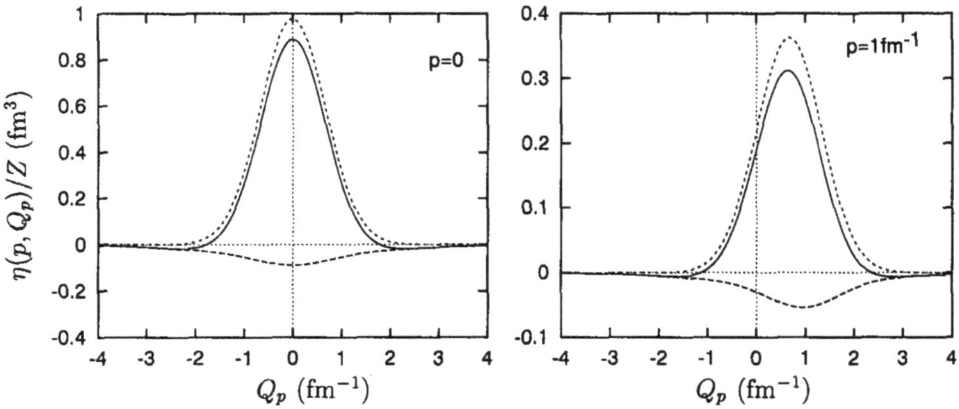

Figure 7. Generalized momentum distribution $n(p, Q_p)$ of ^{4}He for $\vec{p}$ parallel to $\vec{Q}$ as a function of Q_p for $p = 0$ and 1 fm^{-1}, calculated with the first two terms of the LOA approximation (Eq. (16)) (continuous line) and in the harmonic oscillator model (short-dashed line), along with their difference, which exhibits the dynamical correlations (long-dashed line).

where $n^{\rm unc}(\vec{p}, \vec{Q})$ is the GMD in the harmonic oscillator model calculated from Eqs. (11) and (14) for the case of the ^{4}He nucleus,

$$n^{\rm unc}(\vec{p}, \vec{Q}) = \frac{2b^3}{\pi^{3/2}} e^{-\frac{p^2b^2}{2}} e^{-\frac{w^2b^2}{2}} e^{-\frac{Q^2b^2}{4}}, \tag{17}$$

and $y = b^2/\beta^2$. One can verify that in the case of ^{4}He ($Z = 2$), $n^{\rm c}(\vec{p}, \vec{Q})$ obeys the sequential relation (5), if use is made of the expression of the momentum distribution calculated within the same approximation. Results have been obtained for the parameters values $b = 1.2195$ fm and $\beta = 0.813$ fm, which yield $y = 2.25$ (Ref. 35). In Fig. 7, $n^{\rm c}(\vec{p}, \vec{Q})$ (continuous line) along with $n^{\rm unc}(\vec{p}, \vec{Q})$ (short-dashed line) are plotted for $\vec{p}$ parallel $\vec{Q}$ ($\vec{Q} = Q_p\vec{p}/p$) as a function of Q_p for $p = 0$ and 1 fm^{-1}. The difference $n^{\rm c}(\vec{p}, \vec{Q}) - n^{\rm unc}(\vec{p}, \vec{Q})$, also plotted in Fig. 7 (long-dashed lines), gives mainly the contribution of short-range correlations. In Fig. 8, a comparison is made of the GMD per particle for $\vec{p}$ parallel $\vec{Q}$ in the cases of ^{4}He and infinite nuclear matter, calculated with the use of Eq. (16) (continuous line) and (8) (diamond chain) respectively, for the values of $p = 0$ and $2k_F^{NM}$ ($k_F = 1.3915$ fm^{-1}), using the correlation function G2 of Eq. (10). The deviation $n^{\rm c}(\vec{p}, \vec{Q}) - n^{\rm unc,d}(\vec{p}, \vec{Q})$ ($n^{\rm unc,d} \equiv n^{\rm d}$ of Eq. (11)) (dashed line), which is primarily due to the effect of dynamical and statistical correlations, shows qualitatively a similar behaviour to $n(\vec{p}, \vec{Q})$ of nuclear matter.

5 Conclusions

In summary, the momentum space transform $n(\vec{p}, \vec{Q})$ of the half-diagonal two-body density matrix of the ground state of model nuclear matter and of finite nuclei has been determined, using different approximative schemes. In the case of uniform, isospin symmetrical, spin-saturated nuclear matter, $n(\vec{p}, \vec{Q})$ has been determined

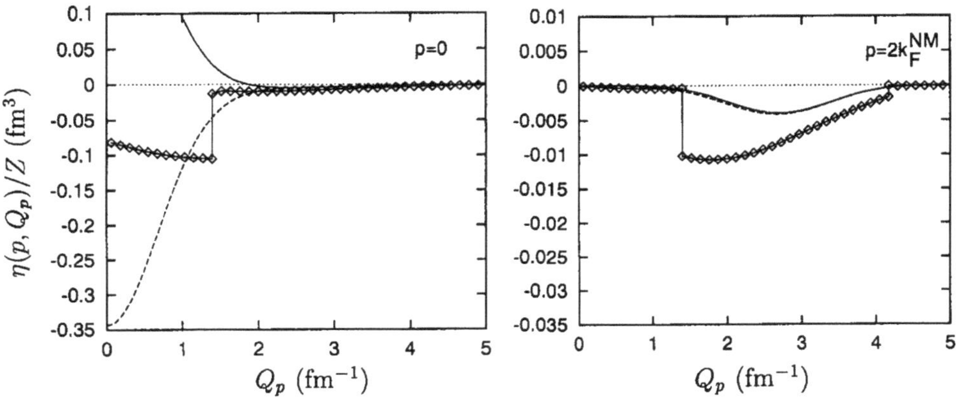

Figure 8. Generalized momentum distribution per particle $n(p, Q_p)/Z$ for $\vec{p}$ parallel to $\vec{Q}$, as a function of Q_p for $p = 0$ and $2k_F^{NM}$ ($k_F^{NM} = 1.3915$ fm^{-1}), calculated for ^{4}He with the first two terms of the LOA approximation (Eq. (16)) (continuous line), and calculated for infinite nuclear matter with Fermi wave number k_F^{NM} in the Fermi hypernetted-chain approximation (FHNC/0) (Eq. (8)) (diamond chain). The difference $n^c(p, Q_p)/Z - n^{unc,d}(p, Q_p)/Z$ is also plotted (dashed line).

microscopically, assuming state-independent, central, two-body correlations, via a Fermi hypernetted-chain calculation. The results exhibit interesting features that reflect the interplay of statistical and dynamical correlations. Regarding $n(\vec{p}, \vec{Q})$ of the ground state of finite nuclei, it has been first determined within the independent-particle model, using a harmonic oscillator basis. Results have been derived for the magic nuclei ^{4}He, ^{16}O, and ^{40}Ca, and exhibit interesting features stemming from the finite size and the Fermi statistics. Since in certain regions of momenta p and Q, dynamical correlations play a significant role, we have next considered Jastrow correlations and evaluated $n(\vec{p}, \vec{Q})$ using the first two terms of the low-order approximation of Dal Ri, Stringari and Bohigas.[30] Further investigations of $n(\vec{p}, \vec{Q})$ in the case of nuclear matter should consider realistic, state-dependent correlations; in the case of finite nuclei, the investigation should be extended to other nuclei and, if possible, consider higher-order terms in the cluster expansion. For this purpose, it might be fruitful to apply a suitable local density approximation based on inputs from the evaluation of $n(\vec{p}, \vec{Q})$ in (uniform) nuclear matter over a range of densities. Another important direction for future work is the determination of other Fourier-transforms of the two-body density matrix, for example $n(\vec{p}, \vec{k}, \vec{Q})$ (Ref. 5). One realizes that a decade after the Argonne Workshop,[2] the agenda for calculations beyond the one-body density matrix and the momentum distribution proposed by Clark and Ristig is still open.

Acknowledgments

Partial financial support from grant 70/4/3309 from the University of Athens is gratefully acknowledged.

References

1. Per-Olov Löwdin, *Phys. Rev.* **97**, 1474 (1955).
2. *Momentum Distributions*, ed. R. N. Silver and P. E. Sokol (Plenum, New York, 1989).
3. M. L. Ristig, P. M. Lam, and J. W. Clark, *Phys. Lett.* A **55**, 101 (1975).
4. J. W. Clark and M. L. Ristig, in *Momentum Distributions*, eds. R. N. Silver and P. E. Sokol (Plenum, New York, 1989), p. 39.
5. M. L. Ristig and J. W. Clark, *Phys. Rev.* B **40**, 4355 (1989).
6. M. L. Ristig and J. W. Clark, *Phys. Rev.* B **41**, 8811 (1990).
7. J. W. Clark, in *Progress in Particle and Nuclear Physics*, Vol. 2, ed. D. H. Wilkinson, (Pergamon Press, Oxford, 1979), p. 89.
8. V. R. Pandharipande and R. B. Wiringa, *Rev. Mod. Phys.* **51**, 821 (1979).
9. E. Mavrommatis, M. Petraki, and J. W. Clark, *Phys. Rev.* C **51**, 1849 (1995).
10. P. Papakonstantinou, E. Mavrommatis, and T. S. Kosmas, *Nucl. Phys.* A **673**, 171 (2000).
11. I. Sick, *Prog. Part. Nucl. Phys.* **34**, 323 (1995); and references therein.
12. J. Arrington *et al.*, *Phys. Rev. Lett.* **82**, 2056 (1999).
13. A. S. Carroll *et al.*, *Phys. Rev. Lett.* **61**, 1698 (1988).
14. L. Lapikas, *Nucl. Phys.* A **553**, 297c (1993).
15. D. Abbott *et al.*, *Phys. Rev. Lett.* **80**, 5072 (1998)
16. I. Mardor *et al.*, *Phys. Rev. Lett.* **81**, 5085 (1998).
17. P. D. Harty *et al.*, *Phys. Rev.* C **47**, 2185 (1993).
18. J. R. Annand *et al.*, *Phys. Rev. Lett.* **71**, 2703 (1993).
19. J. W. Clark and R. N. Silver, in *Proceedings of the 5th International Conference on Nuclear Reaction Mechanisms*, ed. E. Gadioli, (Universita degli Studi di Milano, Milan, 1988), p. 531.
20. O. Benhar *et al.*, *Phys. Rev.* C **44**, 2328 (1991).
21. O. Benhar *et al.*, *Phys. Lett.* B **359**, 8 (1995).
22. A. S. Rinat and M. F. Taragin, *Nucl. Phys.* A **571**, 733 (1994); *Nucl. Phys.* A **620**, 417 (1997).
23. J. Beprosvany, *Nucl. Phys.* A **601**, 269 (1996)
24. M. Petraki, PhD thesis, University of Athens (2000).
25. E. Feenberg, *Theory of Quantum Fluids*, (Academic Press, New York 1969.)
26. S. Stringari, *Phys. Rev.* B **46**, 2974 (1992).
27. R. N. Silver, *Phys. Rev.* B **38**, 2283 (1988).
28. J. W. Clark, E. Mavrommatis, and M. Petraki, *Acta. Phys. Pol.* **24**, 659 (1993).
29. M. Petraki, E. Mavrommatis, and J. W. Clark, submitted for publication.
30. M. Dal Ri, S. Stringari, and O. Bohigas, *Nucl. Phys.* A **376**, 81 (1982).
31. D. M. Ceperley, G. V. Chester, and M. H. Kalos, *Phys. Rev.* B **16**, 3081 (1977).
32. M. F. Flynn, J. W. Clark, R. M. Panoff, O. Bohigas, and S. Stringari, *Nucl. Phys.* A **427**, 253 (1984).
33. T. S. Kosmas and J. D. Vergados, *Nucl. Phys.* A **536**, 72 (1992).
34. P. Papakonstantinou, MSc Thesis, University of Athens, 1998.
35. S. S. Dimitrova, D. N. Kadrev, A. N. Antonov, and M. V. Stoitsov, *Eur. Phys. J.* A **7**, 335 (2000).

THE TRANSLATIONALLY INVARIANT COUPLED CLUSTER METHOD WITH APPLICATIONS TO NUCLEAR SYSTEMS

I. MOLINER
Department of Physics, UMIST, PO Box 88, Manchester M60 1QD, United Kingdom
E-mail: moliner@dirac.phy.umist.ac.uk

The translationally invariant reformulation of the coupled cluster method (TICCM) is reviewed, showing the results obtained for both bosonic and fermionic nucleonic systems using the different approximations within the method

1 Introduction

The coupled cluster method (CCM)[1] is one of the most successful *ab initio* many-body methods, and it has been used in many different fields. Nevertheless, when dealing with finite systems the correct treatment of the centre-of-mass problem is very important, and it was the motivation for a translationally invariant reformulation of the CCM,[2] which will be reviewed in this article. In this first section the basic concepts of the CCM will be explained, before the translationally invariant reformulation and the results obtained are explained in the next sections.

The basic starting point of the CCM is the exponential form of the wave function, that for the ground state of a N-body closed-shell system has the structure

$$|\Psi\rangle = e^S|\Phi\rangle, \tag{1}$$

where $|\Phi\rangle$ is an uncorrelated reference state and S is an operator that promotes clusters of $1, 2, \ldots, N$ particles out of the reference state, $S = \sum_{i=1}^{N} S_i = \sum_I \mathcal{S}_I C_I^\dagger$, where I stands for the set of indices that labels the cluster excitation, and we have the intermediate normalization $\langle\Phi|\Psi\rangle = 1$. The traditional CCM procedure consists on projecting the Schrödinger equation onto $\langle\Phi|$ to obtain an equation for the energy of the system

$$\langle\Phi|He^S|\Phi\rangle = E, \tag{2}$$

while pre-multiplying by e^{-S} and projecting onto the excitations $\langle\Phi|C_I$, we find equations for the unknown coefficients in the S operator

$$\langle\Phi|C_I\, e^{-S} H\, e^S|\Phi\rangle = 0, \quad \forall\, I. \tag{3}$$

If e^{-S} is not used we have an alternate form of the equations, in which the energy appears explicitly

$$\langle\Phi|C_I H\, e^S|\Phi\rangle = E\,\langle\Phi|C_I\, e^S|\Phi\rangle, \quad \forall\, I. \tag{4}$$

In order to compute observables, one must truncate the wave function in some way, and the most straightforward truncation scheme is the SUB(n) approximation, where all the coefficients $\mathcal{S}_J$ with $J > n$ are set to zero. At the SUB(2) level, the excitation operator is truncated at second order, thus the operator S includes only one- and two-particle excitations

$$|\Psi\rangle = e^{S_1+S_2}|\Phi\rangle. \tag{5}$$

2 The Translationally Invariant Coupled Cluster Method (TICCM)

In order to get translationally invariant excitations we must build a new operator $S^{(1,2)}$ from S_1 and S_2, because neither one- nor two-body excitations are in general translationally invariant. Besides, the reference state $|\Phi\rangle$ must factorise properly the centre-of-mass contribution, thus we will use a harmonic oscillator uncorrelated reference state. For a bosonic system the structure of the new operator is

$$S^{(1,2)} = \sum_{n=1}^{\infty} \mathcal{S}(n) \sum_{n_1 n_2 l_1} \langle n000; 0|n_1 l_1 n_2 l_1\rangle [a^{\dagger}_{n_1 l_1} \times a^{\dagger}_{n_2 l_1}]^0_0 \, a^2_{00}, \tag{6}$$

where the translational (and rotational) invariance is attained by appropriately coupling the single-particle states using Clebsch-Gordan coefficients and Brody-Moshinsky brackets, and we see that the above operator is a mixture of one-body (n_1 or n_2 zero) and two-body (both n_1 and n_2 non-zero) terms. The exponential of the operator produces again non-invariant terms, which are eliminated by using the normal ordering prescription,

$$|\Psi\rangle =: e^{S^{(1,2)}} : |\Phi\rangle. \tag{7}$$

The wave function has a more familiar translation in coordinate space

$$\Psi(\boldsymbol{r}_1,\ldots,\boldsymbol{r}_N) = \Big(1+\sum_{i<j} f(r_{ij}) + \tfrac{1}{2!}\sum_{i<j}\sum_{k<l}{}' f(r_{ij})f(r_{kl}) + \ldots\Big)\Phi(\boldsymbol{r}_1,\ldots,\boldsymbol{r}_N), \tag{8}$$

where the primed sum means that there are no repeated indices because the pair excitations are independent.

If we only take the linear terms into account we have the TICI2 method (translationally-invariant configuration-interaction method). This linear wave function has been used, together with gaussian expansions, to describe bosonic and fermionic systems, but within a variational approach, i.e., computing $E = \langle\Psi|H|\Psi\rangle/\langle\Psi|\Psi\rangle$.

3 TICI 2 results for nuclei

Although the above wave functions were developed for bosonic systems one can use them as well with fermionic ones, finite nuclei for example. The TICI2 wave function in coordinate space was used to describe closed-shell nuclei in the p-shell,[3] using scalar and state-dependent correlations. In Table 1 we see some results for two semi-realistic $N-N$ interactions, the Brink-Boeker BB1[4] and the modified Afnan-Tang MS3[5] potentials. The results with these interactions were good, especially when introducing V4 state-dependent correlations, but when dealing with realistic interactions and tensor correlations the linear *ansatz* proved to be insufficient.[6] This was the reason for both the introduction of a mixed Jastrow-TICI2 scheme, and for the study of the effect of the higher orders of the wave function.

As mentioned above, when dealing with realistic interactions, the linear approach is not flexible enough to cope with the repulsive core. On the other hand, it has been known for a while that Jastrow factors give a fairly good representation

Table 1. Binding energies (in MeV)[3] of p-shell nuclei for two V4 interactions, using the TICI2 wave function with state-independent (SI) and state-dependent (SD) correlations. In the state-dependent case, the correlation function f is given an operatorial structure similar to that of the interaction.

		^{4}He	^{8}Be	^{12}C	^{16}O
BB1	SI	37.86	49.18	84.91	145.94
	SD	37.86	61.30	103.93	167.30
MS3	SI	25.41	26.26	46.22	85.56
	SD	27.99	37.30	62.99	105.64

Table 2. Comparison of ^{4}He binding energies (in MeV) using several V4 interactions (Brink-Boeker BB1,[4] the Wigner part of the Afnang-Tang S3,[9] and the Malfliet-Tjon MT V[10]). The Jastrow part in J-TICI2 contains only one gaussian.

	TICI2[3]	J-TICI2[6]	DMC[11]
BB1	37.86	38.28	38.32 ± 0.01
S3	25.41	27.20	27.35 ± 0.02
MT V	29.45	31.21	31.32 ± 0.02

of short-range correlations, and one could think of incorporating a Jastrow scalar correlation, keeping the TICI2 structure (scalar or state dependent),

$$\Psi(r_1,\ldots,r_N) = \left(\sum_p \sum_{i<j} f^p(r_{ij})\Theta^p_{ij}\right)\left(\prod_{k<l} g(r_{kl})\right)\Phi(r_1,\ldots,r_N). \tag{9}$$

The Jastrow correlation function can be parametrised by a sum of gaussians, $g(r) = 1+\sum_i a_i\, e^{-b_i r^2}$. The idea behind this mixed scheme is similar to the correlated basis functions approach,[7] in which we build the correlations upon a *correlated* reference state. In this way we let the Jastrow correlations to take care of the short-range repulsion of the interaction, and we build the medium- and long-range correlations following the TICI2 prescriptions. The results for ^{4}He shown in Table 2 are good, as we can see by comparing them to the Diffusion Monte Carlo (DMC) numbers, and similar wave functions have been used to treat other systems successfully.[8] The second point to be addressed, the role of the higher order terms of the wave function, is more directly related to the study of the TICCM method and will be dealt with in the next sections.

4 Full TICC2 equations for bosonic systems

For a N-boson system, the full TICC2 wave function is given by Eq. (7). In order to obtain the energy, the Schrödinger equation is projected onto the reference state,

$$E = \langle\Phi|H : e^{S^{(1,2)}} : |\Phi\rangle \tag{10}$$

and the unknown amplitudes $\mathcal{S}(n)$ in Eq. 6) are obtained by projecting onto the generalised $2p$-$2h$ excitations, given by the structure of $S^{(1,2)}$

$$|2p\text{-}2h\rangle = \sum_{n_1 n_2 l_1} \langle n_x 000; 0 | n_1 l_1 n_2 l_1 \rangle [a^\dagger_{n_1 l_1} \times a^\dagger_{n_2 l_1}]^0_0 \, a^2_{00}. \tag{11}$$

The energy of the excited pair is $2n_x\hbar\omega$, and the schematic expression for the projection equations is $\langle 2p\text{-}2h | H : e^{S^{(1,2)}} : |\Phi\rangle = E \, \langle 2p\text{-}2h | : e^{S^{(1,2)}} : |\Phi\rangle$ which is a set of n_{max} coupled non-linear equations. The expression obtained for the ground-state energy is

$$\begin{aligned} E = {} & N_1 \, K^0_0 + N_2 \, V^{00}_{00} \\ & + N_2 \left[K^p_0 S_p + 2(N-1) V^{p0}_{00} S_p + 2 V^{pq}_{00} S^q_p \right] + N_4 \, V^{pq}_{00} S_p S_q \end{aligned} \tag{12}$$

where $N_i \equiv N(N-1)\ldots(N-i+1)$, and K^n_m and V^{pq}_{mn} stand for the matrix elements of kinetic and potential energy between harmonic oscillator states, including couplings to relative and centre-of-mass coordinates. The equations for the coefficients are more complicated

$$\begin{aligned} & E \times \left\{ N_2 \left[(N-1) C_m S_m + 2 C^n_m S^n_m \right] + N_4 \, C^n_m S_m S_n \right\} \\ & = N_2 \left\{ C_m K^0_m + 2(N-1) C_m V^{00}_{m0} + 2 C^n_m V^{00}_{mn} \right\} \\ & + N_2 \left\{ (N-1) \left[(N-1) C_m K^0_0 S_m + 2 C^n_m K^0_m S_n + 2 C_m K^p_0 S^p_m + C_m K^p_m S_p \right] \right. \\ & \quad + 4 C^n_m K^p_m S^p_n + (N-2) \left[2 C^n_m K^0_0 S^n_m + (N-1)^2 C_m V^{00}_{00} S_m + 2(N-3) C^n_m V^{00}_{00} S^n_m \right] \\ & \quad + 4(N-1) \left[(N-2) C^n_m V^{00}_{m0} S_n + (N-2) C_m V^{p0}_{00} S^p_m + (N-1) C_m V^{p0}_{m0} S_p \right] \\ & \quad \left. + 16(N-2) C^n_m V^{p0}_{n0} S^p_m + 4(N-1) C_m V^{pq}_{m0} S^q_p + 4(N-1) C^n_m V^{p0}_{mn} S_p + 4 C^n_m V^{pq}_{mn} S^q_p \right\} \\ & + N_4 \left\{ (N-2) C^n_m K^0_0 S_m S_n + 2 C^n_m K^p_0 S^n_m S_p + 4 C^n_m K^p_0 S^p_m S_n + (N-1) C_m K^p_0 S_m S_p \right. \\ & \quad + 2 C^n_m K^p_m S_n S_p + (N-2)(N-3) C^n_m V^{00}_{00} S_m S_n + 4(N-3) C^n_m V^{p0}_{00} S^n_m S_p \\ & \quad + 8(N-3) C^n_m V^{p0}_{00} S^p_m S_n + 2(N-1)(N-2) C_m V^{p0}_{00} S_m S_p + 4 C^n_m V^{pq}_{00} S^n_m S^q_p \\ & \quad + 8(N-2) C^n_m V^{p0}_{n0} S_m S_p + 2(N-1) \left[4 C_m V^{pq}_{00} S^q_m S_p + C_m V^{pq}_{00} S_m S^q_p + C_m V^{pq}_{m0} S_p S_q \right] \\ & \quad \left. + 8 C^n_m V^{pq}_{00} S^p_m S^q_n + 16 C^n_m V^{pq}_{n0} S^q_m S_p + 8 C^n_m V^{pq}_{n0} S_m S^q_p + 2 C^n_m V^{pq}_{mn} S_p S_q \right\} \\ & + N_6 \left\{ C^n_m K^p_0 S_m S_n S_p + 2(N-3) C^n_m V^{p0}_{00} S_m S_n S_p + (N-1) C_m V^{pq}_{00} S_m S_p S_q \right. \\ & \quad \left. + 8 C^n_m V^{pq}_{00} S^p_m S_n S_q + 2 C^n_m V^{pq}_{00} S^n_m S_p S_q + 2 C^n_m V^{pq}_{00} S_m S_n S^q_p + 4 C^n_m V^{pq}_{n0} S_m S_p S_q \right\} \\ & + N_8 \left\{ C^n_m V^{pq}_{00} S_m S_n S_p S_q \right\} \end{aligned} \tag{13}$$

The expressions in Eqs. (12) and (13) are very schematic, with different sums and couplings hidden in the notation, but they give a fair idea of their structure in terms of the $\mathcal{S}$ coefficients (each S_p or S^q_p term contains one of them). See Ref. 12 for details. In Table 3 we can see results for three bosonic systems, using the linear and the complete wave functions. While for the $A = 4$ system the higher order terms give a negligible contribution, the importance of the same terms is much greater in heavier systems. It is worth noticing the comparison between the linear results in configuration and coordinate space. The latter are fully converged, while the former are almost, but not fully converged, when using 30 amplitudes. The convergence problem (and the very time consuming computation in configuration space) led to the transcription of the equations to real space, where the convergence is granted.

Table 3. Binding energies (in MeV) for bosonic systems using the Wigner part of the Afnan-Tang S3 interaction.[9] The first row shows the TICI2 coordinate space results.

	N=4	N=16	N=40
TICI2[13]	25.42	1131.2	7495.3
TICI2[12]	25.31	1130.9	7493.3
TICC2[12]	25.49	1234.9	8456.6

5 Coordinate Space TICC2

In order to avoid the convergence problem, we can use the coordinate space representation of the TICC2 wave function

$$\Psi(\boldsymbol{r}_1,\ldots,\boldsymbol{r}_N)=\left(1+\sum_{i<j} f(r_{ij})+\frac{1}{2!}\sum_{i<j}\sum_{k<l}{}' f(r_{ij})f(r_{kl})+\ldots\right)\Phi(\boldsymbol{r}_1,\ldots,\boldsymbol{r}_N), \quad (14)$$

and proceed as before, projecting onto the reference state and the excitations. The excited state is given by generic function, g, with the same characteristics as the correlation function f, mainly $\langle\Phi|f|\Phi\rangle = 0$. Both functions are expanded in gaussians to solve the equations, that can be expressed diagrammatically in a compact form: a circle represents a particle, solid lines denote the correlation function f, dashed lines stand for the interaction, crosses indicate the action of the kinetic energy operator, and wavy lines denote the function g. To clarify the meaning of the diagrammatic notation, consider the following diagram,

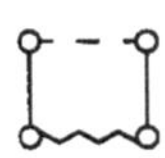

$$\equiv \int \Phi^*(\boldsymbol{r}_1,\ldots,\boldsymbol{r}_N)\, g^*(r_{12})V(r_{34})f(r_{13})f(r_{24})\,\Phi(\boldsymbol{r}_1,\ldots,\boldsymbol{r}_N)\; d\boldsymbol{r}_1\ldots d\boldsymbol{r}_N. \quad (15)$$

The equation corresponding to the projection onto a generic excitation reads

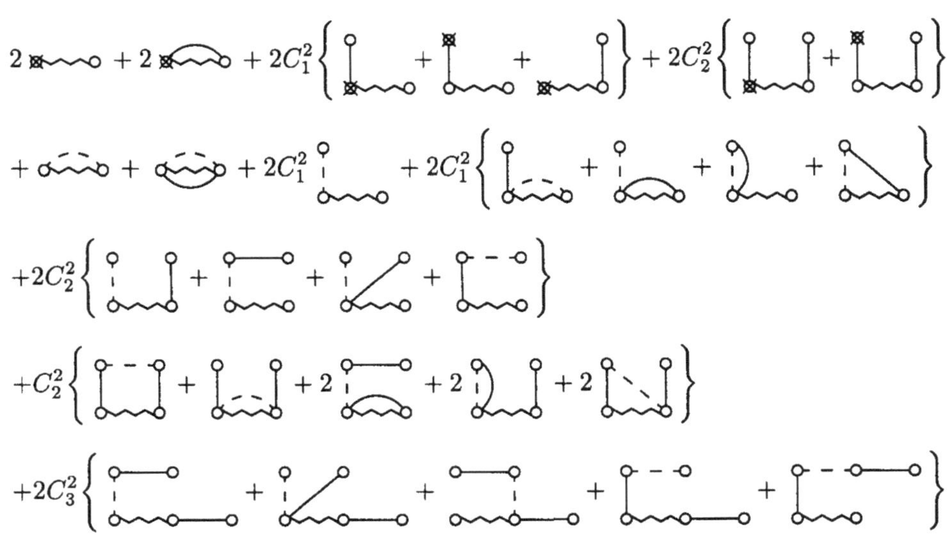

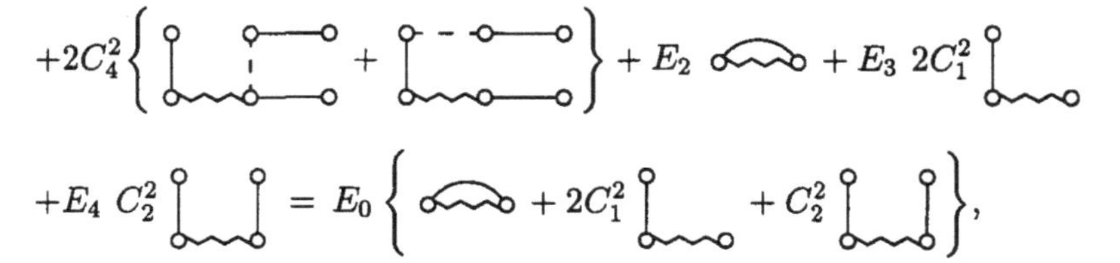

(16)

where $C_k^n \equiv (A-n)(A-n-1)\dots(A-n-k+1)$ are statistical factors. The E_n factors

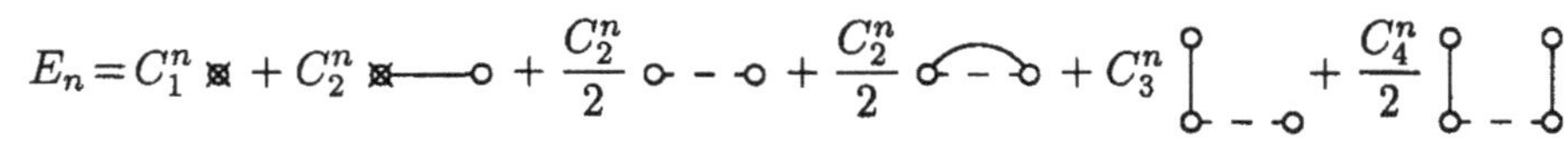

(17)

are related to unlinked diagrams, and E_0 corresponds to the ground-state energy. Eqs. (16) and (17) are the coordinate space equivalent to Eqs. (12) and (13). In Table 4 we see the results obtained for the same boson systems of Table 3. As we supposed by comparing the linear results, we can see that the full results in configuration space were almost but not fully converged. Another interesting feature of these calculations is the relative importance of the different orders in powers of the correlation function f in the wave function. The actual equations are fourth order in f, but we can clearly see that the relevant terms are the linear and quadratic ones. One final remark about the calculation in real space is that the computation for these bosonic systems was roughly 500 times faster than in configuration space.

6 TICC2 Calculations in Nuclei

We have seen in the previous section that, for bosons, the third and fourth order terms in the wave function give an almost negligible contribution to the ground state energy, so for nuclei calculations we are going to disregard them and use the following wave function,

$$\Psi(\boldsymbol{r}_1,\dots,\boldsymbol{r}_N) = \left(1 + \sum_{i<j} f(r_{ij}) + \frac{1}{2!}\sum_{i<j}\sum_{k<l}{}' f(r_{ij})f(r_{kl})\right)\Phi(\boldsymbol{r}_1,\dots,\boldsymbol{r}_N). \quad (18)$$

The reference state, $|\Phi\rangle$, is now a Slater determinant of harmonic oscillator single-particle wave functions, as is appropriate for the fermionic statistics of the wave function. There is another difference with the bosonic case, regarding diagrams containing disconnected correlation functions. The intermediate normalisation imposes $\langle\Phi|f|\Phi\rangle = 0$, and that constraint cancels out all diagrams with unlinked correlation functions in the bosonic case. However, in the fermionic case these diagrams do not cancel because the unlinked f does not factorise. Thus, with nuclei we have to consider diagrams such as, for example,

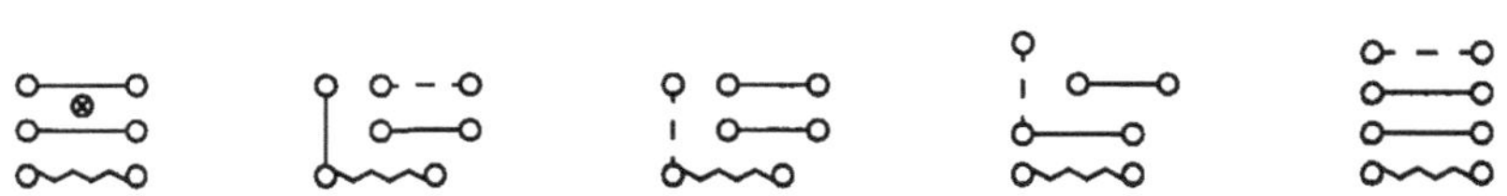

(19)

and the number of diagrams to consider is much larger, but the equations have the same structure. We have performed the calculation both with and without the

Table 4. Binding energies (in MeV)[14] for various "bosonic nuclei" with the Wigner part of the S3 interaction. Order 1 corresponds to the TICI2 case and Order 4 is the full TICC2 result. The last row shows the TICC2 configuration space results from the previous table.

Order	N=4	N=16	N=40
1	25.42	1131.2	7495.3
2	25.60	1235.7	8457.1
3		1235.1	8458.5
4		1235.1	8458.7
4	25.49	1234.9	8456.6

Table 5. Binding energies (in MeV) for $4n$ nuclei in the p-shell, using the BB1 and MS3 interactions. TICC2(b) corresponds to the calculation disregarding diagrams with disconnected f, and TICC2 to the full calculation up to seven-particle diagrams (six in the ^{16}O case).

		^{4}He	^{8}Be	^{12}C	^{16}O
BB1	TICI2	37.86	49.18	84.91	145.92
	TICC2(b)	37.92	50.04	86.16	149.36
	TICC2	37.92	49.96	86.17	149.27
MS3	TICI2	25.41	26.26	46.22	85.56
	TICC2(b)	25.59	28.37	50.53	94.71
	TICC2	25.59	28.35	50.52	94.81

unlinked diagrams, and the results are displayed in Table 5. It is worth noting that of the two interactions used, the MS3 has the stronger core, and the gain in energy is much greater with this interaction than with the milder BB1 potential. On the other hand, we can check the relevance of the unlinked diagrams by comparing the two last rows for each interaction. While for ^{4}He there is no contribution, because is basically a boson system, for the other nuclei the contribution is non-zero, but small.

7 Comments and Prospects

The Translationally Invariant Coupled Cluster Method was introduced to tackle from the very outset the centre-of-mass problem within the framework of the CCM, and was first used to treat ^{4}He. At the linear level of approximation, TICI2, the results for finite nuclei proved the linear wave function to be a good starting point, and a very clever way of reformulating the configuration-interaction wave function taking advantage of the rotational and translational invariance. Nevertheless, the linear wave function is not good enough to deal with realistic nuclear interactions, and a exploration of the higher orders was needed. The results for bosons in configuration space showed that the non-linear terms were indeed very important beyond the four-body system, but the calculation faced problems with the convergence ratio. This problems led to the translation of the TICC2 equations into the real coordinate space, where the convergence is granted and the computation turned out to be much faster. At the same time, the third- and fourth-order terms of the wave function were found to be irrelevant for the ground state energy of bosonic systems, and were thus disregarded, when facing calculations for closed-shell nuclei in the p-shell. These last results show that the contribution of the second-order terms in the wave function is very important and seems to be the more relevant, the harder the core of the interaction is. As future prospects of the method, a state-dependent TICC2 calculation will be very interesting, mainly to check the importance of the quadratic operatorial correlations, which could give us a hint about the results to be expected, when tensor correlations and interactions are used. Once the two-body correlations have been explored fully, higher order ones should be taken into account (which means formulating TICC3). On the other hand, due to the nature of the nuclear repulsive core, realistic calculations in nuclei are likely to need Jastrow correlations, at least at the scalar level, and thus mixed schemes as Jastrow-TICI2 are worth exploring as well.

References

1. F. Coester, *Nucl. Phys.* **7**, 421 (1958); F. Coester and H. Kümmel, *Nucl. Phys.* **17**, 477 (1960).
2. R. F. Bishop, M. F. Flynn, M. C. Boscá, E. Buendía, and R. Guardiola, *Phys. Rev.* C **42**, 1341 (1990).
3. R. Guardiola, P. I. Moliner, J. Navarro, R. F. Bishop, A. Puente, and N. R. Walet, *Nucl. Phys.* A **609**, 218 (1996).
4. D. M. Brink and E. Boeker, *Nucl. Phys.* A **91**, 1 (1967).
5. R. Guardiola, A. Faessler, H. Müther and A. Polls, *Nucl. Phys.* A **371**, 79 (1981).
6. R. F. Bishop, R. Guardiola, I. Moliner, J. Navarro, M. Portesi, A. Puente, and N. R. Walet, *Nucl. Phys.* A **643**, 243 (1998).
7. E. Feenberg and J. W. Clark, *Phys. Rev.* **113**, 388 (1959); E. Feenberg and C. W. Woo, *Phys. Rev.* **137**, 391 (1965); E. Krotscheck and J. W. Clark, *Nucl. Phys.* A **328**, 73 (1979).
8. R. Guardiola and J. Navarro, *Phys. Rev. Lett.* **84**, 1144 (2000).
9. I. R. Afnag and Y. C. Tang, *Phys. Rev.* **175**, 1337 (1968).
10. R. A. Malfliet and J. A. Tjon, *Nucl. Phys.* A **127**, 161 (1969).
11. R. F. Bishop, E. Buendía, M. F. Flynn and R. Guardiola, *J. Phys. G (Nucl. Part. Phys.)* **18**, L21 (1992).
12. R. Guardiola, I. Moliner, J. Navarro, and M. Portesi, *Nucl. Phys.* A **628**, 187 (1998).
13. R. Guardiola, P. I. Moliner and J. Navarro, *Phys. Lett.* B **383**, 243 (1996).
14. I. Moliner, R. F. Bishop, N. R. Walet, R. Guardiola, J. Navarro, and M. Portesi, *Phys. Lett.* B **480**, 61 (2000).

MEAN FIELD APPROACH TO QUARK MATTER IN THE NJL MODEL

STEVEN A. MOSZKOWSKI
Physics Department, UCLA, Los Angeles, CA 90095, USA
E-mail: stevemos@ucla.edu

CONSTANÇA PROVIDÊNCIA AND JOÃO DA PROVIDÊNCIA
Departamento de Física, Universidade de Coimbra,
3000 Coimbra, Portugal
E-mail: constanca@teor.fis.uc.pt, providencia@teor.fis.uc.pt

The NJL model yields a coupling for scalar meson exchange which is in quite good agreement with what is needed to fit the properties of nuclear matter. Unfortunately, according to the NJL model, quark matter does not saturate until the density has reached the critical value at which chiral symmetry restoration is reached. However, it may be shown that quark matter may saturate at a density for which the effective mass is still finite if the scalar coupling constant is allowed to increase with density. So, for a conveniently generalized NJL model, saturation is reached before chiral symmetry is restored. In the generalized model, restoration of chiral symmetry may occur at a density an order of magnitude higher than nuclear matter density. The properties of hadronic matter in the framework of the generalized model are here discussed.

1 Introduction

In the nuclear medium, the same underlying Hamiltonian leads to slightly different masses than in free space. This can be considered as due to effective interactions via scalar and vector meson exchange. It turns out that the NJL model yields a coupling for scalar meson exchange which is in quite good agreement with what is needed to fit the properties of nuclear matter. However, according to the NJL model, quark matter does not saturate until the density has reached the critical value at which the effective mass vanishes, i.e. saturation does not occur before chiral symmetry is restored, which is not satisfactory. Nevertheless, if the scalar coupling constant is allowed to increase with density, then quark matter may saturate at a density where the effective mass is still finite. We use a generalization of the NJL model to generate an effective density dependent coupling constant. The generalized model yields an expression for the energy of quark matter as function of density predicting saturation to be reached, before chiral symmetry is restored. The properties of hadronic matter in the framework of the generalized model are here computed and discussed. For the model parameters, the critical density for chiral symmetry restoration is an order of magnitude higher than nuclear matter density.

In Section 2 the Generalized NJL model is introduced. In Section 3, results are presented and conclusions are drawn. In Section 4, some open problems are listed.

2 Generalized NJL model

The NJL model[1] is defined by the Lagrangian density

$$\mathcal{L} = \bar{\psi}(i\gamma^\mu\partial_\mu)\psi + G_S[(\bar{\psi}\psi)^2 + (\bar{\psi}i\gamma_5\boldsymbol{\tau}\psi)^2]. \tag{1}$$

A regularizing momentum cutt-off Λ is part of the model. The Lagrangian is equivalent to the Hamiltonian

$$\mathcal{H}_{NJL} = \sum_{k=1}^{N} \boldsymbol{p}_k \cdot \boldsymbol{\alpha}_k + G_S \sum_{k,l=1}^{N} \delta(\boldsymbol{r}_k - \boldsymbol{r}_l)\beta_k\beta_l(1 - \gamma_k^5\gamma_l^5\boldsymbol{\tau}_k \cdot \boldsymbol{\tau}_l). \tag{2}$$

The vacuum is described by a Slater Determinant $|\Phi_0\rangle$ constructed from plane waves which are negative energy eigenfunctions of the single particle Hamiltonian $h = \boldsymbol{p} \cdot \boldsymbol{\alpha} + \beta m$. The "constituent mass" m is a variational parameter.

If positive energy eigenfunctions with momentum $\boldsymbol{p}$ satisfying $|\boldsymbol{p}| < p_F$ are occupied, so that p_F is the Fermi momentum, the energy expectation value $\mathcal{E} = \langle\Phi_0|\mathcal{H}_{NJL}|\Phi_0\rangle$ reads

$$\begin{aligned}\mathcal{E} &= -\nu \sum_{p=p_F}^{\Lambda} \frac{p^2}{E_p} - \frac{G_S}{V}\left[\nu \sum_{p=p_F}^{\Lambda} \frac{m}{E_p}\right]^2 \\ &= -\frac{\nu V}{2\pi^2}\int_{p_F}^{\Lambda} \mathrm{d}p \frac{p^4}{E_p} - G_S V \left[\frac{\nu}{2\pi^2}\int_{p_F}^{\Lambda} \mathrm{d}p \frac{mp^2}{E_p}\right]^2,\end{aligned} \tag{3}$$

where $E_p = \sqrt{p^2 + m^2}$, V is the normalization volume, and $\nu = 2N_cN_f$ is the degeneracy. The condition $\partial\mathcal{E}/\partial m$=0 leads to the gap equation

$$1 = \frac{2\nu G_S}{V}\sum_{p=p_F}^{\Lambda} \frac{1}{E_p} = \frac{\nu G_S}{\pi^2}\int_{p_F}^{\Lambda} \mathrm{d}p \frac{p^2}{E_p}, \tag{4}$$

which fixes m.

Strong interaction dynamics of mesons and baryons is believed to be described by QCD, which exhibits a non perturbative behaviour at low energies. This circumstance renders the analytic study of the theory rather difficult. The NJL model is a popular substitute which has in common with QCD important symmetries of the quark-flavour dynamics. This model has been very successful in the description of the meson sector. The question arises: does the model allow also for soliton solutions (non-homogeneous solutions) which are of interest for the description of the baryon sector and of hadronic matter? It is found that the model leads to saturation of hadronic matter[3] providing $\Lambda/m_0 \leq 1.8$, where m_0 is the vacuum value of m (for $p_F = 0$). We observe, in passing, that the existence of saturation is not enough to ensure the occurrence of inhomogeneous solutions describing droplets of hadronic matter. In Ref. 3, the surface tension has been computed in the framework of a semi-classical approach based on the Wigner-Kirkwood expansion, with second order gradient terms included. It was found in Ref. 3 that the NJL model leads to a rather high value for the surface tension, so that solitons exist only if the number A of quarks in positive energy states is sufficiently large. For instance, if

$\Lambda/m_0 = 1.5$, solitons exist only if there are more than 12 quarks in positive energy states. For higher values of Λ/m_0 the situation is still worse. A high value for the surface tension is probably due to the high incompressibility which is expected from the NJL model, and is apparent, for instance, in Fig. 1 of Ref. 2.

The NJL model can be generalized to yield more reasonable saturation of quark matter. All we need is a modified gap equation

$$\sum_{p=p_F}^{\Lambda} \frac{1}{E_p} = \frac{V}{2\nu G(\rho)}. \tag{5}$$

where ρ denotes the nucleon density.

An effective density dependent coupling constant is obtained, if the following generalized NJL Lagrangian density, which actually pushes chiral symmetry restoration to higher densities, is considered

$$\begin{aligned}\mathcal{L} = &\bar{\psi}(i\gamma^\mu \partial_\mu)\psi + G_S[(\bar{\psi}\psi)^2 + (\bar{\psi}i\gamma_5\boldsymbol{\tau}\psi)^2] - G_V(\bar{\psi}\gamma^\mu\psi)^2 \\ &- G_{VS}[(\bar{\psi}\psi)^2 + (\bar{\psi}i\gamma_5\boldsymbol{\tau}\psi)^2](\bar{\psi}\gamma^\mu\psi)^2.\end{aligned} \tag{6}$$

The resulting thermodynamic potential per volume is

$$\omega(\omega, p_F, \mu) = \langle\bar{\psi}(\boldsymbol{\gamma}\cdot\mathbf{p})\psi\rangle - G_S\langle\bar{\psi}\psi\rangle^2 + G_V\langle\psi^\dagger\psi\rangle^2 + G_{SV}\langle\bar{\psi}\psi\rangle^2\langle\psi^\dagger\psi\rangle^2 - \mu\langle\psi^\dagger\psi\rangle, \tag{7}$$

where exchange terms have been neglected. By $\langle\bar{\psi}\Gamma\psi\rangle$ we denote the following expectation value per volume

$$\langle\bar{\psi}\Gamma\psi\rangle = \frac{1}{V}\langle\Phi_0|\sum_k \beta_k\Gamma_k|\Phi_0\rangle.$$

We find

$$\langle\bar{\psi}(\boldsymbol{\gamma}\cdot\mathbf{p})\psi\rangle = -2N_cN_f\int\frac{d^3p}{(2\pi)^3}\frac{p^2}{E_p}(\theta_\Lambda - \theta_F), \tag{9}$$

$$\langle\bar{\psi}\psi\rangle = -2N_cN_f\int\frac{d^3p}{(2\pi)^3}\frac{m}{E_p}(\theta_\Lambda - \theta_F), \tag{10}$$

$$\langle\psi^\dagger\psi\rangle = 2N_cN_f\int\frac{d^3p}{(2\pi)^3}\theta_F, \tag{11}$$

with $E_p = \sqrt{m^2+p^2}$, $\theta_\Lambda = \theta(\Lambda^2 - p^2)$, $\theta_F = \theta(p_F^2 - p^2)$. The condition $\partial\omega/\partial m = 0$ implies

$$m = -2G_S\langle\bar{\psi}\psi\rangle + 2G_{VS}\langle\bar{\psi}\psi\rangle\langle\psi^\dagger\psi\rangle^2.$$

The condition $\partial\omega/\partial p_F = 0$ implies

$$E_{p_F} = \mu - 2G_V\langle\psi^\dagger\psi\rangle - 2G_{VS}\langle\psi^\dagger\psi\rangle\langle\bar{\psi}\psi\rangle^2.$$

These conditions fix the values of p_F, m for given μ.

In terms of the following functions:

$$F_0(m,p) = \int dp\frac{p^2}{E_p} = \frac{1}{2}pE_p - \frac{1}{2}m^2\log(p+E_p), \tag{12}$$

$$F_2(m,p) = \int dp\frac{p^4}{E_p} = \frac{1}{8}(-3m^2p + 2p^3)E_p + \frac{3}{8}m^4\log(p+E_p), \tag{13}$$

we have

$$\langle\bar{\psi}(\boldsymbol{\gamma}\cdot\boldsymbol{p})\psi\rangle = -2N_cN_f\frac{4\pi}{(2\pi)^3}(F_2(m,\Lambda)-F_2(m,p_F)), \tag{14}$$

$$\langle\bar{\psi}\psi\rangle = -2N_cN_f\frac{4\pi m}{(2\pi)^3}(F_0(m,\Lambda)-F_0(m,p_F)), \tag{15}$$

$$\langle\psi^\dagger\psi\rangle = 2N_cN_f\frac{4\pi}{(2\pi)^3}\frac{p_F^3}{3}. \tag{16}$$

The properties of the generalized NJL model are now easily computed.

3 Results and Conclusions

The value of G_S is determined by the choice of the vacuum constituent quark mass m_0 and of the cut-off momentum Λ. We take $m_0 = 322\,\mathrm{MeV}$ and $\Lambda = 2m_0 = 644\,\mathrm{MeV}$. Then, we find $G_S\Lambda^2 = 2.17$. We also obtain important properties of the vacuum, namely, the pion decay constant f_π, the order parameter (quark condensate) $\langle\bar{q}q\rangle$ and the slope of the constituent quark mass $\mathrm{d}\,m/\mathrm{d}\,\rho$ at the vacuum, that is, for $\rho = 0$, where ρ denotes the baryonic density. We find $f_\pi = 93$, $(\rho_0/m_0)\mathrm{d}\,m/\mathrm{d}\,\rho = -0.350$, $\langle\bar{q}q\rangle = -(313)^3\mathrm{MeV}^3$. Here ρ_0 denotes the nuclear matter density.

The numerical results displayed in Table 1 were obtained as follows. Having fixed G_S by the vacuum properties, several values were given to G_V, as displayed in the first column, and for each value of G_V the corresponding value of G_{VS} was determined by the requirement that the energy per nucleon, at saturation, should be $E/A = -15.80\,\mathrm{MeV}$.

At saturation, corresponding to the chosen energy per nucleon E/A, we show the constituent quark mass m^*, the nucleon density ρ, the Fermi momentum p_F, the chemical potential μ, and the incompressibility of hadronic matter per nucleon K. At restoration of chiral symmetry we show the nucleon density ρ_c.

The term in G_{VS}, which is responsible for the density dependence of the effective coupling constant, plays an important role in pushing to higher energies the restoration of chiral symmetry and in lowering the incompressibility. Chiral symmetry is restored at a density about 11.5 times the nuclear matter density ρ_0.

The resulting equation of state can be put into a form similar to that found some years ago by Glendenning and Moszkowski,[7] which is intermediate between the Walecka model[6] and the derivative coupling model.[8] As shown in Fig. 1, the energy per particle increases less steeply for the GNJL model than for the Walecka model. Concerning the predicted incompressibilty, the present model is in reasonable agreement with these relativistic models (Refs. 7,8).

4 Open Problems

The NJL model is essentially a quark shell model.[4] What may be said about confinement, i.e. quark clustering, which is not included in the model? It is known that the neglect of quark clustering in the quark shell model leads to serious deviations in the calculation of some nuclear properties.[5] Another unsolved problem of great

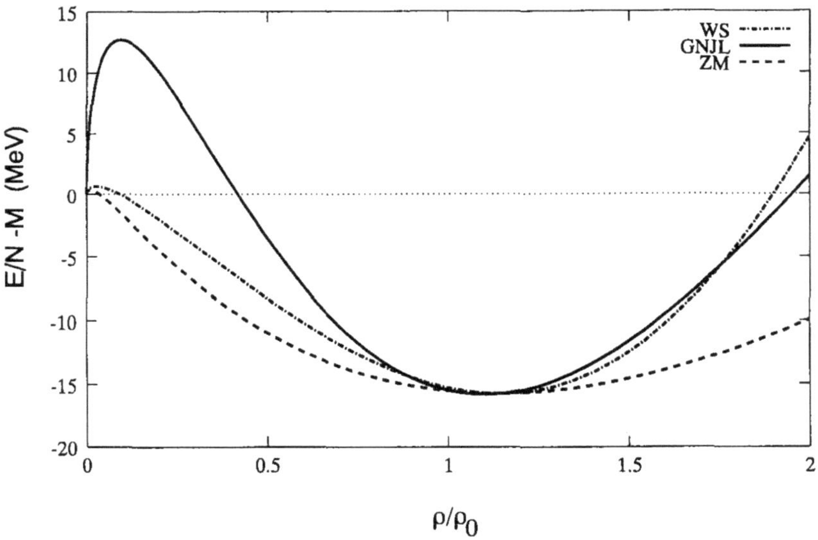

Figure 1. Comparison of energyi per nucleon as a function of ρ for the GNJL, Walecka and Zimani-Moszkowski model

Table 1. Numerical results. Having fixed G_S by the vacuum properties, for each value of G_V, the corresponding value of G_{VS} was determined by requiring that the energy per nucleon, at saturation, is $E/A = -15.80\,\text{MeV}$. Here ρ_0 denotes the nuclear matter density.

$G_V\,\Lambda^2$	$G_{VS}\,\Lambda^8$	E/A (MeV)	m^* (MeV)	ρ (fm^{-3})	p_F (MeV/c)	μ (MeV)	K (MeV)	ρ_c/ρ_0
0.00	-358.60	-15.80	227.64	0.206	285.94	316.49	908.37	10.95
0.43	-426.80	-15.80	242.29	0.185	275.63	316.44	776.04	11.60
0.87	-488.33	-15.80	251.64	0.172	269.19	316.68	696.51	11.98
1.30	-545.76	-15.80	258.94	0.160	262.75	316.52	632.64	12.13
1.74	-600.40	-15.80	264.57	0.151	257.60	316.43	584.64	12.36

interest is the microscopic basis for the mechanism that removes the deeply bound states which would appear, if only the scalar meson exchange occurred. In other words, how are our effective vector and scalar-vector terms in the generalized NJL model related to correlations between nucleons?

Acknowledgments

One of us (S.A.M.) is very grateful to Dr. T. Goldman for a helpful discussion. Two of us (J.P. and C.P.) are grateful to Dr. Célia Sousa and Dr. Yasuhiko Tsue for valuable comments.

References

1. S. P. Klevansky, *Rev. Mod. Phys.* **64**, 649 (1992).

2. M. Fiolhais, J. da Providência, M. Rosina, and C. A. Sousa, *Phys. Rev.* C **56**, 3311 (1997).
3. J. da Providência, Hans Walliser and Herbert Weigel, *Nucl. Phys.* A **671**, 547 (2000).
4. H. R. Petry, H. Hofestadt, S. Merk, K. Bleuler, H. Bohr and K. S. Narain, *Phys. Lett.* B **159**, 363 (1985).
5. I. Talmi, *Phys. Lett.* B **205**, 140 (1988).
6. B. D. Serot and J. W. Walecka, *Adv. Nucl. Phys.* **16**, 1 (1986).
7. N. K. Glendenning, F. Weber, and S. A. Moszkowski, *Phys. Rev.* C **45**, 844 (1992).
8. J. Zimanyi and S. A. Moszkowski, *Phys. Rev.* C **42**, 1416 (1990).

EFFECTIVE FIELD THEORY IN NUCLEAR MANY-BODY PHYSICS

BRIAN D. SEROT
Physics Department and Nuclear Theory Center, Indiana University
Bloomington, IN 47405, USA

JOHN DIRK WALECKA
Department of Physics, The College of William and Mary
Williamsburg, VA 23187, USA

Recent progress in Lorentz-covariant quantum field theories of the nuclear many-body problem(*quantum hadrodynamics*, or QHD) is discussed. The importance of modern perspectives in effective field theory and density functional theory for understanding the successes of QHD is emphasized.

1 Overview

Reference 1 is a presentation entitled *Relativistic Nuclear Many-Body Theory* given at the Seventh International Conference on Recent Progress in Many-Body Theories held in Minneapolis, Minnesota, in August, 1991. This was a report on a long-term effort to understand the nuclear many-body system in terms of relativistic quantum field theories based on hadronic degrees of freedom,[2,3] a topic we refer to as *quantum hadrodynamics* (or QHD). An extensive, more recent review of work in this area is contained in Ref. 4, and a text now exists[5] that provides background material.[a] There has been significant recent progress in this area,[4,6–8] and the goal of this contribution is to summarize briefly what has transpired since the presentation in Ref. 1.

The only consistent framework we have for discussing the relativistic many-body system is relativistic quantum field theory based on a local Lagrangian density. In any Lagrangian approach, one must first decide on the generalized coordinates, and hadronic degrees of freedom—baryons and mesons—are the most appropriate for ordinary nuclear systems (QHD). Early attempts involved simple renormalizable models, which reproduced some basic features of the nuclear interaction.[2] The advantage of such models is that in principle, one can consistently investigate and relate all aspects of nuclear structure to a small number of renormalized coupling constants and masses. The disadvantage, in addition to the strong coupling constants that make reliable approximation schemes difficult to come by, is that limiting the discussion to renormalizable Lagrangians is too restrictive. Despite these drawbacks, the simple models led to interesting insights. In relativistic mean-field theory (MFT), nuclear densities, the level structure of the nuclear shell model, and the spin dependence of nucleon–nucleus scattering are reproduced.[2] The simplest model (QHD–I) consists of baryons and isoscalar scalar and vector mesons. A basic feature of all these models is that there are strong scalar and vector mean fields present in the nucleus, which cancel in the binding energy but which add to give

[a]Extensive references to other work in this field are contained in Refs. 1 through 5.

the large spin-orbit interaction.[7]

There is now overwhelming evidence that the underlying theory of the strong interaction is *quantum chromodynamics* (QCD), a Yang–Mills non-abelian gauge theory built on an internal color symmetry of a system of quarks and gluons. If mass terms for the u and d quarks are absent in the Lagrangian, QCD possesses *chiral symmetry* in the nuclear domain; although spontaneously broken in manifestation, this symmetry should play an essential role in nuclear dynamics. The challenge[1] was to understand the theoretical basis of QHD, the successes that it had, and its limitations, in terms of QCD. Indeed, as we say in our summary in Ref. 1:

> More generally, it is probable that at low energies and large distances, QCD can be represented by an *effective field theory* formulated in terms of a few hadronic degrees of freedom. All possible couplings must be included in the low-energy effective Lagrangian, which is then to be used at tree level. The underlying assumption of QHD is that of a local relativistic theory formulated in terms of baryons and the lightest mesons. The theory is assumed to be renormalizable, and one then attempts to extract predictions for long-range phenomena by computing both tree-level diagrams and *renormalized quantum loop corrections.* In the end, it may turn out that this assumption is untenable, and that the only meaningful interpretation of QHD is as an *effective* theory, to be used at the tree or one-loop level. The limitation to renormalizable couplings may then be too restrictive. Nevertheless, the phenomenological success of the MFT of QHD–I in the nuclear domain implies that *whatever the effective field theory for low-energy, large-distance QCD, it must be dominated by linear, isoscalar, scalar, and vector interactions.*

The major progress since Ref. 1, in addition to the multitude of applications discussed in Ref. 4, has been the following:[4,6–8]

- The understanding of QHD as a low-energy, effective Lagrangian for QCD, which can be used to improve MFT calculations systematically;
- The understanding of the way spontaneously broken chiral symmetry is realized in QHD;
- The development of a consistent, controlled expansion and approximation scheme that allows one to compute reliable results for bulk nuclear properties;
- The relation of relativistic MFT to density functional theory and Kohn–Sham potentials, placing it on a sounder theoretical basis;
- The understanding of the robustness of many of the QHD–I results.

In Sec. 2 we discuss the relation to density functional theory[9,10] and Kohn–Sham potentials.[11] Section 3 contains a brief presentation of the effective Lagrangian, and Sec. 4 summarizes some recent results.

2 Density Functional Theory

We begin with a discussion of nonrelativistic density functional theory (DFT) and generalize later to include relativity. The basic idea behind DFT is to compute the energy E of the many-fermion system (or, at finite temperature, the grand potential Ω) as a functional of the particle density. DFT is therefore a successor to Thomas–Fermi theory, which uses a crude energy functional, but eliminates the need to calculate the many-fermion wave function.

The strategy behind DFT can be seen most easily by working in analogy to thermodynamics.[12] For a uniform system in a box of volume V at temperature T, one first computes the grand potential $\Omega(\mu, T, V)$, where μ is the chemical potential. It then follows that the number of particles N is determined by

$$N = \langle \hat{N} \rangle = -\partial\Omega/\partial\mu \, . \tag{1}$$

The convexity of Ω implies that N is a monotonically increasing function of μ, so this relation can be inverted for $\mu(N)$. Finally, one makes a Legendre transformation to the Helmholtz free energy $F(N,T,V) = \Omega(\mu(N),T,V) + \mu(N)N$ to discuss systems with a fixed density $n = N/V$.

For a finite system, we replace the chemical potential with an external, single-particle potential[b] $\sum_i v(\mathbf{r}_i)$. The grand potential is now a *functional*: $\Omega([v(\mathbf{r})], T)$, and a functional derivative with respect to v gives the particle density:[c]

$$n(\mathbf{r}) = \langle \hat{n}(\mathbf{r}) \rangle = \delta\Omega/\delta v(\mathbf{r}) \, . \tag{2}$$

The convexity of Ω allows us (in principle) to invert this relation and find $v(\mathbf{r})$ as a (complicated) functional of $n(\mathbf{r})$. Finally, we make a functional Legendre transformation to define the Hohenberg–Kohn free energy, which is a functional of $n(\mathbf{r})$:

$$F_{\mathrm{HK}}[n(\mathbf{r})] = \Omega[v(\mathbf{r})] - \int d\mathbf{r} \; n(\mathbf{r})v(\mathbf{r}) \, . \tag{3}$$

(T is suppressed.) The variational derivative of this free energy functional with respect to n now gives

$$\delta F_{\mathrm{HK}}/\delta n(\mathbf{r}) = -v(\mathbf{r}) \, . \tag{4}$$

If we now restrict consideration to $T = 0$ and $v(\mathbf{r}) = 0$, then the *Hohenberg–Kohn theorem* follows:[4,10] If the functional form of $F_{\mathrm{HK}}[n(\mathbf{r})]$ is known exactly, the ground-state expectation value of any observable is a *unique* functional of the exact ground-state density. Moreover, it follows immediately from Eq. (4) that the exact ground-state density can be found by minimizing the energy functional. Although we have assumed here that the ground state is non-degenerate, this assumption can be easily relaxed.[10]

The generalization of DFT to relativistic systems is straightforward.[13] The energy functional F_{HK} now becomes a functional of *both* scalar and vector densities (or more precisely, vector four-currents). Extremization of the functional gives rise to variational equations that determine the ground-state densities.

[b] In fact, one can absorb μ into the definition of v. We suppress all spin-dependence at this point.
[c] Higher variational derivatives yield various correlation functions.

Significant progress in solving these equations was made by Kohn and Sham,[11] who introduced a complete set of single-particle wave functions. In our case, these wave functions allow us to recast the variational equations as Dirac equations for occupied orbitals. The single-particle Hamiltonian contains *local*, density-dependent, scalar and vector potentials, even when the exact energy functional is used. Moreover, one can introduce auxiliary (scalar and vector) fields corresponding to the local potentials, so that the resulting equations resemble those in a relativistic MFT calculation.[4,6]

The strength of the approach rests on the following theorem:

> The exact ground-state scalar and vector densities, energy, and chemical potential for the fully interacting many-fermion system can be reproduced by a collection of (quasi)fermions moving in appropriately defined, self-consistent, local, classical fields.

The proof is straightforward.[10] Start with a collection of noninteracting fermions moving in an externally specified, local, one-body potential. The exact ground state for this system is known: just calculate the lowest-energy orbitals and fill them up.[d] Therefore, if one can find a suitable local, one-body potential based on an *exact* energy functional, the exact ground state of that system can be determined. But this potential is precisely what one obtains by differentiating the interaction parts of F_{HK} with respect to $n(\mathbf{r})$.[10] The resulting one-body potential will generally be density dependent and thus must be determined self-consistently.

Several points are noteworthy. As noted by Kohn,[10] the single-particle basis constructed as described above can be considered "density optimal", in contrast to the Hartree (or Hartree–Fock) basis, which is "total-energy optimal". Thus the *exact* scalar and vector densities are given by sums over the squares of the Dirac wave functions, with unit occupation probability. Moreover, since these densities are guaranteed to make the energy functional stationary [the external $v(\mathbf{r}) = 0$], the exact ground-state energy is also obtained. The proof that the eigenvalue of the least-bound state is exactly the Fermi energy is given in Ref. 14. Note, however, that aside from this association, the exact Kohn–Sham wave functions (and remaining eigenvalues) have no known, directly observable meaning.

If one knows the exact functional form of the energy on the density, one can describe the observables noted in the theorem exactly (and easily) in terms of the Kohn–Sham basis. Observables of this type are typically the ones calculated in relativistic MFT. Moreover, it has been known for many years[2] that the mean-field contributions dominate the single-particle potentials at ordinary densities. Thus, by *parametrizing* the energy functional in a mean-field (or "factorized") form, and by fitting the parameters to empirical bulk and single-particle nuclear data, one should obtain an excellent approximation to the exact energy functional in the relevant density regime. This is the key to the success of relativistic MFT calculations, as we will verify below, using the effective Lagrangian constructed in the next section.

---

[d]For simplicity, we assume that the least-bound orbital is completely filled, so the ground state is non-degenerate.

3 Effective Lagrangian

We cannot give a detailed derivation and discussion of the effective Lagrangian of QHD in this short article, but we can illustrate the basic principles. To exhibit how spontaneously broken chiral symmetry is incorporated quite generally into the hadronic theory, consider the linear σ-model with an additional linear coupling of an isoscalar V^μ to the baryon current, the so-called "chiral (σ, ω) model".[5] Define right- and left-handed nucleon fields by $\psi_{R,L} \equiv (1 \pm \gamma_5)\psi/2$ and the $SU(2)$ matrix $\underset{\sim}{U} \equiv \exp(i\boldsymbol{\tau} \cdot \boldsymbol{\pi}/s_0)$, where $\boldsymbol{\pi}$ is the isovector pion field. If M is the nucleon mass, determined by the spontaneous breaking of chiral symmetry, and $s_0 = M/g_\pi$, then the Lagrangian for the chiral (σ, ω) model can be written as the $s_0 \to \infty$ limit of the following generalized Lagrangian[4]

$$\begin{aligned}
\mathcal{L} = {} & i\left[\overline{\psi}_R \gamma_\mu (\partial^\mu + i g_v V^\mu)\psi_R + \overline{\psi}_L \gamma_\mu (\partial^\mu + i g_v V^\mu)\psi_L\right] \\
& - g_\pi s_0 \left(1 - \frac{\sigma}{s_0}\right)\left[\overline{\psi}_R \underset{\sim}{U}^\dagger \psi_L + \overline{\psi}_L \underset{\sim}{U} \psi_R\right] + \frac{1}{2}(\partial_\mu \sigma \partial^\mu \sigma) + \frac{1}{4}s_0^2 \operatorname{tr}(\partial_\mu \underset{\sim}{U} \partial^\mu \underset{\sim}{U}^\dagger) \\
& - \mathcal{V}(\underset{\sim}{U}, \partial_\mu \underset{\sim}{U}; \sigma) + \frac{1}{4}m_\pi^2 s_0^2 \operatorname{tr}(\underset{\sim}{U} + \underset{\sim}{U}^\dagger - 2) - \frac{1}{4}F_{\mu\nu}F^{\mu\nu} + \frac{1}{2}m_v^2 V_\mu V^\mu \; . \qquad (5)
\end{aligned}$$

For $m_\pi^2 = 0$, this Lagrangian is evidently invariant under chiral $SU(2)_L \times SU(2)_R$ transformations of the form (σ and V^μ are unchanged)

$$\psi_L \to \underset{\sim}{L}\psi_L \; , \qquad \psi_R \to \underset{\sim}{R}\psi_R \; , \qquad \underset{\sim}{U} \to \underset{\sim}{L}\,\underset{\sim}{U}\,\underset{\sim}{R}^\dagger \; . \qquad (6)$$

Here $\underset{\sim}{L}$ and $\underset{\sim}{R}$ are independent, global $SU(2)$ matrices, and the generalized potential $\mathcal{V}$ is chosen to be invariant, with the limit $\mathcal{V} \to m_\sigma^2\sigma^2/2 + O(1/s_0)$. Conventional notation is recovered with the identification

$$s_0 = M/g_\pi \equiv f_\pi \; . \qquad (7)$$

The change of variables $\underset{\sim}{U} \equiv \underset{\sim}{\xi}\underset{\sim}{\xi}$, $N_L \equiv \underset{\sim}{\xi}^\dagger \psi_L$, $N_R \equiv \underset{\sim}{\xi}\psi_R$ reduces the fermion terms in the preceding Lagrangian to

$$\begin{aligned}
\mathcal{L}_{\text{fermion}} &= \overline{N}\left[i\gamma^\mu(\partial_\mu + i\underset{\sim}{v}_\mu + i g_v V_\mu) + \gamma^\mu\gamma_5\, \underset{\sim}{a}_\mu - M + g_\pi \sigma\right] N \; , \\
\underset{\sim}{v}_\mu &\equiv -\frac{i}{2}(\underset{\sim}{\xi}^\dagger \partial_\mu \underset{\sim}{\xi} + \underset{\sim}{\xi}\partial_\mu \underset{\sim}{\xi}^\dagger) \; , \qquad \underset{\sim}{a}_\mu \equiv -\frac{i}{2}(\underset{\sim}{\xi}^\dagger \partial_\mu \underset{\sim}{\xi} - \underset{\sim}{\xi}\partial_\mu \underset{\sim}{\xi}^\dagger) \; . \qquad (8)
\end{aligned}$$

This Lagrangian is invariant under the following *nonlinear* chiral transformation:

$$\underset{\sim}{\xi}(x) \to \underset{\sim}{L}\,\underset{\sim}{\xi}(x)\underset{\sim}{h}^\dagger(x) \equiv \underset{\sim}{h}(x)\underset{\sim}{\xi}(x)\underset{\sim}{R}^\dagger \; , \qquad N(x) \to \underset{\sim}{h}(x)N(x) \; , \qquad (9)$$

where $\underset{\sim}{h}(x)$ is a local $SU(2)$ matrix. It follows that $\underset{\sim}{U}$ still transforms globally according to Eq. (6). Additional mesons and interactions can now be introduced requiring only invariance under the *local* isospin transformations of Eq. (9). While illustrated within the framework of a simple model, this nonlinear realization of $SU(2)_L \times SU(2)_R$ is, in fact, quite general, and can be used as a basis for constructing the most general QHD Lagrangian.[4]

The effective Lagrangian, which reflects the underlying spontaneously broken chiral symmetry of QCD, and from which the energy functional of the previous section is obtained, is constructed from the following series of steps:[4,6]

1. A baryon field and low-mass meson fields that concisely describe the important interaction channels, namely, $\pi(0^-,1), \phi(0^+,0), V_\mu(1^-,0)$, and $\rho_\mu(1^-,1)$, are the generalized coordinates of choice. The pion, a Goldstone boson, is treated as in the example above. Higher mass meson fields are assumed to be "integrated out" and their contributions contained in the effective coupling constants.

2. Dimensional analysis is first used to characterize the various terms in the effective Lagrangian. Briefly, this is done as follows. The initial couplings of the meson fields to the baryon fields are linear, with a strong coupling constant g. The dimensionless form of this combination is $g\phi/M = \phi/f_\pi$ [see Eq. (7)]; non-Goldstone boson fields are assumed to enter in this dimensionless form. From the mass term of the meson fields $\propto m^2\phi^2$, with $m^2 \approx M^2$, one then deduces an overall scale factor in the Lagrangian density of $f_\pi^2 M^2$. From the baryon mass term $M\overline{\psi}\psi$ one concludes that the appropriate dimensionless form of the baryon densities is $\overline{\psi}\psi/Mf_\pi^2$. This "naive" dimensionless analysis (NDA) then implies that, after appropriate combinatorial factors are included, the various terms in the effective Lagrangian enter with dimensionless coefficients of order unity.

3. The various interaction terms allowed by the $SU(2)_L \times SU(2)_R$ symmetry of QCD are then constructed using the nonlinear realization of chiral symmetry illustrated above. Simply writing down all possible terms does not get one very far unless there is an *organizational principle*, and the following provides the crucial insight:

4. Although the mean scalar and vector field energies are *large* compared to the nuclear binding energy, the dimensionless combinations $g_s\phi_0/M \approx \phi_0/f_\pi$ and $g_v V_0/M \approx V_0/f_\pi$ are roughly 1/3 and *thus provide convenient expansion parameters*. Furthermore, spatial variations of the meson fields and of the baryon densities in the nucleus are observed to occur over the scale of the nuclear surface region, and hence the dimensionless ratio ∇/M also provides a useful expansion quantity (as does the characterization of chiral symmetry violation at the Lagrangian level, m_π/M).

5. A combination of these observations allows one to construct a hierarchy of decreasing contributions to the effective Lagrangian for the nuclear many-body system characterized by an integer ν defined by[6]

$$\nu = d + \frac{n}{2} + b \,, \tag{10}$$

where d is the number of derivatives, n is the number of nucleon fields, and b is the number of non-Goldstone boson fields present in the interaction term. The effective Lagrangian at various levels of ν is given in Refs. 4, 6.[e]

[e] The extension of the effective Lagrangian to include electromagnetic interactions as an expansion in powers of derivatives is also discussed in these references.

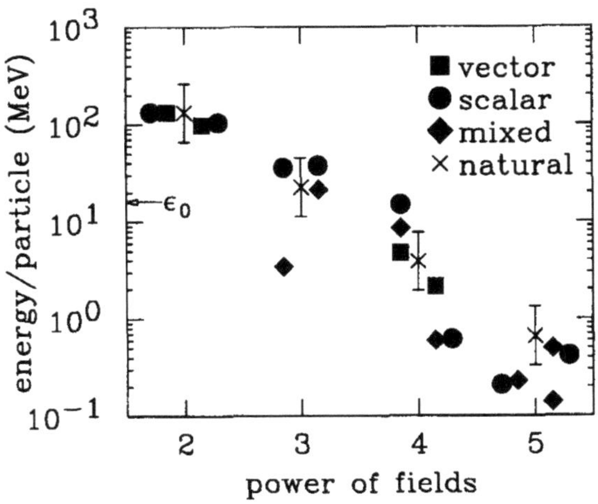

Figure 1. Nuclear matter energy/particle for two QHD parameter sets, one on the left and one on the right of the error bars. The power of fields is $b \equiv j + \ell$ for a term of the form $(g_s\phi_0)^j(g_v V_0)^\ell$ (ℓ is even). The arrow indicates the total binding energy, $\epsilon_0 = 16.1\,\text{MeV}$. Absolute values are shown.

The effective Lagrangian with mean meson fields then determines the energy functional of the previous section, and a representation in terms of Dirac–Hartree orbitals leads to local, nonlinear Hartree equations, which can be solved numerically. The extent to which nuclei exhibit this hierarchy of interactions, and to which this effective Lagrangian indeed describes the nucleus, is discussed in the next section.

4 Results and Summary

The QHD mean-field energy functional discussed above is given in several different forms in Refs. 4, 6, 8, 15, as are the field equations that result from extremization. The equations are solved self-consistently for the closed-shell nuclei ^{16}O, ^{40}Ca, ^{48}Ca, ^{88}Sr, and ^{208}Pb, and also in the nuclear matter limit. The parameters are then best-fit to empirical properties of the charge densities, the binding energies, and various splittings between energy levels near the Fermi surface using a figure of merit (χ^2) defined by a weighted, squared deviation between the 29 calculated and empirical values. When working at the highest order of truncation (essentially $\nu = 4$), the calculated results are very accurate, as we illustrate shortly, but they are too numerous to reproduce here.[6,8,15]

The critical question is whether the hierarchal organization of interaction terms is actually observed. This is illustrated in Fig. 1, where the nuclear matter energy/particle is shown as a function of the power of the mean fields, which is called b in Eq. (10). (There are no gradient contributions in nuclear matter and $\langle\hat{\pi}\rangle = 0$.) The crosses and error bars are estimates based on NDA and naturalness, that is, overall coefficients are of order unity. It is clear that each successive term in the hierarchy is reduced by roughly a factor of five, and thus for any reasonable desired accuracy, the Lagrangian can be truncated at a low value of ν. Derivative terms

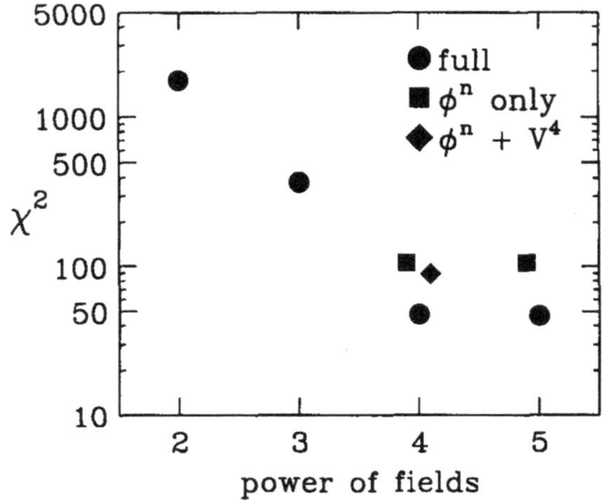

Figure 2. χ^2 values for QHD parameter sets, as a function of the level of truncation.

and other coupling terms are analyzed in Ref. 8, with similar conclusions.

The quality of the fits to finite nuclei and the appropriate level of truncation is illustrated in Fig. 2, where the figure of merit is plotted as a function of truncation order and of various combinations of terms retained in $\mathcal{L}$. The full calculations (●) retain all allowed terms at a given level of ν, while the other two choices keep only the indicated subset. There is clearly a great improvement in the fit (more than a factor of 35) in going from $\nu = 2$ to $\nu = 4$, but there is no further improvement in going to $\nu = 5$. Speaking chronologically, the $\nu = 2$ results show the level of accuracy obtained almost 20 years ago, while the $\nu = 4$ results were obtained four years ago.[6] Moreover, the ϕ^n only results at $\nu = 4$ (■) show the state of the situation in the late 1980s, as discussed in Ref. 3. Recent work[8] shows that the full complement of parameters at order $\nu = 4$ is underdetermined, and that only six or seven are determined by this data set, which explains the great success of these earlier models with a restricted set of parameters.

In *summary*, the hadronic theory of QHD is truly a manifestation of low-energy, strong-coupling QCD. The modern viewpoint of QHD based on effective field theory and density functional theory explains the accurate description of bulk and single-particle nuclear properties. Corrections to the mean-field parametrization of the energy functional can be calculated systematically using the effective hadronic Lagrangian.[16] An important goal for the future is finding an efficient, tractable, nonperturbative way to match the low-energy, strong-coupling, effective field theory of QHD to the underlying QCD Lagrangian.

Acknowledgements

We thank Dick Furnstahl for useful comments. This work was supported in part by the US Department of Energy under contracts DE–FG02–87ER40365 and DE–FG02–97ER41023.

References

1. B. D. Serot and J. D. Walecka, in *Recent Progress in Many-Body Theories*, vol. 3, Eds. T. L. Ainsworth, C. E. Campbell, B. E. Clements, and E. Krotscheck (Plenum, New York, 1992), p. 49.
2. B. D. Serot and J. D. Walecka, *Adv. Nucl. Phys.* **16**, 1 (1986).
3. B. D. Serot, *Rep. Prog. Phys.* **55**, 1855 (1992).
4. B. D. Serot and J. D. Walecka, *Int. J. Mod. Phys.* E **6**, 515 (1997).
5. J. D. Walecka, *Theoretical Nuclear and Subnuclear Physics* (Oxford Univ. Press, New York, 1995).
6. R. J. Furnstahl, B. D. Serot, and H.-B. Tang, *Nucl. Phys.* A **615**, 441 (1997); **640**, 505 (1998) (E).
7. R. J. Furnstahl and B. D. Serot, *Comments on Modern Physics* (2000), in press.
8. R. J. Furnstahl and B. D. Serot, *Nucl. Phys.* A **671**, 447 (2000).
9. R. M. Dreizler and E. K. U. Gross, *Density Functional Theory* (Springer, New York, 1990).
10. W. Kohn, *Rev. Mod. Phys.* **71**, 1253 (1999).
11. W. Kohn and L. J. Sham, *Phys. Rev.* A **140**, 1133 (1965).
12. N. Argaman and G. Makov, *Am. J. Phys.* **68**, 69 (2000).
13. A. H. MacDonald and S. H. Vosko, *J. Phys. C (Cond. Matt.)* **12**, 2977 (1979).
14. C. O. Almbladh and U. von Barth, *Phys. Rev.* B **31**, 3231 (1985).
15. J. J. Rusnak and R. J. Furnstahl, *Nucl. Phys.* A **627**, 495 (1997).
16. Y. Hu, Ph.D. thesis, Indiana University (2000).

HYPERSPHERICAL METHODS FOR $A \geq 4$ SYSTEMS

M. VIVIANI
INFN and Phys. Dept., Via Buonarroti, 2, I-56100 Pisa, Italia
E-mail: michele.viviani@pi.infn.it

In this contribution, the application of the Correlated Hyperspherical Harmonic technique to the calculation of bound and continuum states of $A = 4$ systems is reviewed. The extension of the Hyperspherical technique to larger systems solving pseudo Faddeev-Yakubovsky equations is also outlined.

1 Introduction

A sophisticated variational technique to describe the bound and scattering states of three-nucleon (3N) systems has been developed in Refs. 1,2. The wave function (WF) is expanded in channels, as in the Faddeev technique, and the radial amplitude of each channel is expanded in terms of correlated functions. Two different types of correlations have been investigated. The pair-correlated hyperspherical harmonic (PHH) basis includes a pair correlation function per channel, while the correlated hyperspherical harmonic (CHH) basis uses a correlation factor of the product (Jastrow) form. Such bases result to be well suited for taking into account the correlations induced by the large repulsive terms of the nucleon-nucleon (NN) potential. The technique was used in Ref. 1 to calculate the bound and continuum state WF of 3N systems, with realistic models of nucleon-nucleon (NN) and 3N forces. The obtained results are in complete agreement with those given by the best available methods.[3–6]

The generalization of the CHH expansion method to the four-nucleon (4N) systems is more involved. The method is outlined in Sec. 2, where a review of the results for the bound state of the α-particles and for some observables of $p - {}^3$He scattering are also presented.

The extension of this technique to larger systems has not been attempted so far. However, an interesting technique which can be applied to a generic A-body systems has been proposed recently.[7] It consists in solving pseudo Faddeev-Yakubovsky (PFY) equations derived from a trial wave function including only two-, three- and special four-body correlations. That technique is reviewed in Sec. 3, where some preliminary results obtained in the case of bosonic systems are also reported.

Finally, in the last section we discuss the future applications of these Hyperspherical techniques to $A \geq 4$ systems.

2 The CHH expansion for $A = 4$ systems

The WF of a four-nucleon system with total angular momentum J, J_z and total isospin T, T_z can be written as a sum of amplitudes

$$\Psi = \sum_{p=1}^{12} \Big[\psi_A(\boldsymbol{x}_{1p}, \boldsymbol{x}_{2p}, \boldsymbol{x}_{3p}) + \psi_B(\boldsymbol{y}_{1p}, \boldsymbol{y}_{2p}, \boldsymbol{y}_{3p}) \Big] \ , \tag{1}$$

where p denotes an even permutation of the particles. Through this section the permutation p corresponds to the order i, j, k, m of the particles. The dependence of ψ_A and ψ_B on the nucleon spin-isospin variables is understood. The vectors $\boldsymbol{x}_{1,2,3}$ and $\boldsymbol{y}_{1,2,3}$ are the two possible sets of Jacobi vectors which can be constructed for a system of four particles of equal masses, and they are defined as

$$\begin{array}{ll} \underline{\text{set A}} & \underline{\text{set B}} \\ \boldsymbol{x}_{1p} = \sqrt{\tfrac{3}{2}}(\boldsymbol{r}_m - \frac{\boldsymbol{r}_i + \boldsymbol{r}_j + \boldsymbol{r}_k}{3}) \ , & \boldsymbol{y}_{1p} = \boldsymbol{r}_m - \boldsymbol{r}_k \ , \\ \boldsymbol{x}_{2p} = \sqrt{\tfrac{4}{3}}(\boldsymbol{r}_k - \frac{\boldsymbol{r}_i + \boldsymbol{r}_j}{2}) \ , & \boldsymbol{y}_{2p} = \sqrt{2}(\frac{\boldsymbol{r}_m + \boldsymbol{r}_k}{2} - \frac{\boldsymbol{r}_i + \boldsymbol{r}_j}{2}) \ , \\ \boldsymbol{x}_{3p} = \boldsymbol{r}_j - \boldsymbol{r}_i \ , & \boldsymbol{y}_{3p} = \boldsymbol{r}_j - \boldsymbol{r}_i \ . \end{array}$$

In the L-S coupling scheme the amplitudes ψ_A and ψ_B are written as

$$\psi_A(\boldsymbol{x}_{1p}, \boldsymbol{x}_{2p}, \boldsymbol{x}_{3p}) = \sum_\alpha F_{\alpha p} \Phi_{A\alpha}(x_{1p}, x_{2p}, x_{3p}) \mathcal{Y}^A_{\alpha p} \ , \tag{3}$$

$$\mathcal{Y}^A_{\alpha p} = \Big\{ \Big[[Y_{\ell_{1\alpha}}(\hat{x}_{1p}) Y_{\ell_{2\alpha}}(\hat{x}_{2p})]_{\ell_{12\alpha}} Y_{\ell_{3\alpha}}(\hat{x}_{3p}) \Big]_{L_\alpha} \times \\ \Big[\Big[[s_i s_j]_{S_{a\alpha}} s_k \Big]_{S_{b\alpha}} s_m \Big]_{S_\alpha} \Big\}_{JJ_z} \Big[\Big[[t_i t_j]_{T_{a\alpha}} t_k \Big]_{T_{b\alpha}} t_m \Big]_{TT_z} \ , \tag{4}$$

and

$$\psi_B(\boldsymbol{y}_{1p}, \boldsymbol{y}_{2p}, \boldsymbol{y}_{3p}) = \sum_\alpha F_{\alpha p} \Phi_{B\alpha}(y_{1p}, y_{2p}, y_{3p}) \mathcal{Y}^B_{\alpha p} \ , \tag{5}$$

$$\mathcal{Y}^B_{\alpha p} = \Big\{ \Big[[Y_{\ell_{1\alpha}}(\hat{y}_{1p}) Y_{\ell_{2\alpha}}(\hat{y}_{2p})]_{\ell_{12\alpha}} Y_{\ell_{3\alpha}}(\hat{y}_{3p}) \Big]_{L_\alpha} \times \\ \Big[[s_i s_j]_{S_{a\alpha}} [s_k s_m]_{S_{b\alpha}} \Big]_{S_\alpha} \Big\}_{JJ_z} \Big[[t_i t_j]_{T_{a\alpha}} [t_k t_m]_{T_{b\alpha}} \Big]_{TT_z} \ , \tag{6}$$

respectively. In order to ensure the antisymmetry of the WF the amplitudes $\psi_{A,B}$ must change sign under the exchange of particles i and j. Therefore, if the function $F_{\alpha p}$ is even (odd) when $i \leftrightarrow j$, then the integer $l_{3\alpha} + S_{a\alpha} + T_{a\alpha}$ must be odd (even); moreover, $\ell_{1\alpha} + \ell_{2\alpha} + \ell_{3\alpha}$ must be an even or odd number depending on whether the parity of the state considered is either even or odd.

The correlation functions $F_{\alpha p}$ in Eqs. (3) and (5) are taken to be of the form

$$F_{\alpha p} = f_{a\alpha}(r_{ij}) f_{b\alpha}(r_{ik}) f_{b\alpha}(r_{jk}) f_{c\alpha}(r_{im}) f_{c\alpha}(r_{jm}) f_{d\alpha}(r_{km}) \ , \tag{7}$$

where $f_{a\alpha}$, $f_{b\alpha}$, $f_{c\alpha}$ and $f_{d\alpha}$ are one-dimensional functions of the interparticle distances. The inclusion of such (non-optimized) Jastrow factors significantly improves the convergence rate of the usual hyperspherical harmonic (HH) expansion. The choice of these functions and the α-channels selected in our calculation have been discussed in Ref. 8.

One can replace the magnitudes of the Jacobi variables with the hyperspherical coordinates, given by the hyperradius[9]

$$\rho = \sqrt{(x_{1p})^2 + (x_{2p})^2 + (x_{3p})^2} = \sqrt{(y_{1p})^2 + (y_{2p})^2 + (y_{3p})^2} \,, \tag{8}$$

which turns out to be independent on p, and the "hyperangular" variables,

$$\begin{aligned} \cos\phi_{3p} &= x_{3p}/\rho = y_{3p}/\rho \,, \\ \cos\phi_{2p}^{A} &= x_{2p}/(\rho\sin\phi_{3p}) \,, \\ \cos\phi_{2p}^{B} &= y_{2p}/(\rho\sin\phi_{3p}) \,. \end{aligned} \tag{9}$$

Each Φ_α function in Eqs. (3) and (5) is then expanded in terms of the HH basis as

$$\Phi_{A\alpha}(x_{1p}, x_{2p}, x_{3p}) = \sum_{n,m} \frac{u^{\alpha}_{nm}(\rho)}{\rho^4} x_{1p}^{\ell_{1\alpha}} x_{2p}^{\ell_{2\alpha}} x_{3p}^{\ell_{3\alpha}} Y^{\alpha}_{nm}(\phi^{A}_{2p}, \phi_{3p}) \,, \tag{10}$$

$$\Phi_{B\alpha}(y_{1p}, y_{2p}, y_{3p}) = \sum_{n,m} \frac{w^{\alpha}_{nm}(\rho)}{\rho^4} y_{1p}^{\ell_{1\alpha}} y_{2p}^{\ell_{2\alpha}} y_{3p}^{\ell_{3\alpha}} Y^{\alpha}_{nm}(\phi^{B}_{2p}, \phi_{3p}) \,, \tag{11}$$

where

$$Y^{\alpha}_{nm}(\beta,\gamma) = N^{\alpha}_{nm}\,(\sin\gamma)^m P_n^{K_{2\alpha},\ell_{3\alpha}+\frac{1}{2}}(\cos 2\gamma) P_m^{\ell_{1\alpha}+\frac{1}{2},\ell_{2\alpha}+\frac{1}{2}}(\cos 2\beta) \,. \tag{12}$$

In the last expression, $P^{a,b}$ are Jacobi polynomials, the integers n and m range from zero to infinity, $K_{2\alpha} = \ell_{1\alpha} + \ell_{2\alpha} + 2m + 2$ and N^{α}_{nm} are normalization factors. When the functions Y^{α}_{nm} are multiplied with the product of the spherical harmonics $Y_{\ell_1} Y_{\ell_2} Y_{\ell_3}$, as in Eqs. (3) and (5), the standard expression[9] of the HH functions is recovered.

2.1 Results for the α-particle

The Rayleigh-Ritz variational principle is used to determine the functions $u^{\alpha}_{nm}(\rho)$ and $w^{\alpha}_{nm}(\rho)$ and the eigenvalue E. The following nuclear potential models have been considered: 1) the Argonne v_{18} NN potential[11] (the AV18 model), and 2) the Argonne v_{18} NN potential plus the Urbana IX 3N potential[10] (the AV18UR model). For these realistic potentials, the results obtained with the CHH technique with $N_c = 22$ are compared in Table 2.1 to those calculated by means of the Green's Function Monte Carlo (GFMC) method[10] and the Faddeev-Yakubovsky equations[6] (FYE). In the case of the AV18 interaction, the GFMC, FYE, and CHH estimates for the binding energy are rather close. For AV18UR potential model, the B value obtained differs from the FYE one by approximately 0.5 MeV. A larger number of channels (perhaps also of HH components per channel) should be presumably considered in order to resolve this discrepancy.

Table 1. α-particle energy for the AV18 and AV18UR potential models obtained with the CHH expansion with $N_c = 22$. Energies are in MeV and radii in fm. T is the mean value of the kinetic energy, R is the mass radius and P_P, P_D are the P- and D-wave percentages. For the sake of comparison also the results obtained with the GFMC[10] and FYE[6] techniques have been reported. The numbers in parentheses give the statistical errors in the last digit.

Potential	Method	B	T	R	P_P	P_D
	FYE	24.28	97.83		0.36	13.74
AV18	CHH	24.18	97.79	1.511	0.341	13.69
	GFMC	24.1(1)				
	FYE	28.50	113.21		0.75	16.04
AV18UR	CHH	28.0	111.72	1.437	0.646	15.78
	GFMC	28.3(2)	112.1(8)			

2.2 *Results for $p - {}^3$He scattering*

The WF of a $p - {}^3$He state with relative orbital angular momentum L and channel spin S can be written as[12]

$$\Psi_{LS} = \Psi_C + \Phi_{LS} \,. \tag{13}$$

The first term Ψ_C of Eq. (13) must be sufficiently flexible to guarantee a detailed description in the "internal region", where all the particles are close to each other and the mutual interaction is large; Ψ_C goes to zero when the distance d_i between the ^{3}He and the unbound proton i increases. This term in the WF is expanded in terms of CHH basis functions as described in Sect.2.

The second term Φ_{LS} describes the asymptotic configuration of the system, for large d_i values, where the nuclear p-^{3}He interaction is negligible. Let us introduce the surface functions

$$\Omega_{LS}^{(\lambda)} = \sum_{i=1}^{4} \left\{ Y_L(\hat{d}_i) \left[\Psi_{jk\ell}^{{}^3\mathrm{He}} \chi_i \right]_S \right\}_{JJ_z} \mathcal{R}_L^{(\lambda)}(d_i) \,, \tag{14}$$

where χ_i is the spin function of the unbound nucleon i and $\Psi_{jk\ell}^{{}^3\mathrm{He}}$ is the ^{3}He bound state WF. This latter function is normalized to unity and is antisymmetric under the exchange of any pair of particles j, k, and ℓ. $\Psi^{{}^3\mathrm{He}}$ has been determined as discussed in Ref. 2 by using the CHH expansion for a three-body system.

The functions $\mathcal{R}_L^{(\lambda)}(r_i)$ of Eq. (14) are the ingoing ($\lambda \equiv -$) and outgoing ($\lambda \equiv +$) radial solutions of the two-body Schrödinger equation without nuclear interaction.[13] The asymptotic WF is then written as

$$\Phi_{LS} = \Omega_{LS}^{(-)} - \sum_{L'S'} \mathcal{S}_{LL'}^{SS'} \Omega_{L'S'}^{(+)} \,, \tag{15}$$

where the quantities $\mathcal{S}_{LL'}^{SS'}$ give the amplitude of the outgoing $(L'S')$ component relative to the the ingoing (LS) wave. The elastic S-matrix, whose dimensionality is 1 (2) for the $J = 0$ $(J > 0)$ states, should be unitary since there are no open

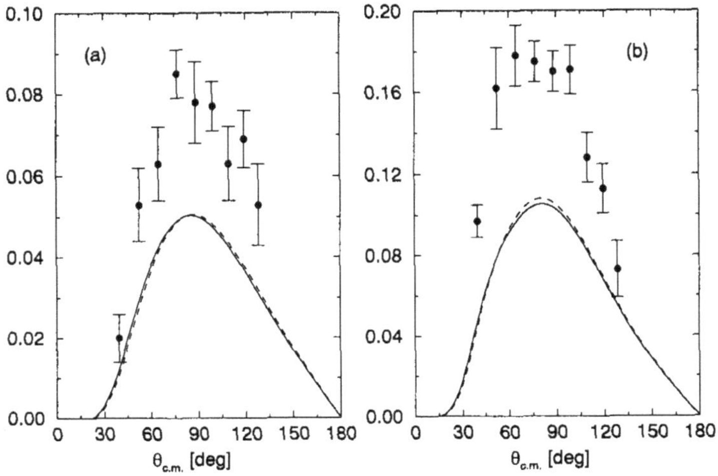

Figure 1. Measurements[14] of the proton analyzing power A_y for $p - {}^3\mathrm{He}$ scattering as a function of the scattering angle at $E_{c.m.} = 1.20$ MeV (panel a) and 1.69 MeV (panel b). The theoretical estimates obtained with the AV18 (solid curves) and the AV18UR (dashed curves) interaction models are also reported.

reaction channels for the energies considered here. It follows that the eigenvalues of the S-matrix may be written as $\exp(2i\delta_{LS})$, where δ_{LS} is the eigenphase shift of the ${}^{2S+1}L$ wave. These quantities are calculated by means of the complex form of the Kohn variational principle with a procedure similar to that one used in the N-d case.[2,13]

The convergence (in terms of the number of the channels included in Ψ_C) of the $L = 0$ waves ($J^\Pi = 0^+, 1^+$) at $E_{c.m.} = 0$ was studied previously in Ref. 12. At $E_{c.m.} = 0$ and at the energies considered here, a rather small number of channels is sufficient to provide a good convergence. This is due mainly to the Pauli principle which limits overlaps between the four nucleons. As a consequence, the internal part is rather small and does not require a large number of channels.

On the other hand, for the $L = 1$ waves ($J^\Pi = 0^-$, 1^- and 2^-) the convergence rate is slow and many channels ($N_c \approx 100$) have to be included. In these cases the interaction between the p and ^{3}He clusters is very attractive (it has been speculated that some 4N resonant states exist) and the construction of the internal wave function is more delicate. At the energies considered here, the scattering in the $L = 2$ waves ($J^\Pi = 1^+$, 2^+ and 3^+) is very peripheral and the corresponding phase shifts are small. They can be calculated with good precision by considering only the asymptotic part in Eq. (13). The contribution of the waves with $L > 2$ is very tiny and has been disregarded.

The predicted analyzing powers are compared with the measurements[14] in Fig. 2.2. The solid (dashed) curves correspond to the AV18 (AV18UR) interaction model. The main aspect to be seen in Fig. 2.2 is that the calculations are significantly smaller in magnitude than the measurements at both energies. This is precisely the same result that has previously been seen and well documented for

N-d scattering. We also see in Fig. 2.2 that the 3N interaction has almost no effect on A_y.

3 The PFY method

Let us now consider a systems with A identical particles. The system will be described in terms of $N = A-1$ Jacobi vectors. For a given order p of the particles $p \equiv (i,j,k,m,\ldots)$, the usual choice for $\boldsymbol{x}_N$ is

$$\boldsymbol{x}_N = \boldsymbol{r}_j - \boldsymbol{r}_i \; . \tag{16}$$

As discussed in the previous section, there are two possible choices for $\boldsymbol{x}_{N-1}$, namely

$$\boldsymbol{x}^K_{N-1} = \sqrt{\frac{4}{3}}(\boldsymbol{r}_k - \frac{\boldsymbol{r}_i+\boldsymbol{r}_j}{2}) \, , \qquad \boldsymbol{x}^H_{N-1} = \sqrt{2}(\frac{\boldsymbol{r}_m+\boldsymbol{r}_k}{2} - \frac{\boldsymbol{r}_i+\boldsymbol{r}_j}{2}) \, , \tag{17}$$

and so on. Moreover, the hyperspherical coordinates are defined by the hyperradius ρ, defined as

$$\rho^2 = \sum_{i=1}^{N} x_i^2 = \tfrac{2}{A}\sum_{i>j} r_{ij}^2 \, , \tag{18}$$

and the $3N-1$ angular coordinates $\Omega \equiv \{\hat{x}_1, \cdots, \hat{x}_N, \theta_2, \cdots, \theta_N\}$, where the hyperangles θ_j are defined by

$$\cos\theta_j = \frac{x_j}{\sqrt{x_1^2+\ldots+x_j^2}} \; . \tag{19}$$

Now, let $\Phi_0(\Omega)$ be a symmetry-carrying hyperspherical function, i.e. an antisymmetric functions for fermions and a constant for bosons, and consider the following set of approximations for the WF of the system:

$$\Psi^{(0)} = \Phi_0 \psi^{(0)}(\rho) \, , \tag{20}$$

$$\Psi^{(1)} = \Phi_0 \sum_{\text{pairs } ij} \psi^{(1)}(\rho, \boldsymbol{x}_N) \, , \tag{21}$$

$$\Psi^{(2)} = \Phi_0 \Big[\sum_{\text{triplets } ijk} \psi^{(2)}_K(\rho, \boldsymbol{x}_N, \boldsymbol{x}^K_{N-1}) + \sum_{\text{quartets } ijkm} \psi^{(2)}_H(\rho, \boldsymbol{x}_N, \boldsymbol{x}^H_{N-1}) \Big] \, , \tag{22}$$

and so on. The 0th-order approximation (20) is usually too crude, whereas the 1st-order approximation (21) was used with some success[15,16] since it includes in the WF two-body correlations. However, for $A > 3$, also the first approximation leads to unsatisfactory results. The successive approximation (2nd-order) was considered in Ref. 7. It incorporates also triplet/quartet correlations and leads to better results. Some of the results obtained[7] for bosonic systems interacting via the Afnan-Tang S3 potential[17] are reported in Table 3. Using such an approximation, the problem

Table 2. Binding energy for A-bosons interacting via the S3 potential.[17] Calculations were done with 0th, 1st and 2nd order expansion. Also shown calculations done with the SVM.[18] Results are given in MeV.

A	$E^{(0)}$	$E^{(1)}$	$E^{(2)}$	SVM
4	-7.19	-27.23	-27.37	-27.42
6	-47.66	-111.17	-114.98	-115.06
8	-124.67	-247.41	-260.20	-
10	-238.30	-430.70	-457.17	-

reduces to the solution of pseudo Faddeev-Yakubovsky equations. In practice, the $\psi^{(2)}$ have been determined by expanding them in HH functions.[7]

As can be seen by inspection of the Table, the use of the 0th order WF, Eq. (20), is very poor. The energies $E^{(1)}$, obtained by considering Eq. (21), show the importance of the two-body correlations. The triplet/quartet correlations give smaller (but still sizeable) contributions. As expected, for Bose systems the importance of the second-order terms increases with increasing number of particles. For the 4- and 6-body systems we have seen that the estimated $E^{(2)}$ binding energies agree with the results obtained using the stochastic variational method, SVM,[18] to within 0.05 MeV in the 4-body case and within 0.1 MeV in the 6-body case.

From the results obtained, it seems clear that the PFY approach provides a rather convenient way of studying the A-body problem. The method now is being applied to fermion systems. In this case the relative weight of the 2nd order terms might decrease behind the α particle as its density is larger than the density of the heavier systems.

4 Perspectives

The increase of computer power and the development of optimal methods are rapidly bringing the theory of nuclear systems with $A \geq 4$ particles to a degree of accuracy comparable to that achieved for $A = 3$. The very accurate calculations performed by means of the FYE[6] (where the α-particle wave function is expanded in approximately 2000 channels) is defining the state-of-the-art for the $A = 4$ calculation. The convergence is estimated to have been reached within 50 keV, namely at the level of 0.2 %. Moreover, the GFMC calculations have been extended to treat bound states of nuclei with $A \leq 10$. In the meanwhile, new techniques as those devised by by Varga *et al*[18] (the SVM) and by Barnea *et al*[19] seem to be very promising for calculating binding energies of $A \geq 4$ systems.

In this contribution, we have presented some of the results which can be achieved by using (correlated) Hyperspherical methods. For the α-particle the precision reached is somewhat smaller than that obtained by other techniques. However, the results presented here have been obtained using 22 channels. New calculations with larger number of channels are now underway. One of the positive aspects of the method is that it can be applied to scattering states as well. In the last years, such a study has been brought forth for the $p-{}^3$He elastic scattering. Large calculations (of approximately 100 channels) have been performed, reaching a degree of convergence of 1%. Such a precision is sufficient to allow new stringent tests of the current models of NN and 3N interaction. The study of this reaction, together with the $p - {}^3$H, $n - {}^3$H and $d - d$ ones, will be one of the most interesting areas of research in the next years.

For larger system, we have presented a convenient method to calculate binding energies for bosonic systems. The extension to fermionic systems interacting via realistic potentials is currently underway. The success of such an application would allow accurate calculations of reactions involving $A = 7 \div 8$ nucleons, which are of noticeable interest for the astrophysical models of the sun and the stars.

Acknowledgments

The author would like to give to John W. Clark, Alpo J. Kallio, Manfred L. Ristig, and especially to Sergio Rosati his best wishes for their 65th birthdays.

References

1. A. Kievsky, S. Rosati and M. Viviani, *Nucl. Phys.* A **551**, 241 (1993);
2. A. Kievsky, M. Viviani, S. Rosati, *Nucl. Phys.* A **577**, 511 (1994); A. Kievsky, M. Viviani, S. Rosati, *Phys. Rev.* C **52**, R15 (1995).
3. H. Kameyama, M. Kamimura and Y. Fukushima, *Phys. Rev.* C **40**, 974 (1989); H. Kameyama, M. Kamimura and Y. Fukushima, *Nucl. Phys.* A **508**, 17c (1990).
4. C. R. Chen, G. L. Payne, J. L. Friar and B. F. Gibson, *Phys. Rev.* C **31**, 266 (1985); J. L. Friar, B. F. Gibson and G. L. Payne, *Phys. Rev.* C **36**, 1138 (1987).
5. S. Ishikawa, T. Sasakawa, T. Sawada and T. Ueda, *Phys. Rev. Lett.* **53**, 1877 (1984); T. Sasakawa and S. Ishikawa, *Few-Body Sys.* **1**, 3 (1986).
6. A. Nogga, H. Kamada and W. Glöckle, *Phys. Rev. Lett.* **85**, 944 (2000).
7. N. Barnea and M. Viviani, *Phys. Rev.* C **61**, 034003 (2000).
8. M. Viviani, A. Kievsky and S. Rosati, *Few-Body Sys.* **18**, 25 (1995).
9. M. Fabre de la Ripelle, *Ann. Phys. (NY)* **147**, 281 (1983).
10. B.S. Pudliner *et al.*, *Phys. Rev.* C **56**, 1720 (1997).
11. R.B. Wiringa, V.G.J. Stoks, and R. Schiavilla, *Phys. Rev.* C **51**, 38 (1995).
12. M. Viviani, S. Rosati, A. Kievsky, *Phys. Rev. Lett.* **81**, 1580 (1998).
13. A. Kievsky, *Nucl. Phys.* A **624**, 125 (1997).
14. E.A. George and L.D. Knutson, submitted to *Phys. Rev. Lett.*.
15. M. Fabre de la Ripelle, Y. J. Jee, A. D. Klemm and S. Y. Larsen, *Ann. Phys. (NY)* **212**, 195 (1991).
16. R. Brizzi, M. Fabre de la Ripelle and M. Lassaut, *Nucl. Phys.* A **596**, 199 (1996).
17. L.R. Afnan and Y.C. Tang, *Phys. Rev.* **175**, 1337 (1968).
18. K. Varga, Y. Suzuki, *Comp. Phys. Com.* **106**, 157 (1997).
19. N. Barnea, W. Leidemann and G. Orlandini, nucl-th/9910062.

Spin Systems

AB INITIO CALCULATIONS OF THE SPIN-HALF *XY* MODEL

D. J. J. FARNELL
Department of Physics, UMIST, P.O. Box 88, Manchester M60 1QD, UK

M. L. RISTIG
Institut für Theoretische Physik, Universität zu Köln, Zülpicher Str., 50937 Köln, Germany

In this article we perform CBF calculations for the spin-half quantum XY model on the linear chain, the square lattice, and the simple cubic lattice for the first time. The results for the ground-state energy and sublattice magnetisation are compared to previous CCM results for this model and excellent quantitative and qualitative agreement is observed. A brief comparison of the relative strengths and weaknesses of the CBF and CCM techniques may thus be made, and it is seen that the application, in parallel, of two such techniques of quantum many-body theory to a given system leads not only to insights into this system but also into the techniques themselves.

In this paper we consider the $T = 0$ properties of the quantum spin system known as the spin-half XY model, described by the Hamiltonian

$$H = \frac{1}{8}\sum_{i,j}\left[(1+\gamma)\sigma_i^x\sigma_j^x + (1-\gamma)\sigma_i^y\sigma_j^y\right] \quad , \tag{1}$$

in the regime $0 < \gamma \leq 1$. Note that the index i runs over all N lattice sites and that the index j runs over the z nearest-neighbour to i on the linear chain ($z = 2$), the square lattice ($z = 4$), and the cubic lattice ($z = 6$).

In the regime $0 < \gamma \leq 1$ the ground state is believed to exhibit Néel ordering in the x-direction, and for $-1 < \gamma \leq 0$ the ground state is again believed to possess Néel ordering in the y-direction. We note that a phase transition point occurs for the linear chain model at exactly $\gamma = 0$, and that Néel ordering is found to disappear at this point. For spatial dimensionality greater than one, the phase transition point of the anisotropic model is also believed to be at (or very near to) $\gamma = 0$ from approximate calculations. The ground state of the spin-half XY model on the square and cubic lattices at $\gamma = 0$ is also believed to be Néel-ordered in the xy-plane.

The spin-half XY model was solved exactly by Lieb, Schultz and Mattis[1] for the linear chain using the Jordan-Wigner transformation. Since then the ground- and excited-state properties have been extensively studied by many authors (see, for examples, Refs. 2,3). However, no exact results exist for higher spatial dimensionality, although approximate results such as those from spin-wave theory,[4] Monte-Carlo (QMC) methods,[5,6] series expansions,[7] and the coupled cluster method (CCM)[8] has proven to be highly successful. Extrapolated finite size calculations[9] have also been performed for $\gamma = 0$.

The correlated basis function (CBF) method[10–20] is a widely applied and accurate method of modern-day quantum many-body theory. Recently, this method has been applied with great success to the Ising model in a transverse magnetic field at zero temperature.[21–24] In this article we wish to apply the CBF method to the spin-half XY model. We begin this process by firstly performing a number of unitary transformations on the local spin axes on two sublattices $\{A, B\}$ in order to simplify the problem. The first such transformation on the A-sublattice is given by,

$$\sigma^x \to \sigma^z \ ; \ \sigma^y \to \sigma^x \ ; \ \sigma^z \to \sigma^y \ , \tag{2}$$

and the second transformation on the B-sublattice is given by,

$$\sigma^x \to -\sigma^z \ ; \ \sigma^y \to -\sigma^x \ ; \ \sigma^z \to \sigma^y \ . \tag{3}$$

Note that both of these transformations are simply rotations of the local spin-axes of the spins, and that the eigenvalue spectrum of the problem is left unchanged because these transformations are unitary. The Hamiltonian may now be rewritten in terms of these new spin-axes as

$$H = -\frac{1}{8}\sum_{i,j}\Big[(1-\gamma)\sigma_i^x\sigma_j^x + (1+\gamma)\sigma_i^z\sigma_j^z\Big] \ . \tag{4}$$

We may now define a ground-state trial wave function, given by

$$|\psi\rangle = \exp\{U\}\ |0\rangle \ ; \ U = \frac{1}{2}\sum_{i<j}^{N} u(\mathbf{r}_{ij})\sigma_i^x\sigma_j^x \ , \tag{5}$$

where $u(\mathbf{r}_{ij})$ is the pseudo-potential. The reference state $|0\rangle$ is given by a tensor product of spin states which have eigenvalues of $+1$ with respect to σ^z, and this state is an exact ground eigenstate of the Hamiltonian Eq. (4) when $\gamma = 1$. Translational invariance also implies that the pseudo-potential, $u(\mathbf{r}_{ij})$, depends only on the relative distance, $\mathbf{n} = \mathbf{r}_i - \mathbf{r}_j \equiv \mathbf{r}_{ij}$.

The treatment of the spin-half XY model by the CBF method is continued by defining the lattice magnetisation (i.e., again the magnetisation in the z-direction in terms of the rotated local spin-axes), given by

$$M = \frac{\langle\psi \mid \sigma_i^z \mid \psi\rangle}{\langle\psi|\psi\rangle} \ , \tag{6}$$

for a ground-state trial wave function, $|\psi\rangle$. Furthermore, the 'transverse' magnetisation (in terms of the rotated local spin-axes) is given by,

$$A = \frac{\langle\psi \mid \sigma_i^x \mid \psi\rangle}{\langle\psi|\psi\rangle} \ . \tag{7}$$

We may now define a spatial distribution function (which plays a crucial part in any CBF calculation) in the following manner,

$$G(\mathbf{n}) = \frac{\langle\psi \mid \sigma_i^x\sigma_j^x \mid \psi\rangle}{\langle\psi|\psi\rangle} \ . \tag{8}$$

Furthermore, we may also determine an expression for the expectation value of the ground-state energy of the spin-half XY Hamiltonian of Eq. (4), where

$$\frac{E}{N} = \frac{\langle\psi|H|\psi\rangle}{N\langle\psi|\psi\rangle} \ . \tag{9}$$

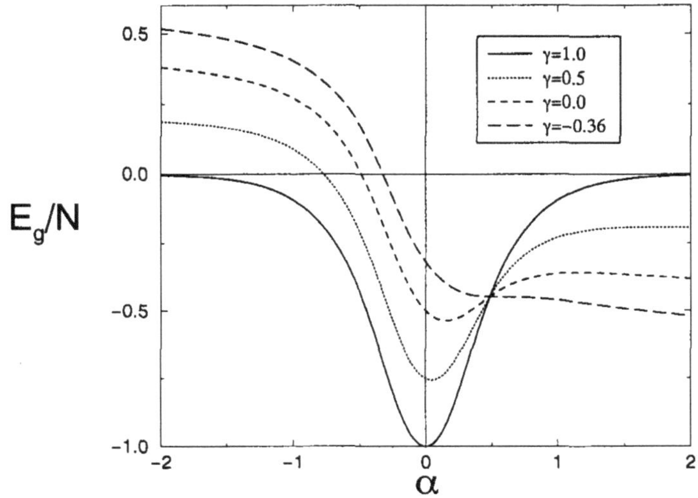

Figure 1. Results for the ground-state energy of the spin-half XY model for the square lattice plotted as a function of the strength of the nearest-neighbour pseudo-potential, α, for varying γ. At $\gamma = -0.36$ we see that the minimal solution that we have tracked from $\gamma = 0$ is lost.

In the region $\gamma \geq 0$ we now make the explicit assumption that $A = 0$, which is in agreement with our Ansatz for the trial wave function of Eq. (2). This assumption furthermore implies that $A/N \equiv \langle P^{\rightarrow}\rangle - \langle P^{\leftarrow}\rangle = \rho^{\rightarrow} - \rho^{\leftarrow} = 0$, where $P^{\rightarrow}$ and $P^{\leftarrow}$ are spin projection operators in the positive and negative x-directions respectively. However, we note that $\rho^{\rightarrow} + \rho^{\leftarrow} = 1$ must also be correct, which therefore implies that $\rho^{\rightarrow} = \rho^{\leftarrow} = 1/2$. Hence, we may treat this problem completely analogously to a binary-mixture of two types of bosons[25] each with a density equal to one-half.

The expression in Eq. (9) may be determined via a hypernetted chain (HNC) cluster expansion, and it is readily found using this procedure that the ground-state energy is given in terms of a functional with respect to the pseudo-potential, $u(\mathbf{n})$, where

$$\frac{E}{N} = -\frac{1}{8}\sum_{\mathbf{n}} \Delta(\mathbf{n}) \Big[(1-\gamma)G(\mathbf{n}) + (1+\gamma)\; M^2 \cosh[u(\mathbf{n})] \Big] \quad . \tag{10}$$

Note that $\Delta(\mathbf{n})$ is unity if $\mathbf{n}$ is a nearest-neighbour vector and is zero otherwise.

Self-consistent HNC equations may also be determined. These equations may be then iteratively solved, and thus $G(\mathbf{n})$ (and so the ground-state energy) may be also obtained. The first method of finding the pseudo-potential has a "variational" flavour, and we parametrise $u(\mathbf{n})$ in the following way,

$$u(\mathbf{n}) = \alpha\; \Delta(\mathbf{n}) \quad . \tag{11}$$

$\Delta(\mathbf{n})$ is unity if $\mathbf{n}$ is a nearest-neighbour vector and is zero otherwise. We now minimise the ground-state energy with respect to α at a given value of γ. Indeed, at $\gamma = 1$ we already know that all correlations have zero strength as our reference state $|0\rangle$ is an exact ground eigenstate of Eq. (4), and this implies that $\alpha = 0$. We thus track this solution at $\gamma = 1$ in the regime $\gamma < 1$, and the ground-state energy

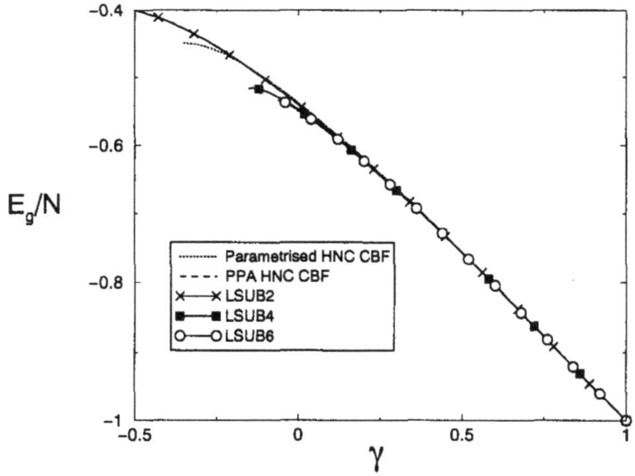

Figure 2. CBF results for the ground-state energy of the spin-half XY model on the square lattice compared to results of high-order CCM results of Ref. 8.

as a function of α for various values of γ is plotted in Fig. 1 for the square lattice. We may see that at $\gamma = -0.36$ the minima that we have tracked from $\gamma = 1$ become a point of inflection.

The second such method of determining the pseudo-potential is to determine the optimal value for the function $u(\mathbf{n})$ with respect to the ground-state energy, E/N. This is stated as,

$$\frac{\delta E}{\delta u(\mathbf{n})} = 0 \quad , \tag{12}$$

which may be determined analytically from Eq. (10). In the context of this article, this approach shall be referred to as the paired-phonon approximation (PPA), in analogy with a binary mixture of two types of bosons, for example. Note that we do not explicitly state here the resulting PPA equations for this model, although the treatment is fully analogous to that performed for the transverse Ising model and the interested reader is referred to Ref. 22 for a full account of this calculation.

For details of the specific application of the CCM to the spin-half XY model the interested reader is referred to Ref. 8. We note however that two types of approximations are made, namely, the SUB2 approximation which retains all two-body correlations in the approximate CCM ground-state wave function, and the LSUBm which retains all correlations in a locale defined by m.

Results for the CBF ground-state energy of the spin-half XY model on the square lattice compared to results of high-order CCM results are given in Fig. 2 and, for the isotropic point ($\gamma = 0$) only, in Table 2. We may see from Fig. 2 that both sets of results are in excellent qualitative agreement over a wide range of γ. It is furthermore seen from Tables 1-3 that CBF results are in excellent quantitative agreement with LSUB2 CCM results at $\gamma = 0$. This is a perfectly reasonable result, because both the CBF and CCM LSUB2 results only utilise two-body correlations. It is, however, expected that the inclusion of higher-order correlations in the CBF

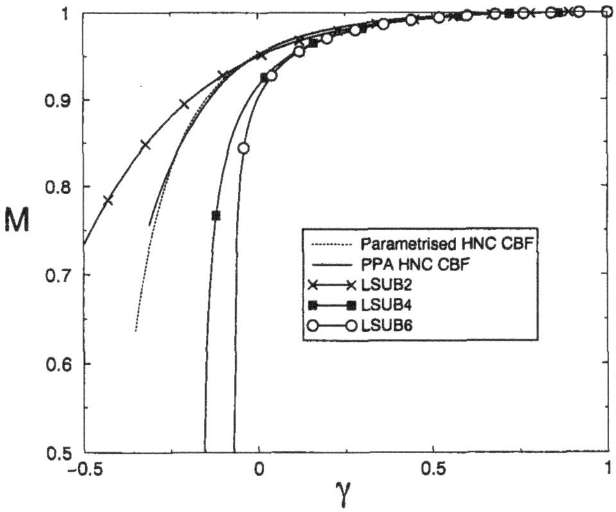

Figure 3. CBF results for the sublattice magnetisation of the spin-half XY model on the square lattice compared to results of high-order CCM results of Ref. 8.

trial wave function would produce more accurate results for the energy, as is seen for the CCM. Thus, from Tables 1-3, we see that the CBF results at $\gamma = 0$ capture about 59% of the correlation energy for the linear chain, 76% of the correlation energy for the square lattice and 85% of the correlation energy for the cubic lattice (in comparison with exact and extrapolated CCM results). Indeed, the extrapolated CCM results present some of the most accurate results yet seen for the isotropic XY model on the square and cubic lattices. (Results for the linear chain and cubic lattice are qualitatively similar to the results presented for the square lattice in Fig. 2 and so are not plotted here.)

Results for the sublattice magnetisation of the spin-half XY model on the square lattice are presented in Fig. 3 and in Table 2 for the isotropic point, $\gamma = 0$. Again, it is seen from Fig. 3 that the CBF results are in good qualitative agreement with the known results of this model. However, the CBF result for the sublattice magnetisation at the isotropic point ($\gamma = 0$) is slightly too high, although it is again expected that higher accuracy would be achieved with the inclusion of higher-order correlations in the approximate CBF ground-state wave function. Again, results for the linear chain and cubic lattice are fully analogous to the square lattice case and so are presented only for the isotropic model in Tables 1 and 3.

Results for the phase transitions points predicted by the CCM method are also given in Tables 1-3, although no such results are explicitly given for the CBF method in these tables. It is however noted here that the loss of "minima" within the parametrized HNC CBF approach (at $\gamma = -0.36$ for the square lattice) may be associated with a phase transition within this system. This constitutes a powerful result for such a simple variational-style calculation. Note that similar behaviour is also seen for both the linear chain and cubic lattices. An analogous change in the energy surface with respect to $u(\mathbf{n})$ for the CBF PPA approach seems to occur

Table 1. Ground-state energy and sublattice magnetisation for the one-dimensional *XY* model at $\gamma = 0$ compared to exact results of Ref. 3 and CCM results of Ref. 8. The critical values of γ for the anisotropic model are also given.

LSUBn	E_g/N	M	$\gamma_c(n)$
Parametrised CBF	−0.29025	0.8919	–
PPA CBF	−0.29030	0.8904	–
LSUB2	−0.30381	0.8373	–
SUB2	−0.31038	0.7795	−0.10789
LSUB∞	−0.31829	–	–
Exact	−0.318310	0.0	0

for varying values of γ. However, in this case, the situation is much less clear-cut because, near to this point, convergence of the PPA equations becomes very difficult.

In this article, the CBF method has been applied with much success to the quantum spin-half *XY* model on the linear chain, the square lattice, and the cubic lattice in order to obtain accurate results for the ground-state energy and the sublattice magnetisation. These results were found to be in excellent qualitative agreement with previous CCM calculations,[8] although more quantitatively accurate CBF results would be possible with the inclusion of higher-order correlations (than Jastrow correlations) in the trial ground-state wave function. A strength of the CBF method is that it is not limited by the presence of frustration, in contrast with QMC methods for example, and a direct extension of this work would be to include next-neighbour-neighbour interactions in our model. Indeed, this presents the possibility that this calculation might be utilised to provide a trial or guiding wave function for these QMC techniques in the presence of such frustrating next-nearest-neighbour bonds. Also, it is possible to see that an extension of this work to Heisenberg antiferromagnetic (HAF) models could follow a similar path to that outlined in this article. One would perform a similar set of rotations of the local spin-axes and then perform HNC re-summations of the relevant quantities that one is interested in. For example, one might consider the HAF on the triangular lattice, and in this case one would perform a rotation on three sublattices – such as that utilised by Singh and Huse for this model.[27] Previous CCM results for the spin-half *XY* model quoted in this article were also seen to provide excellent results for this model, and they are furthermore a valuable yardstick with which to compare our new CBF results with.

We note that the CBF approach utilises a Jastrow wave function and its bra states are always the explicit Hermitian adjoint of the corresponding ket state. Hence, for the CBF approach, an upper bound to the true ground-state energy is, in principle, obtainable, although the approximations made in calculating the energy may destroy it. By contrast, the CCM uses a bi-variational approach in which the bra and ket states are not manifestly constrained to be Hermitian adjoints and hence an upper bound to the true ground-state energy is not necessarily obtained. Also, the CCM uses creation operators with respect to some suitably nor-

Table 2. Ground-state energy and sublattice magnetisation for the square lattice XY model at $\gamma = 0$ compared to CCM calculations of Ref. 8 and series expansion calculations of Ref. 7. The critical values of γ for the anisotropic model are also given, where the value in parentheses is the estimated error in the final decimal place shown.

LSUBn	E_g/N	M	$\gamma_c(n)$
Parametrised CBF	−0.53738	0.9524	–
PPA CBF	−0.53774	0.9515	–
LSUB2	−0.54031	0.9496	–
SUB2	−0.54633	0.9190	−0.030(1)
LSUB∞	−0.54892	0.869	0.00(1)
Series Expansion	−0.5488	0.872	–

Table 3. Ground-state energy and sublattice magnetisation for the cubic lattice XY model at $\gamma = 0$ compared to CCM results of Ref. 8. The critical values of γ for the anisotropic model are also given, where the value in parentheses is the estimated error in the final decimal place shown.

LSUBn	E_g/N	M	$\gamma_c(n)$
Parametrised CBF	−0.78572	0.9710	–
PPA CBF	−0.78625	0.9695	–
LSUB2	−0.78687	0.9715	–
SUB2	−0.79090	0.9583	−0.01666(1)
LSUB∞	−0.79201	0.948	0.01(1)

malised model state in order to span the complete set of (here) Ising states. The CBF method, in essence, uses projection operators to form the Jastrow correlations with respect to a reference state, $|0\rangle$. In some sense, the CCM is found to contain *less* correlations than the others at 'equivalent' levels of approximation (e.g., the CCM LSUB2 approximation versus Hartree and nearest-neighbour Jastrow correlations). A strength of the CCM is that it is well-suited to the inclusion of high-order correlations in the approximate ground-state wavefunction (for example, via computational techniques). Furthermore, the CCM requires no information other than the approximation in S and $\tilde{S}$ in order to determine an approximate ground state of a given system. The CBF method, however, may require that only a certain subset of all possible diagrams are summed over (e.g., the HNC/0 approximation).

We finally note that the results of one method reinforce and sometimes elucidate the results of the other, and the application, in parallel, of two such methods to the same model can lead to a deeper understanding of the behaviour of it.

References

1. E. Lieb, T. Schultz and D. Mattis, *Ann. Phys. (NY)* **16**, 407 (1961).
2. Th. Niemeyer, *Physica* **36**, 377 (1967).

3. E. Barouch and B. M. McCoy, *Phys. Rev.* A **3**, 786 (1971).
4. W. Zheng, J. Oitmaa and C. J. Hamer, *Phys. Rev.* B **44**, 11869 (1991).
5. H-Q. Ding, 1992, *Phys. Rev.* B **45**, 230 (1992).
6. S. Zhang and K. J. Runge, *Phys. Rev.* B **45**, 1052 (1992).
7. C. J. Hamer, J. Oitmaa and W. Zheng, *Phys. Rev.* B **43**, 10789 (1991).
8. D. J. J. Farnell, S. A. Krueger, and J. B. Parkinson, *J. Phys. C (Cond. Matt.)* **9**, 7601 (1997).
9. D. D. Betts, S. Matsui, N. Vats and G. E. Stewart, *Czech. J. Phys.* **74**, 54 (1996).
10. J. W. Clark and E. Feenberg, *Phys. Rev.* **113**, 388 (1959).
11. H. W. Jackson and E. Feenberg, *Rev. Mod. Phys.* **34**, 686 (1962).
12. E. Feenberg, in *Theory of Quantum Liquids*, Ed. K. Binder, (Springer, New York, 1969).
13. J. W. Clark, in *Progress in Particle and Nuclear Physics*, Ed. D. H. Wilkinson, Vol. **2** (Pergamon, Oxford, 1979), p. 89.
14. V. R. Pandharipande and R. B. Wiringa, *Rev. Mod. Phys.* **51**, 821 (1979).
15. E. Krotscheck and J. W. Clark, *Nucl. Phys.* A **333**, 77 (1980).
16. J. W. Clark, in *The Many-Body Problem: Jastrow Correlations Versus Brueckner Theory*, Eds. R. Guardiola and J. Ros, Lecture Notes in Physics, Vol. **138** (Springer-Verlag, Berlin, 1981), p. 184.
17. S. Rosati, in *International School of Physics Enrico Fermi, Course LXXIX*, Ed. A. Molinari (North-Holland, Amsterdam, 1981), p. 73.
18. A. Fabrocini and S. Fantoni, in *First International Course on Condensed Matter*, ACIF Series, Eds. D. Prosperi, S. Rosati and S. Violini, Vol. **8** (World Scientific, Singapore, 1987), p. 87.
19. S. Fantoni and V. R. Pandharipande, *Phys. Rev.* C **37**, 1687 (1988).
20. S. Fantoni and A. Fabrocini, in *Microscopic Quantum Many-Body Theories and Their Applications*, Eds. J. Navarro and A. Polls, Lecture Notes in Physics, Vol. **510** (Springer-Verlag, Berlin, 1998), p. 119.
21. M. L. Ristig and J. W. Kim, *Phys. Rev.* B **53**, 6665 (1996).
22. M. L. Ristig, J. W. Kim, and J. W. Clark, in *Theory of Spin Lattices and Lattice Gauge Models*, Eds. J. W. Clark and M. L. Ristig, Lecture Notes in Physics, Vol. **494** (Springer-Verlag, Berlin 1997), p. 62.
23. M. L. Ristig, J. W. Kim, and J. W. Clark, *Phys. Rev.* B **57**, 56 (1998).
24. R. F. Bishop, D. J. J. Farnell, and M. L. Ristig, in *Condensed Matter Theories*, Vol. **14**, (2000) – in press.
25. M. L. Ristig, S. Fantoni, and K. E. Kürten, *Z. Phys.* B **51**, 1 (1983).
26. R. P. Feynman, *Phys. Rev.* **94**, 262 (1954).
27. R. R. P. Singh and D. A. Huse, *Phys. Rev. Lett.* **68**, 1766 (1992).

QUANTUM ANTIFERROMAGNETS WITH EASY-PLANE ANISOTROPY

P. GIANINETTI

Laboratoire de Physique Theorique des Liquides, Université Pierre and Marie Curie
4, place Jussieu, 75252 Paris Cedex 05, France

A. PAROLA

Dipartmento di Scienze and INFM, Universitá dell'Insubria
Via Lucini 3, 23100 Como, Italy

L. REATTO

Dipartmento di Fisica and INFM, Universitá di Milano
Via Celoria 16, 20133 Milano, Italy

By means of a recently developed formalism for the description of the phase diagram of quantum systems, we have investigated the effects of easy-plane anisotropy in three dimensional quantum antiferromagnets. We show that the cross-over between Heisenberg and XY universality classes can be located at reduced temperatures proportional to the easy-plane anisotropy parameter, which usually falls in the experimentally accessible range.

1 Introduction

The study of phase transitions in quantum models is a challenging subject characterized by the interplay of thermal and quantum fluctuations which both deeply affect the physics of the system. The most celebrated example of an inherently quantum phase transition is probably the onset of superfluidity in liquid ^{4}He, but also lattice models, like quantum antiferromagnets, provide a very rich variety of symmetry breaking phenomena with multi-component order parameters. Actually, there is a close relationship between the two cases: the Heisenberg model with easy plane anisotropy is in fact isomorphic to a lattice gas of hard core bosons with nearest neighbor coupling and its magnetic phase transition maps into the superfluid transition of the bose gas, thereby sharing the universal critical properties. The magnetic properties of many compounds are often described by nearest neighbor Heisenberg models with suitable anisotropy terms but reliable theoretical treatments of these systems are limited to the low temperature regime, where spin wave theory (SWT) is applicable or to the high temperature paramagnetic region, where interaction effects play a marginal role. This leaves out the phase transition region, where thermal and quantum fluctuations are entangled in a subtle and non-intuitive way. Semiclassical methods have been proposed to tackle this problem[1] and have been quite useful in the interpretation of experimental results, but the low temperature regime cannot be accessed by this formalism. Numerical simulations and high temperature expansions are sometime available and provide a valuable check to test new theoretical development in the regions where the numerics can be trusted.

A novel approach to map the phase diagram of quantum models has been recently proposed and applied to Heisenberg quantum antiferromagnets. The quantum hierarchical reference theory (QHRT), which extends previous works in the context of classical fluids,[2] is a microscopic implementation of the momentum shell integration renormalization group approach. The main advantages of this method are: (i) it keeps full information about the non universal properties of the model, avoiding mapping onto long wavelength effective hamiltonians; (ii) it reproduces the correct renormalization group structure in the critical region, with scaling laws and non classical critical exponents; (iii) it provides a consistent description of the first order transition characterized by convex free energy and rigorously flat isotherms in the two phase regions.

In this note we examine the phase diagram of the Heisenberg quantum antiferromagnet with easy plane anisotropy by the QHRT method, with special emphasis on the possibility to observe the crossover between the planar ($O(2)$) and $O(3)$ universality class introduced by the anisotropy. The range of parameters studied here is typical for a class of transition metal compounds.[3]

After a brief outline of the theoretical foundations of the QHRT approach, we will present the results for the two limiting cases of spin one half (quantum) and infinite spin (classical) systems on cubic lattices. Comparisons with available numerical results in the planar and isotropic case testify the accuracy of the method. Analysis of the critical region for several values of the anisotropy will clearly show the presence of a well defined crossover region at experimentally reachable reduced temperatures.

2 QHRT equations

The system we are going to study is the easy plane Heisenberg antiferromagnet defined by the Hamiltonian

$$H = \frac{1}{2} \sum_{\boldsymbol{R},\boldsymbol{R}'} J_{\boldsymbol{R},\boldsymbol{R}'} \left[S^y_{\boldsymbol{R}} S^y_{\boldsymbol{R}'} + S^z_{\boldsymbol{R}} S^z_{\boldsymbol{R}'} + \lambda S^x_{\boldsymbol{R}} S^x_{\boldsymbol{R}'} \right] + h \sum_{\boldsymbol{R}} e^{i\boldsymbol{g}\cdot\boldsymbol{R}} S^z_{\boldsymbol{R}}. \tag{1}$$

The model is defined on a hypercubic lattice in D dimensions and the coupling $J_{\boldsymbol{R},\boldsymbol{R}'}$ is assumed to be antiferromagnetic ($= J > 0$) at the nearest neighbor sites. The vector $\boldsymbol{g}$ is the antiferromagnetic wavevector, $0 \leq \Delta = (1 - \lambda) \leq 1$ is the anisotropy parameter and for future convenience we have included a staggered external magnetic field (ordering field) h along the z axis (which lies in the easy plane).

The basic idea underlying the QHRT method is to calculate the partition function of the model $Z = \mathrm{Tr} \exp(-\beta H)$ by a gradual introduction of spin fluctuations, starting from the shortest wavelengths. A convenient way to implement this RG procedure is to introduce a "cut-off" parameter $0 \leq Q \leq D$ which defines a sequence of systems (Q-systems) in which fluctuation over wavelengths larger than Q^{-1} are inhibited. This can be achieved by defining a sequence of interactions $J^Q_{\boldsymbol{R},\boldsymbol{R}'}$:

$$J^Q_{\boldsymbol{R},\boldsymbol{R}'} = \int \frac{d^D k}{(2\pi)^D} \, e^{i\boldsymbol{k}\cdot(\boldsymbol{R}-\boldsymbol{R}')} \, 2J\gamma_{\boldsymbol{k}} \, \Theta\left(\sqrt{D^2 - \gamma^2_{\boldsymbol{k}}} - Q\right), \tag{2}$$

where the integration domain is the first Brillouin zone of the lattice, $\Theta(x)$ is the usual step function and $\gamma_{\boldsymbol{k}} = \sum_{\alpha=1}^{D} \cos k_\alpha$. When $Q = D$ the interaction $J^Q_{\boldsymbol{R},\boldsymbol{R}'}$ vanishes identically while in the $Q \to 0$ limit $J^Q_{\boldsymbol{R},\boldsymbol{R}'}$ coincides with the physical nearest neighbor coupling of the original hamiltonian (1). Following Ref. 2 it is possible to prove that the properties of a system interacting through $J^Q_{\boldsymbol{R},\boldsymbol{R}'}$ coincide with the properties of the Q-system apart from trivial mean field contributions to the free energy and in the two body correlations which are included in the definition of Q-system. Remarkably, it is possible to obtain the exact "evolution" equations describing the change in the Helmholtz free energy per lattice site of the Q-system $a_Q = -N^{-1} k_B T \ln Z$ when the cut-off Q is varied. The details, together with a more thorough discussion of the QHRT method, can be found in Ref. 4. Here we just recall that the evolution equation for the Q-dependent free energy does depend on the imaginary-time two-body dynamical correlation functions of the Q-system, $F^{ij}_Q(\boldsymbol{k},\omega)$, and that therefore the equation for the free energy, although formally exact, is not closed. As usual ω represents Matsubara frequencies, which are integer multiples of $2\pi/\beta$. It is also possible to derive analogous differential equations for the "evolution" of the two point functions, but those turn out to depend on the three and four body correlations thereby defining an infinite hierarchy of differential equations which represents a microscopic implementation of the momentum shell integration RG in quantum models.

Here we will not pursue the detailed analysis of the exact hierarchy but rather we will study a simple but accurate closure to the first equation of the hierarchy obtained by choosing an approximate form for the two point functions. The closure we have adopted is a straightforward generalization of the mean-field form of the dynamical structure factor which is also the exact initial condition for the evolution equation of the two point function. The explicit form we have considered is also known as the *simple mode approximation* because it amounts to neglecting the incoherent part of the dynamical structure factor. Due to the symmetries of the problem, the only non vanishing elements of the two-point correlation tensor are the diagonal terms

$$\begin{aligned} F^{xx}_Q(\boldsymbol{k},\omega) &= \frac{\mu_{\perp Q} - 2J\gamma_{\boldsymbol{k}}}{m^{-2}\omega^2 + \mu^2_{\perp Q} - \lambda\,(2J\gamma_{\boldsymbol{k}})^2 - 2J\Delta\gamma_{\boldsymbol{k}}\mu_{\perp Q}}, \\ F^{yy}_Q(\boldsymbol{k},\omega) &= \frac{\mu_{\perp Q} - 2J\lambda\gamma_{\boldsymbol{k}}}{m^{-2}\omega^2 + \mu^2_{\perp Q} - \lambda\,(2J\gamma_{\boldsymbol{k}})^2 + 2J\Delta\gamma_{\boldsymbol{k}}\mu_{\perp Q}}, \\ F^{zz}_Q(\boldsymbol{k},\omega) &= \frac{\delta_{\omega,0}}{\mu_{||Q} + 2J\gamma_{\boldsymbol{k}}}, \end{aligned} \tag{3}$$

and the off-diagonal xy components

$$F^{xy}_Q(\boldsymbol{k},\omega) = -F^{yx}_Q(\boldsymbol{k},\omega) = \frac{m^{-1}\omega}{m^{-2}\omega^2 + \mu^2_{\perp Q} - (2J\gamma_{\boldsymbol{k}})^2}. \tag{4}$$

Here m is the staggered magnetization and the two parameters $\mu_{\perp Q}$ and $\mu_{||Q}$ are related, via the susceptibility sum rule, to the derivatives of the free energy density a_Q with respect to the staggered magnetization m,

$$\chi_\perp^{-1} = m\left(\frac{\partial a_Q}{\partial m}\right)^{-1} = F_Q^{yy}(\boldsymbol{g},0) = (\mu_{\perp Q} - 2JD)^{-1},$$
$$\chi_{||}^{-1} = \left(\frac{\partial^2 a_Q}{\partial m^2}\right)^{-1} = F_Q^{zz}(\boldsymbol{g},0) = (\mu_{||Q} - 2JD)^{-1}. \qquad (5)$$

Within this approximation for the two-point dynamical correlation functions the evolution equation for the free energy becomes a closed partial differential equation for $a_Q(m)$ which should be numerically solved at fixed temperature T and for given spatial dimension D and easy plane anisotropy Δ,

$$\frac{da_Q}{dQ} = \frac{1}{2\beta}\int \frac{d^Dk}{(2\pi)^D}\delta\left(\sqrt{D^2-\gamma_{\boldsymbol{k}}^2} - Q\right)\left[\ln\left(\frac{\mu_{||Q}}{\mu_{||Q}+2J\gamma_{\boldsymbol{k}}}\right) + \right.$$
$$\ln\left(\frac{\sinh\left(\frac{m\beta\mu_{\perp Q}}{2}\right)}{\sinh\left(\frac{\beta m}{2}\sqrt{\mu_{\perp Q}^2 - \lambda(2J\gamma_{\boldsymbol{k}})^2 - 2J\Delta\gamma_{\boldsymbol{k}}\mu_{\perp Q}}\right)}\right) +$$
$$\left.\ln\left(\frac{\sinh\left(\frac{m\beta\mu_{\perp Q}}{2}\right)}{\sinh\left(\frac{\beta m}{2}\sqrt{\mu_{\perp Q}^2 - \lambda(2J\gamma_{\boldsymbol{k}})^2 + 2J\Delta\gamma_{\boldsymbol{k}}\mu_{\perp Q}}\right)}\right)\right]. \qquad (6)$$

The momentum integration is actually trivial because the integrand depends on $\boldsymbol{k}$ only through the function $\gamma_{\boldsymbol{k}}$ which, due to the presence of the delta function, can be set $\gamma_{\boldsymbol{k}} = \pm\sqrt{D^2 - Q^2}$. Notice that the value of the spin S just enters the equation through the initial condition at $Q = D$ which, for $S = 1/2$, explicitly reads

$$a_{Q=D} = k_B T\left[(\frac{1}{2}+m)\ln(\frac{1}{2}+m) + (\frac{1}{2}-m)\ln(\frac{1}{2}-m)\right] - JDm^2. \qquad (7)$$

3 Critical properties

As previously mentioned, in the neighborhood of a critical point and at long wavelength (i.e. for $Q \to 0$) our evolution equation (6) reduces to a known RG form derived within the so called *local potential approximation*.[5] Such a RG equation can be shown to reproduce the correct leading order in the $\epsilon = 4 - D$ expansion for the universal properties in the critical region and then it correctly predicts non classical exponents in the physical case $D = 3$ which *depend* on the dimensionality of the order parameter. For the easy plane Heisenberg antiferromagnet, the asymptotic equation reads:[4]

$$\frac{da_Q}{dQ} = -\frac{\Omega_D Q^{D-1}}{2\beta(2\pi\sqrt{D})^D}\left\{(n-1)\ln\left[m^{-1}\frac{\partial a_Q}{\partial m} + J\frac{Q^2}{D}\right] + \ln\left[\frac{\partial^2 a_Q}{\partial m^2} + J\frac{Q^2}{D}\right]\right\}, \quad (8)$$

where Ω_D is the surface of the unit D-dimensional hypersphere and n is just the dimensionality of the order parameter: $n = 2$ for $\Delta \neq 0$ while $n = 3$ in the isotropic $\Delta = 0$ limit. The asymptotic form does depend on n because of the presence of a "mass" term in the xx two point function, which vanishes just for $\Delta = 0$.

Table 1. Critical temperatures for $S = 1/2$ ($T_c(1/2)$) and for $S = \infty$ ($T_c(\infty)$)

Δ	$T_c(1/2)$	$T_c(\infty)$
0	0.902	1.421
0.001	0.909	1.426
0.01	0.926	1.440
0.1	0.971	1.480
1	0.978	1.541

Therefore the isotropic case is actually a singular point in parameter space where fluctuations along the x direction do affect the long wavelength physics, thereby changing the universality class of the model, as expected on the base of symmetry considerations. By suitable rescaling of the variables in Eq. (8) we obtain the RG equation of Ref. 5 which can be studied by standard methods: the associated fixed point equation is first numerically solved for the interesting cases $n = 2$ and $n = 3$ in three dimensions. Due to the analytic form of the two point functions, even at the critical point, the correlation critical exponent η identically vanishes within this approximation. The most relevant eigenvalues of the linearized form of the evolution equation are then obtained and the susceptibility critical exponent γ predicted by QHRT turn out to be $\gamma = 1.65$ for $n = 3$ and $\gamma = 1.53$ for $n = 2$, which are about 15% larger than the accepted ones.[6] By inspection of the denominators of the two point functions (3) it is easy to show that the crossover between the isotropic ($n = 3$) and planar ($n = 2$) regime is expected when the zero field staggered susceptibility χ is larger than $(2D\Delta)^{-1}$, i.e. when the easy plane anisotropy becomes effective in providing a cut-off to the divergence of $F^{xx}(\boldsymbol{g}, 0)$.

4 Results

The QHRT evolution equation has been studied in the isotropic limit $\Delta = 0$ both in two and three dimensions and several values of S: the numerical results have been extensively discussed in Ref. 4 and show a good agreement with available simulation and high temperature expansion data. Here we present some results for the three dimensional spin one half model at non zero Δ in order to assess the observability of the crossover between the XY and Heisenberg universality classes. Comparison with the QHRT results in the classical $S \to \infty$ limit will help to understand the effects of quantum fluctuations.

The partial differential equation (6) has been solved by means of an implicit finite difference method after having reduced it to a quasi-linear form. The mesh size we use consists of several thousand points on the staggered magnetization axis while the evolution in the variable $\ln Q$ has been carried on up to convergence. The critical temperature has been located by a power law fit of the numerical data and our results are reported in Table 1.

The numerical results clearly show a divergence of the staggered susceptibility on approaching T_c. The critical exponent as obtained by the power law fit is always

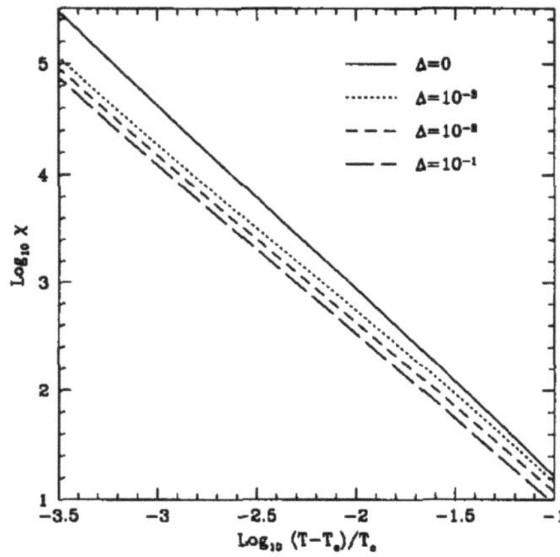

Figure 1. Staggered susceptibility for $S = 1/2$.

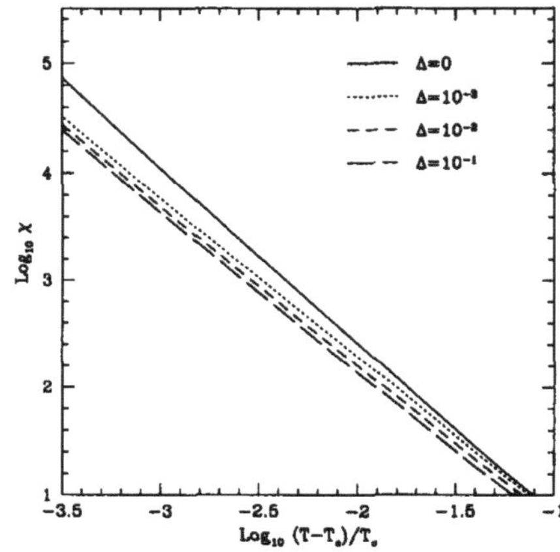

Figure 2. Staggered susceptibility for $S = \infty$.

quite close to the values predicted by the fixed point analysis of the asymptotic equation, showing that QHRT correctly describes the RG structure in the critical region. The power law behavior of χ is clearly demonstrated in Figs. 1 and 2 for $S = 1/2$ and $S = \infty$ respectively.

The different slope of the $\Delta = 0$ lines in Figs. 1 and 2 with respect to the $\Delta \neq 0$ case can be appreciated in the plots and confirms the different universality class of the two cases, irrespective of the value of the spin S. The weak anisotropy case $\Delta = 10^{-3}$ is particularly interesting because it shows a well defined crossover region whose extent is not severely affected by quantum fluctuations and can be located in the range $10^{-2.5} < (T - T_c)/T_c < 10^{-1}$ for both the examined values of the spin.

We have also determined the spontaneous magnetization curves which are plotted for the limiting cases $\Delta = 0$ and $\Delta = 1$ in Fig. 3. The effects of anisotropy on the shape of the coexistence curve are more clearly seen in the quantum case ($S = 1/2$) where also the zero temperature magnetization is affected by the change in dimensionality of the order parameter. The critical exponent β governing the shape of the coexistence region in the neighborhood of the critical point has a non classical value within the QHRT approach: $\beta = 0.41$ for the $O(3)$ model and $\beta = 0.38$ in the $O(2)$ case.

In conclusion, we have shown how QHRT can be applied to the investigation of the phase diagrams of quantum antiferromagnets where it proves to be a valuable tool for determining the location of phase boundaries and the extension of the crossover regions. These problems are usually addressed either by numerical methods or through the mapping onto effective long wavelength hamiltonians[7] whose reliability for studying the non universal properties is questionable. QHRT can also be generalized to frustrated antiferromagnets and to fermionic models where simulations are affected by numerical instabilities. The only approximation of the theory is the possibility to represent the momentum and frequency dependence of

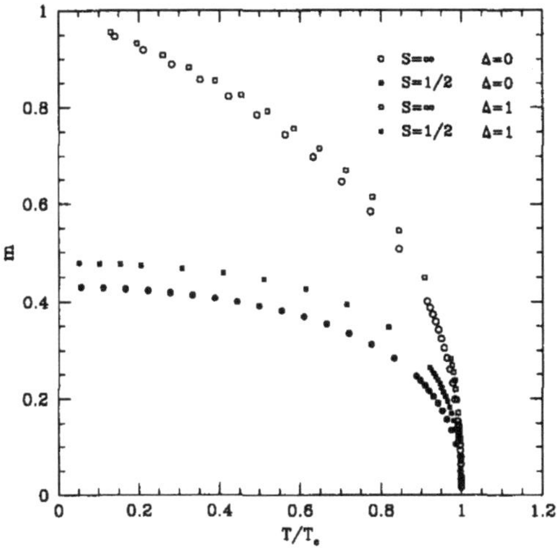

Figure 3. Spontaneous staggered magnetization

the dynamic structure factor in a simple parametric form, like the single mode approximation adopted in this work.

References

1. A. Cuccoli, V. Tognetti, R. Vaia, P. Verrucchi, *Phys. Rev. Lett.* **77**, 3439 (1996).
2. A. Parola, L. Reatto, *Adv. Phys.* **44**, 211 (1995).
3. L. J. de Jongh, *Magnetic Properties of Layered Transition Metal Compounds* (Kluwer, Dordrecht, 1989).
4. P. Gianinetti, A. Parola, *Phys. Lett.* A **268**, 424 (2000); cond-mat/004235.
5. J. F. Nicoll, T. S. Chang, and H. E. Stanley, *Phys. Rev.* A **13**, 1251 (1976).
6. J. Zinn-Justin, *Quantum Field Theory and Critical Phenomena* (Clarendon Press, Oxford. 1989).
7. S. Chakravarty, B. I. Halperin, and D. R. Nelson, *Phys. Rev.* B **39**, 2344 (1989).

QUANTUM PHASE TRANSITIONS IN SPIN SYSTEMS

J. RICHTER AND S. E. KRÜGER
Institute for Theoretical Physics, University of Magdeburg, P.O.Box 4120, D-39016 Magdeburg, Germany
E-mail: johannes.richter@physik.uni-magdeburg.de

D. J. J. FARNELL AND R. F. BISHOP
Department of Physics, UMIST, P.O. Box 88, Manchester M60 1QD, UK

We discuss the influence of strong quantum fluctuations on zero-temperature phase transitions in a two-dimensional spin-half Heisenberg system. Using a high-order coupled cluster treatment, we study competition of magnetic bonds with and without frustration. We find that the coupled cluster treatment is able to describe the zero-temperature transitions in a qualitatively correct way, even if frustration is present and other methods such as quantum Monte Carlo fail.

1 Introduction

Phase transitions have been a subject of great interest to physicists over many decades. Besides classical or thermal phase transitions, the so-called quantum phase transitions (or zero-temperature transitions) have started to attract a lot of attention (e.g., see Ref. 1). The study of a variety of simple models allows us to understand which aspects of thermal and zero-temperature phase transitions are common to classes of models and which are more special. For continuous order-disorder transitions we basically need the interplay between the interparticle interactions and fluctuations. For thermal transitions the Ising model may serve as the simplest model. The equilibrium state corresponds to a minimum of the free energy, and we have competition between energy and entropy controlled by the temperature. For zero-temperature transitions no thermal fluctuations are present, and the fluctuations arise due to Heisenberg's uncertainty principle. A corresponding basic model which has strong quantum fluctuations is the spin-half Heisenberg antiferromagnet (HAFM), particularly in low dimensions.

The subject of quantum spin-half Heisenberg antiferromagnetism in low-dimensional systems has attracted a great deal of interest in connection with the magnetic properties of the high-temperature superconductors. Although we know from the Mermin-Wagner theorem[2] that thermal fluctuations are strong enough to destroy magnetic long-range order (LRO) at any finite temperature in 1D and 2D, the role of quantum fluctuations is less understood. It is now clear that the ground-state of the HAFM in 1D is not long-range ordered, whereas the HAFM on the square lattice is long-range ordered (e.g., see Ref. 3). However, in 2D there are many other lattices with different coordination numbers and topologies, and there is no general statement concerning zero-temperature Néel-like LRO.

Anderson and Fazekas have suggested[4] that additional competition between magnetic bonds may increase quantum fluctuations and can suppress the Néel-like LRO in 2D. Indeed, the strength of this competition may serve as the control parameter of a zero-temperature order-disorder transition. The competition between magnetic bonds in quantum spin systems can be caused in various ways. As for classical spin systems, frustration can affect the magnetic ordering in quantum spin systems. In the classical HAFM the frustration often leads to canted (e.g., spiral) spin states which may or may not have counterparts in the quantum HAFM. Furthermore, due to frustration Marshall's sign rule need not be fulfilled.[5] The violation of the sign rule in frustrated systems makes their theoretical investigation particularly difficult. For example, the quantum Monte Carlo (QMC) method suffers from the minus sign problem in frustrated spin systems.

A generic model of a frustrated HAFM is the spin-half J_1–J_2 model on the square lattice, where the frustrating J_2 bonds plus quantum fluctuations yield a second-order transition from a Néel-ordered state to a disordered quantum spin liquid (see, e.g., Refs. 6–9). On the other hand, there are examples where frustration leads to a first-order transition in quantum spin systems in contrast to a second-order transition in the corresponding classical model (see, e.g., Refs. 10–13).

Besides frustration there is a second type of competition between bonds which favours a Néel-like distribution of spin correlations over the lattice and other bonds which favour the formation of local spin singlets. By contrast to frustration, which yields competition in quantum as well as in classical systems, this type of competition is present only in quantum systems. The formation of local singlets is accompanied by the 'melting' of the magnetic LRO. This mechanism for breaking magnetic LRO may be relevant for the quantum disordered state in bilayer systems[14,15] as well as in CaV_4O_9 (see, e.g., Refs. 16,17). Of course both mechanisms can be mixed as, for instance, in $SrCu_2(BO_3)_2$ (see, e.g., Refs. 12,13,18,19).

In this paper we discuss the quantum order-disorder transition driven by local singlet formation as well as the influence of quantum fluctuations on zero-temperature transitions driven by frustration. To that end we study a spin-half model on the square lattice in which both mechanisms, frustration and singlet formation, are observed in different regions. High-order implementations of the coupled cluster method (CCM) are used to obtain a consistent description of both types of competition for this model. The CCM (see, e.g., Ref. 20) is one of the most powerful and most universal techniques in quantum many-body theory, and has previously been applied to quantum spin systems[8,21–25] with great success. In particular, we shall study to what extent the CCM is able to describe zero-temperature transitions in spin systems.

2 The Model

We consider a spin-half Heisenberg model on a square lattice with two kinds of nearest-neighbour bonds J and J', as shown in Fig. 1,

$$H = J \sum_{\langle ij \rangle_1} \mathbf{S}_i \cdot \mathbf{S}_j + J' \sum_{\langle ij \rangle_2} \mathbf{S}_i \cdot \mathbf{S}_j. \tag{1}$$

The expressions $\langle ij \rangle_1$, and $\langle ij \rangle_2$ indicate nearest-neighbour bonds arranged in a regular zigzag pattern, as shown in Fig. 1 by the dotted and solid lines, respectively. Each square-lattice plaquette consists of three J bonds and one J' bond. If J' and J have different signs, then the plaquettes are frustrated, whereas competition

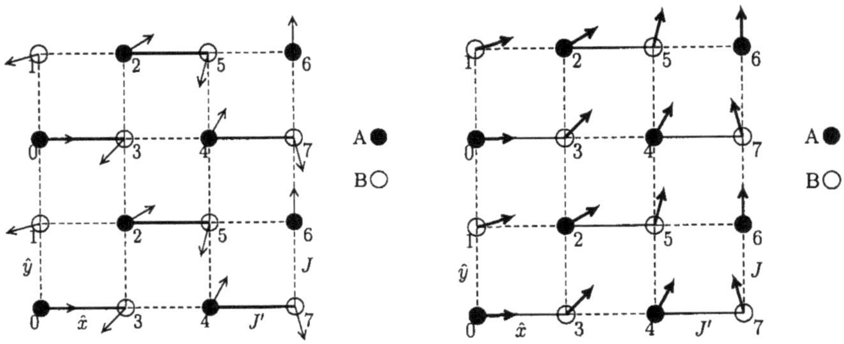

Figure 1. Illustration of the model (J-bonds correspond to dotted lines and J'-bonds to solid lines) and of the classical spiral state for antiferromagnetic $J = +1$ and ferromagnetic $J' < -1/3$ (left graph) and vice versa ($J = -1$, $J' > 1/3$) (right graph). As discussed in the text, both spiral states can be transformed into each other by reversing all of the spins on the B sublattice.

without frustration is realized for antiferromagnetic bonds $J' > J > 0$.

3 The Classical Ground State

To discuss the influence of quantum fluctuations on the ground-state (GS) properties of the model we need to know the classical GS of Eq. (1). We set $J = 1$. Without frustration ($J' > 0$), the classical GS is the Néel state, independent of the strength of J'. Frustration appears for (ferromagnetic) $J' < 0$. One finds that the Néel state remains the GS for small amounts of frustration, $0 \geq J' \geq -1/3$. At the critical point $J' = -1/3$, a second-order transition takes place from the Néel state to a spiral state characterized by a pitch angle $\Phi_{\rm cl} = \arccos(\sqrt{1 - 1/J'}/2)$ (see left graph of Fig. 1). We note that $\Phi_{\rm cl} = 0$ (for $J' > -1/3$) corresponds to the Néel state. The spin directions belonging to the A and B sublattices respectively, are given by $\mathbf{S}_A(\mathbf{R}) = \mathbf{e}_x \cos \mathbf{Q} \cdot \mathbf{R} + \mathbf{e}_y \sin \mathbf{Q} \cdot \mathbf{R}$ and $\mathbf{S}_B(\mathbf{R} + \hat{x}) = \mathbf{e}_x \cos(\mathbf{Q} \cdot \mathbf{R} + \pi + 3\Phi) + \mathbf{e}_y \sin(\mathbf{Q} \cdot \mathbf{R} + \pi + 3\Phi)$, where $\mathbf{R}$ runs over the sites of the sublattice A, and $\mathbf{Q} = (2\Phi, 0)$. We note that this spiral state is incommensurate in the x-direction. By contrast to the quantum case, the classical model with $J = -1$ can be transformed into the model with $J = 1$ considered above by the simultaneous substitution $J \to -J$, $J' \to -J'$, $\mathbf{S}_{i \in B} \to -\mathbf{S}_{i \in B}$. Hence the physics for $J = -1$ is classically the same as for $J = +1$ (c.f., Fig.1).

4 Competition Without Frustration

In this section we restrict our attention to the region where $J = +1$ and $J' > 0$. The terms within the Hamiltonian now compete, because the first term favours Néel LRO on the "honeycomb" lattice, whereas the second term favours an uncorrelated product state of local pair singlets (see below). Again, we emphasize that there is no competition in the classical model.

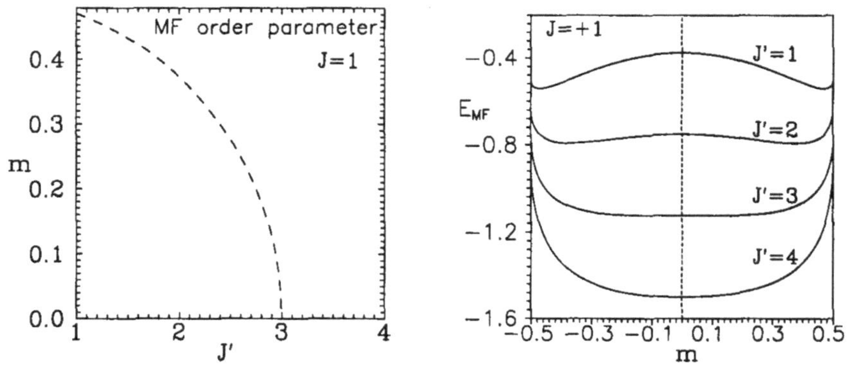

Figure 2. Sublattice magnetization versus J' (for $J = 1$) (left graph) and energy versus sublattice magnetization (right graph) using a mean field approach (2).

Mean Field Approach: We start with a simple mean-field (MF) like description of the order-disorder transition. The corresponding uncorrelated MF state for Néel LRO is the Néel state $|\phi_{\mathrm{MF}_1}\rangle = |\uparrow\downarrow\uparrow\ldots\rangle$, and for the dimerized singlet state it is the rotationally-invariant product state of local pair singlets $|\phi_{\mathrm{MF}_2}\rangle = \prod_{i\in A}[|\uparrow_i\downarrow_{i+\hat{x}}\rangle - |\downarrow_i\uparrow_{i+\hat{x}}\rangle]/\sqrt{2}$, where i and $i+\hat{x}$ correspond to those sites which cover the J' bonds. In order to describe the transition between both states, we consider an uncorrelated product state interpolating between $|\phi_{\mathrm{MF}_1}\rangle$ and $|\phi_{\mathrm{MF}_2}\rangle$ of the form[15,25]

$$|\Psi_{\mathrm{MF}}(t)\rangle = \prod_{i\in A} \frac{1}{\sqrt{1+t^2}}\left[|\uparrow_i\downarrow_{i+\hat{x}}\rangle - t|\downarrow_i\uparrow_{i+\hat{x}}\rangle\right]. \qquad (2)$$

We have $|\Psi_{\mathrm{MF}}(t=0)\rangle = |\phi_{\mathrm{MF}_1}\rangle$ and $|\Psi_{\mathrm{MF}}(t=1)\rangle = |\phi_{\mathrm{MF}_2}\rangle$. We minimize $\langle\Psi_{\mathrm{MF}}|H|\Psi_{\mathrm{MF}}\rangle$ with respect to t and obtain

$$\frac{E_{\mathrm{MF}}}{N} = \frac{\langle\Psi_{\mathrm{MF}}|H|\Psi_{\mathrm{MF}}\rangle}{N} = \begin{cases} -\frac{3J'}{8} - \frac{1}{24J}(3J-J')^2 & J' \le 3J \\ -\frac{3J'}{8} & J' > 3J, \end{cases} \qquad (3)$$

for the energy per site. For the sublattice magnetization $m \equiv \langle\Psi_{\mathrm{MF}}|S^z_{i\in A}|\Psi_{\mathrm{MF}}\rangle$ we get $m = \sqrt{(3J-J')(3J+J')}/(6J)$ for $J' \le 3J$ and $m = 0$ otherwise. Note that m vanishes at a critical point $J'_c = 3J$, and that the critical index is the MF index 1/2. Eq. (3) may be rewritten in terms of m as $E_{\mathrm{MF}}/N = -\frac{1}{8}J' - \frac{1}{4}J'\sqrt{1-4m^2} - \frac{3}{2}J\,m^2$, and Fig. 2 illustrates that the dependence of E_{MF} on m corresponds to a typical scenario of a second-order transition. We can expand E_{MF} up to the fourth order in m near the critical point and find a Landau-type expression, given by $E_{\mathrm{MF}}/N = -\frac{3}{8}J' + \frac{1}{2}(J'-3J)\,m^2 + \frac{1}{2}J'\,m^4$. However, as discussed elsewhere for a similar magnetic model for CaV_4O_9,[17] MF theory probably does not describe the critical behaviour correctly.

CCM: Let us now apply a high-order CCM approach (for details see Refs. 23,25) to this model. We set the classical collinear Néel state to be the reference state $|\Phi\rangle$. We calculate the GS wave function, $|\Psi\rangle = e^S|\Phi\rangle$ within the LSUBn approximation scheme up to $n = 8$ and extrapolate to $n \to \infty$. The CCM results for the order

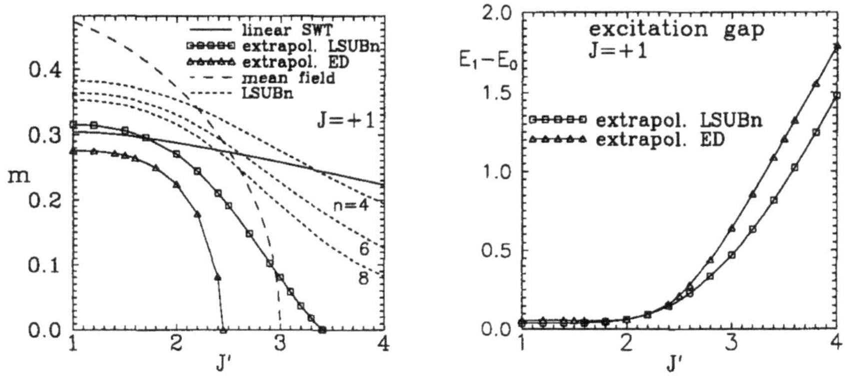

Figure 3. Sublattice magnetization (left graph) and excitation gap (right graph) versus J'.

parameter are shown in the left graph of Fig. 3 and they are compared to results of linear spin wave theory (SWT), exact diagonalization (ED) of N=16,18,20,26,32 sites, and the MF theory. The CCM is able to describe correctly the order-disorder transition, whereas conventional SWT cannot (for more details concerning the SWT and ED results see Ref. 25). The critical value predicted by extrapolation of the LSUBn results is, however, found to be slightly too large. We may also consider the inflection points of m versus J' for the LSUBn approximations. It is assumed that the true $m(J')$-curve will have a negative curvature up to the critical point. Thus we might expect that (for increasing n) the inflection point approaches the critical point. We find the corresponding inflection points at $J' = 3.1$ (n=2), $J' = 3.0$ (n=4), $J' = 2.9$ (n=6), and $J' = 2.85$ (n=8), indicating a critical value J'_c somewhere between $2.5J$ and $3J$. Notice, that the estimation of $2.5 \lesssim J'_c/J \lesssim 3$ is consistent with results of series expansions and exact diagonalizations.[25,26] The breakdown of Néel LRO due to singlet formation is also accompanied by the opening of an excitation gap between the singlet GS and the first triplet excitation. This behaviour is well described by the CCM (right graph of Fig. 3) which predicts that the gap opens in the range $2J < J' < 3J$ (and notice that the non-zero gap below $2J$ is a result of the limited accuracy of the extrapolation).

5 Competition with Frustration

We now consider the frustrated model (where J and J' have different signs). Due to the incommensurate classical spiral state the ED technique for finite-size systems is less appropriate. The CCM intrinsically considers the limit $N \to \infty$ from the outset and thus has no problems in dealing with incommensurate states. Hence the CCM appears to be particularly suitable to attack the frustrated quantum model. We choose the classical state to be our CCM model state, although quantum fluctuations may change the pitch angle of the spiral phase. Hence, we determine the 'quantum pitch angle' Φ by minimizing $E_{\mathrm{LSUB}n}(\Phi)$ with respect to Φ.

Néel versus Spiral: We consider (antiferromagnetic) $J = +1$ and (ferromagnetic) $J' < 0$. Results for $E(\Phi)$ and $\Phi(J')$ are shown in Fig. 4. The main results are

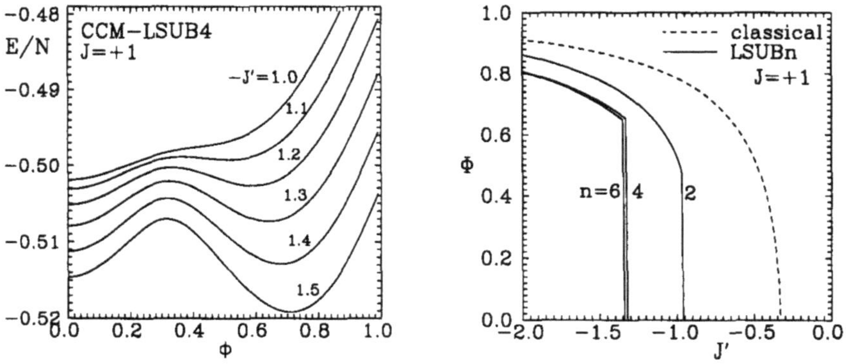

Figure 4. Energy versus quantum pitch angle for LSUB4 (left graph) and quantum pitch angle versus J' (right graph). Note, that $\Phi = 0$ corresponds to the Néel state.

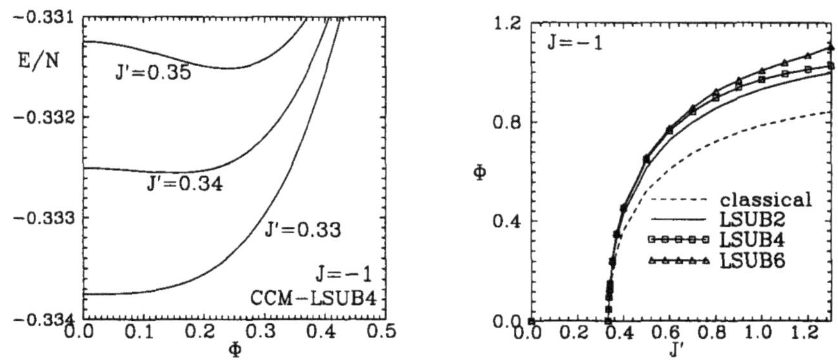

Figure 5. Energy versus quantum pitch angle for LSUB4 (left graph) and quantum pitch angle versus J' (right graph). Note, that $\Phi = 0$ corresponds to the fully polarized ferromagnetic state.

that: (i) In the quantum case the quantum Néel state remains the GS up to much stronger frustration than in the classical case. Indeed, it is generally found for spin systems, that quantum fluctuations favour collinear spin structures as opposed to noncollinear ones. (ii) The quantum fluctuations change the phase transition from second order to first order. (iii) The CCM yields a consistent description of the collinear and the spiral phases.

Ferro versus Spiral: We now consider the model for (ferromagnetic) $J = -1$ and (antiferromagnetic) $J' > 0$. In the classical model we again have a second-order transition from a collinear to a spiral state (see section 3). In the quantum model the situation is quite different. Although the collinear antiferromagnetic state possesses strong quantum fluctuations, the collinear, fully polarized, ferromagnetic state possesses no such quantum fluctuations. The corresponding results for $E(\Phi)$ and $\Phi(J')$ are shown in Fig. 5. By contrast to the situation at $J = +1$, the transition from the ferromagnetic to the noncollinear spiral is now of second order, the same as for the classical model. Furthermore, the classical critical point $J'/J = -1/3$

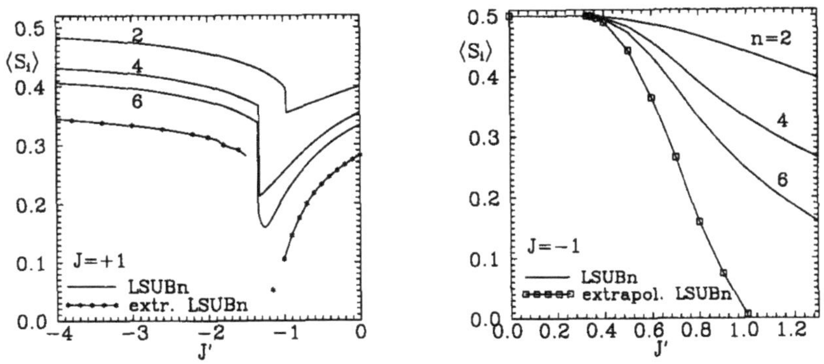

Figure 6. On-site magnetic moment versus J' for $J = +1$ (left graph) and $J = -1$ (right graph).

also holds for the quantum case. The difference between both cases also becomes evident when the order parameter $\langle S_i \rangle$ is considered (Fig. 6). For $J = +1$ there is a discontinuity in $\langle S_i \rangle$ at every level of LSUBn approximation. However, the extrapolation to $n \to \infty$ becomes imprecise close to the phase transition point. We cannot therefore decide whether the order parameter vanishes near to the transition point. For $J = -1$ there is a smooth change in $\langle S_i \rangle$ at the critical point. Increasing the antiferromagnetic J' the spiral magnetic order becomes weaker and vanishes at $J' \approx 1$. The underlying reason for that is local singlet formation, as discussed in section 4, and the continuous vanishing of the spiral order is therefore very similar to this second-order transition. However, the strength of J' needed for local singlet formation is much smaller due to the assisting effects of frustration (c.f., Ref. 15).

6 Conclusions

In this article we have investigated the zero-temperature phase transitions of a spin-half Heisenberg system on the square lattice. The main results of our treatment are: (i) Quantum fluctuations plus competition without frustration are able to destroy Néel LRO by local singlet formation. This is a pure quantum effect and has no classical counterpart. The control parameter is the strength of the competition and the breakdown of Néel ordering is accompanied by the opening of a spin gap. Standard SWT (even to higher orders) fails to describe this transition, whereas the CCM describes both the order parameter and the gap satisfactorily. Since we have no frustration, most standard techniques (e.g., QMC) are applicable and a quantitative description is possible. As was discussed in Ref. 17, the critical properties seem to correspond to the 3D classical Heisenberg model. (ii) Competition due to frustration was found to give more complex magnetic properties. In the model considered we have a second-order transition between collinear (antiferro- or ferro-magnetic) and noncollinear (spiral) states driven by frustration in the classical case. In the quantum spin-half model, standard techniques (e.g., QMC) are not applicable due to the violation of Marshall's sign rule. By contrast, the CCM provides a consistent description of collinear, noncollinear, and disordered phases.

Furthermore, we find a strong influence of quantum fluctuations on the nature of the collinear-noncollinear transition, and quantum fluctuations (which favour collinear ordering) may change the second-order classical transition to a first-order quantum transition. If quantum fluctuations are suppressed in the collinear phase, the transition to the spiral phase is similar for the quantum and classical models.

Acknowledgments

We thank the Deutsche Forschungsgemeinschaft (Ri 615/9-1) for its support.

References

1. S. Sachdev, *Quantum Phase Transitions* (Cambridge University Press 1999).
2. N. Mermin and H. Wagner, *Phys. Rev. Lett.* **17**, 1133 (1966).
3. E. Manousakis, *Rev. Mod. Phys.* **63**, 1 (1991).
4. P. W. Anderson, *Mater. Res. Bull.* **8**, 153 (1973); P. Fazekas and P. W. Anderson, *Phil. Mag.* **30**, 423 (1974).
5. J. Richter, N. B. Ivanov, and R. Retzlaff, *Europhys. Lett.* **25**, 545 (1994).
6. J. Richter, Phys. Rev. B **47**, 5794 (1993).
7. J. Oitmaa and Zheng Weihong, *Phys. Rev.* B **54**, 3022 (1996).
8. R. F. Bishop, D. J. J. Farnell, and J. B. Parkinson, *Phys. Rev.* B **58**, 6394 (1998).
9. L. Capriotti and S. Sorella, *Phys. Rev. Lett.* **84**, 3173 (2000).
10. Y. Xian, *Phys. Rev.* B **52**, 12485 (1995).
11. H. Niggemann, G. Uimin, and J. Zittartz, *J. Phys. C (Cond. Matt.)* **9**, 9031 (1997).
12. J . Richter, N. B. Ivanov, and J. Schulenburg, *J. Phys. C (Cond. Matt.)* **10**, 3635 (1998).
13. A. Koga et al., *Phys. Rev.* B **62**, 5558 (2000).
14. A. W. Sandvik and D. J. Scalapino, *Phys. Rev. Lett.* **72**, 2777 (1994).
15. C. Gros, W. Wenzel, and J. Richter, *Europhys. Lett.* **32**, 747 (1995).
16. M. Troyer, H. Kontani, and K. Ueda, *Phys. Rev. Lett.* **76**, 3822 (1996).
17. M. Troyer, M. Imada, and K. Ueda, *J. Phys. Soc. Jpn.* **66**, 2957 (1997).
18. H. Kageyama et al., *Phys. Rev. Lett.* **82**, 3168 (1999).
19. A. Koga and N. Kawakami, *Phys. Rev. Lett.* **84**, 4467 (2000).
20. R. F. Bishop in *Microscopic Many-Body Theories and Their Applications*, eds. J. Navarro and A. Polls, Lecture Notes in Physics Vol. **510** (Springer 1998).
21. R. F. Bishop, J. B. Parkinson, and Y. Xian, *Phys. Rev.* B **43**, 13782 (1991); *Phys. Rev.* B **44**, 9425 (1991).
22. R. F. Bishop, R. G. Hale, and Y. Xian, *Phys. Rev. Lett.* **73**, 3157 (1994).
23. C. Zeng, D. J. J. Farnell, and R. F. Bishop, J. Stat. Phys. **90**, 327 (1998).
24. R. F. Bishop, D. J. J. Farnell, and C. Zeng, *Phys. Rev.* B **59**, 1000 (1999).
25. S. E. Krüger, J. Richter, J. Schulenburg, D. J. J. Farnell, and R. F. Bishop, *Phys. Rev.* B **61**, 14607 (2000).
26. R. R. P. Singh, M. P. Gelfand, and D. A. Huse, *Phys. Rev. Lett.* **61**, 2484 (1988).

Quantum Fluids and Solids; Bose Condensation

PATH INTEGRAL MONTE CARLO CALCULATIONS OF SYMMETRY-BREAKING IN STRUCTURAL PHASE TRANSITIONS

K. A. GERNOTH
Department of Physics, UMIST, PO Box 88, Manchester M60 1QD, UK
E-mail: mccs7kag@dirac.phy.umist.ac.uk

This paper summarizes recent developments in theoretical Monte Carlo investigations of crystallographic point- and space-group symmetries in solids and of symmetry-breaking in the liquid-solid phase transition. The formal and numerical path integral Monte Carlo treatment of crystallographic symmetries and symmetry-breaking in the one- and two-body densities is outlined and the relation to experiment and to other theories is described.

1 Introduction

Monte Carlo (MC) techniques are one of the most powerful and widely applicable means of computing the microscopic and macroscopic statistical properties of a large variety of materials at zero and at non-zero temperatures from *ab initio* first principles. The physical systems that are being extensively studied with MC methods are as diverse as the liquid and solid phases of rare gas elements,[1–7] adsorbed films,[8–10] superfluid ^{4}He,[4,11] nuclei,[12] or polyethylene,[13,14] to name just a few examples. The path-integral Monte Carlo (PIMC) technique is especially well suited to carry out exact calculations of the behavior of bosonic quantum many-body systems at finite temperatures.[1,4,6,7] In the past the PIMC method has been applied with great success in particular to superfluid ^{4}He and other condensed phases of noble gas elements.[1–7]

It is a rewarding and timely undertaking to vigorously exploit the exact computational means furnished by the path integral Monte Carlo method to calculate also the perplexing systematic point and space group symmetry patterns in the microscopic spatial structure of materials. These symmetries are embodied at the atomistic level in the local one- and two-body densities and are brought to light by means of a thorough group-theoretical analysis of the formal mathematical structure of these quantities.[6] In combination with MC techniques such an analysis enables one to compute the one- and two-body densities also for materials with complex point and space groups and thereby gain valuable insights into the spatial microstructure. Furthermore, the quantities computed in such MC simulations may be easily related to X-ray diffraction and neutron scattering experiments (cf. Sec. 4). The numerical PIMC study of spontaneous crystallization that has been carried out recently[7] and which is described in more detail in Sec. 5 impressively demonstrates that the formal theoretical and the computational methods developed for treatment of point and space group symmetries in the one- and two-body densities provide also ideal means for investigating symmetry-breaking and irreducible order parameter symmetries in structural phase transitions. The exciting novel aspects of this type of analysis of phase transitions are briefly mentioned in Sec. 5. In the near future it is intended to employ the methods sketched in this introduction and described in more detail in the subsequent sections for studying a variety of condensed matter systems of current experimental and theoretical interest, such as, amongst others, condensed phases and structural phase transitions of adsorbed liquid and solid films and electron liquids in crystals. It may be expected that the past and future research in this area will turn out to become a valuable inspiration also for other theories and computational methods, such as, for example, the theoretical treatment of phase transitions in liquid crystals, molecular dynamics simulations of materials, and correlated wave functions and density matrix theories of quantum many-body systems.

2 The path integral Monte Carlo method

The PIMC approach commences with the Hamiltonian $\hat{H}$ of the system under consideration and is limited in its accuracy basically only by how well the interactions between the constituent particles of the system are known. Exploiting repeatedly the convolution property of the exact canonical density operator, the matrix elements $\mathcal{W}(\mathbf{R}, \mathbf{R}'; \beta)$ of the unnormalized canonical density operator $e^{-\beta \hat{H}}$ in coordinate space representation at a given temperature $T = 1/k_B\beta$, where k_B is the Boltzmann constant, may be cast in the form of a multi-dimensional integral of a product of density matrix elements at a higher, effective temperature. The multi-dimensional integral may be interpreted as an integral over all discrete paths, taken at a finite number of discrete imaginary time steps, in $3N$-dimensional configuration space (where N is the number of particles in the system) that connect configurations $\mathbf{R}$ and $\mathbf{R}'$. The art in discrete path integral Monte Carlo (DPIMC) methods is to keep on the one hand the Trotter number of intermediate integrations in the path integral representation of the exact density matrix as small as possible, but on the other to have it large enough, so that known exact expressions for the density matrices at the higher, effective temperature inside the multi-dimensional integral may be used. In DPIMC studies of superfluid ^{4}He,[4] for example, this is accomplished by writing the density matrices at the effective temperature in the form of products of two-body density matrices, which in turn are then computed exactly also by MC techniques.

In contrast to DPIMC methods, where the Trotter number is kept to the minimum that is necessary, the Fourier path integral Monte Carlo (FPIMC) technique[1,6,7] takes the opposite approach. In the limit of infinitely many intermediate integrations imaginary time becomes continuous and the effective temperature approaches infinity, so that the exact classical high-temperature limit can be inserted for the density matrices at the effective temperature in the exact path integral representation of the density matrix at the given temperature T.[15] The integral over all $3N$-dimensional paths from initial configuration $\mathbf{R}$ to final configuration $\mathbf{R}'$ turns into an infinite-dimensional integral over all $3N$-dimensional Fourier coefficients in the Fourier series representation of the paths as a function of continuous imaginary time.[15] In practical numerical applications the number of Fourier coefficients needs to be truncated at a finite number L. The partial averaging technique[1,6,16] allows one to partially take into account the effect of the higher-lying Fourier coefficients and thus achieve convergence of numerical results already with fewer Fourier coefficients than would be needed otherwise. In its final outcome the partial averaging technique amounts to replacing the bare interaction potential by an effective potential, which is of the form of a Gaussian transform of the bare potential, rather than simply cutting off everything in the original Fourier path integral representation of the density matrix after the first L Fourier coefficients.

The microscopic and macroscopic thermodynamic quantities, such as the one-body and two-body densities and the potential, kinetic, and total energy, may now be written in the form of expectation values of the relevant quantum-mechanical operators with respect to the Fourier path integral representation of the exact canonical density operator in the partial averaging approximation.[1,6,7] The multi-dimensional integrals which this procedure yields may be evaluated exactly by means of standard Monte Carlo sampling techniques.[1,6,7,17,18]

3 Path integral Monte Carlo explorations of crystallographic point and space group symmetries in solids

In a recent major work[6] a thorough formal analysis of crystallographic point and space group symmetries in the one- and two-body densities of crystalline quantum many-body systems at zero temperature and at non-zero temperatures has been undertaken. The formal space group-theoretical tools have been applied in exact FPIMC studies of solid argon in the hexagonal close-packed (hcp) structure at the experimental triple point temperature of 83.806 K. Although, with the notable exception of helium, the pure inert gas elements all crystallize in the face-centered cubic (fcc) phase, molecular oxygen and nitrogen impurities cause argon to crystallize in the hcp phase.[19–21] The lower the temperature, the larger an amount of impurities is needed to produce stable hcp Ar crystals. In the vicinity of the triple point temperature already minute impurity amounts of about 2% suffice for argon to form stable hcp crystals. For this reason, the pure hcp Ar solid studied in the above mentioned publication[6] comes at the triple point temperature closest to the experimental situation of hcp argon stabilized by oxygen or nitrogen impurities. Traces of hcp crystals have been found also in freshly frozen argon.[22,23] The hcp structure was observed also in thin neon films condensed on a substrate.[24] A further reason for investigating solid argon in the hcp phase is that the crystallographic space group $P6_3/mmc$ (D^4_{6h} in Schoenflies notation) of the hcp structure[25,26] is, owing to its non-symmorphic nature caused by the hexagonal 6_3 (genuine) screw axis, the more interesting space group to study than the symmorphic space groups $Fm\bar{3}m$ (O^5_h) and $Im\bar{3}m$ (O^9_h) of the fcc and body-centered cubic (bcc) phases, respectively. Furthermore, the numerical FPIMC results for crystalline Ar in the hcp phase provide also valuable guidance for the various two- and three-dimensional rare gas and molecular many-body systems to be investigated in future work.

By means of group-theoretical methods the one-body and two-body densities of a crystal of any crystallographic space group may be cast in the form of series representations in terms of complete sets of suitable symmetry-adapted Hilbert space basis functions.[6] For the two-body density this task is, in particular for non-symmorphic space groups, much more demanding than for the one-body density.

The general group-theoretical formalism has been applied to space group $P6_3/mmc$ of the hcp structure.[6] The one-body density $\varrho(\mathbf{x})$ turns out to be a Fourier series of symmetrized lattice waves and the Fourier coefficients are computed exactly with the FPIMC methods outlined in Sec. 2. The one-body density $\varrho(\mathbf{x})$ is then obtained by inserting the numerical results for the Fourier coefficients in its symmetrized Fourier series representation. The two-body density $\rho_2(\mathbf{x}_1, \mathbf{x}_2)$, which is the joint probability for a particle to be at a position $\mathbf{x}_1$ in three-dimensional real space and another one simultaneously at a position $\mathbf{x}_2$, is analyzed best in terms of the center-of-mass coordinates $\mathbf{S} = (\mathbf{x}_1 + \mathbf{x}_2)/2$ and the relative vector $\mathbf{r} = \mathbf{x}_1 - \mathbf{x}_2$. It may be written in the form of a reciprocal lattice Fourier series in the center-of-mass variable $\mathbf{S}$, where the Fourier coefficients now are actually functions of the relative vector $\mathbf{r}$ which have certain crystallographic symmetries and therefore may be expanded further in complete sets of symmetry-adapted Hilbert space basis functions. This procedure yields coefficient functions which are functions of the relative distance $r = |\mathbf{r}|$ and of the polar angle ϑ that the relative vector $\mathbf{r}$ encloses with the z-axis. These coefficient functions are computed exactly by means of the FPIMC strategy described in Sec. 2.

Although the emphasis in these FPIMC studies[6] of solid hcp argon is on the microscopic one-body and two-body densities, also the macroscopic thermodynamic quantities potential, kinetic, and total energy per particle are calculated. The partial averaging technique sketched in the preceding section is employed in these FPIMC computations. In their present stage FPIMC methods do not incorporate the Bose exchange symmetry, in contrast to DPIMC simulations of superfluid ^{4}He.[4] However, apart from superfluid ^{4}He, Bose exchange may be safely neglected in simulations of the condensed phases of all rare gas elements,[1-4,27] even in case of the solid phases of ^{4}He, unless one wants to calculate specifically Bose exchange rates.[4] Else, all other quantum-mechanical and quantum-statistical effects are exactly and fully included in the FPIMC method. In particular the Debye-Waller thermal broadening of the one-body density with increasing temperature is taken into account in these FPIMC calculations exactly with no further approximations or *a priori* assumptions made about the nature of the broadening. Quantum effects are found to be important even for the heavier inert gas elements up to temperatures of about 100 K.[2,3,6]

4 Relation to experiment (X-ray diffraction and neutron scattering) and to other theories

Beyond being quantities of physical interest in their own right, the one-body and two-body densities may be readily related to experimental analysis of the microscopic structure of materials. The Fourier coefficients of the one-body density $\varrho(\mathbf{x})$ are related to the amplitudes of diffraction of X-rays from crystals[6,28,29] and the lowest-lying Fourier coefficient functions of the two-body density $\rho_2(\mathbf{x}_1, \mathbf{x}_2)$ with reciprocal lattice vector $\mathbf{K} = 0$ to the differential cross section of nearly elastic scattering of neutrons.[6,30] The kinetic energy per particle, one of *the* macroscopic thermodynamic quantities which can be computed exactly with MC methods, may be measured experimentally by means of deep-inelastic scattering of neutrons from intense neutron spallation sources.[27,31,32] Thus the theoretical FPIMC results on the one-body and two-body densities may be directly compared to experimental results, where available. It may be expected that the formal space-group-theoretical tools that have been derived and tested numerically within the FPIMC approach in Refs. [6] and [7] will become valuable also for other Monte Carlo methods and for other microscopic quantum many-body theories, such as correlated basis functions (CBF) and correlated density matrix theories of quantum crystals and liquids.[33] The first attempts of applying these group-theoretical methods to the indications of softening of modes in a CBF treatment (as far as carried out so far) of the liquid-solid transition in ^{4}He and in the charged Bose gas appear to be very promising.[34,35] On the more technical side, fully exploiting the crystallographic point and space group symmetries in the one-body and two-body densities as dictated by the underlying crystalline structure of the material one is investigating significantly reduces the demands on CPU time, RAM memory, and disk space in MC simulations of crystals which calculate these densities.

5 Spontaneous crystallization of liquids

The symmetry-adapted series representations of the one-body and two-body densities[6,36,37] are also an extremely useful tool in analyzing the spontaneous crystallization of liquids.[7] The various (symmetry-breaking) terms in the symmetrized series representations of these quantities may be classified according to which irreducible representations (IR) of the high-symmetry space group of the uniform liquid they belong. The space group of the liquid phase is the extended Euclidean group[38] in three-dimensional space of all proper and improper rotations and translations and combinations thereof and is a continuous six-dimensional linear Lie group,[39] containing all crystallographic space groups as subgroups. When restricted to the low-symmetry crystallographic space group of the solid phase, the IRs of the high-symmetry Euclidean group of the liquid that are present in the symmetrized series representations of the one-body and two-body densities of the solid yield the trivial identity representation of the space group into which crystallization occurs. In this vein the group-theoretical formalism derived in Refs. [6] and [7] may be regarded as an extension of Landau theory of phase transitions[40,41] to 1) transitions from a continuous high-symmetry group and 2) to symmetry-breaking in the two-body density, whereas conventional Landau theory of phase transitions deals mainly with symmetry-breaking in the one-body density in phase transitions between crystalline structures, although generalizations of it to symmetry-breaking in the one-body density $\varrho(\mathbf{x})$ in solidification of liquids and in phase transitions of liquid crystals do exist.[40] It is important to note that it is a particular feature of the continuous nature of the extended Euclidean space group that the vast majority of its IRs that subduce the identity representation of the crystalline phase just do not appear in the Fourier series of the one-body density. The one-body density $\varrho(\mathbf{x})$ can see only certain ones of the possible symmetry-breaking IRs of the high-symmetry space group of the uniform liquid, whereas the two-body density $\rho_2(\mathbf{x}_1, \mathbf{x}_2)$ can see all of them. A study of symmetry-breaking in the two-body density in the spontaneous crystallization process is therefore warranted, since the one-body density might react to the phase transition only after symmetry-breaking has started already in the two-body density. This type of analysis bears the potential to elucidate and cast an entirely new light on the microscopic mechanisms driving spontaneous solidification and other phase transitions.

In recent research[7] the spontaneous crystallization of liquid argon at fixed temperature $T = 83.806\,\mathrm{K}$ into the hcp solid has been investigated by means of the FPIMC methods sketched in Sec. 2. At a bulk particle number density of $0.0170\,\text{Å}^{-3}$ an initial perfect hcp Ar crystal melts in the course of the simulation into the liquid, which is the stable phase of the simulated Ar many-body system at that bulk density and temperature. When one increases the bulk density to $0.0245\,\text{Å}^{-3}$ by suitably adjusting the size of the (hexagonal) simulation box, a single hcp Ar crystal gradually forms in the course of the simulation from the initial liquid, although there was no crystallization seed present in the initial liquid configuration. The liquid and solid phases are analyzed in terms of the space group symmetries in the one-body and two-body densities. Spontaneous crystallization of the Ar liquid into a single hcp crystal is established by monitoring the build-up of symmetry-breaking terms in the one-body and two-body densities and by taking snapshots during the FPIMC simulation of the absolute particle positions and of the positions of the Ar atoms relative to each other. In the final crystal the symmetry-breaking terms in the one-body and two-body densities exhibit exactly the systematic regularities which are characteristic of the space group $P6_3/mmc$ of the hcp structure. The FPIMC simulation of spontaneous crystallization of liquid natural argon[7] demonstrates that the group-theoretical formalism developed in Refs. [6] and [7] enables one to unambiguously establish that the Ar liquid does indeed solidify in a stable hcp crystal with space group $P6_3/mmc$.

Acknowledgments

This work is supported by United Kingdom EPSRC grant GR/M45429 "Microscopic quantum many-body calculations of complex structures" and the Graduiertenkollegs GRK 14 "Klassifizierung von Phasenumwandlungen kristalliner Stoffe aufgrund struktureller und physikalischer Anomalien" and GRK 549 "Azentrische Kristalle" at the University of Cologne in Germany.

References

1. J. D. Doll, D. L. Freeman, and T. L. Beck, *Adv. Chem. Phys.* **78**, 1990 (61).
2. A. Cuccoli, A. Macchi, M. Neumann, V. Tognetti, and R. Vaia, *Phys. Rev.* B **45**, 1992 (2088).
3. A. Cuccoli, A. Macchi, V. Tognetti, and R. Vaia, *Phys. Rev.* B **47**, 1993 (14923).
4. D. M. Ceperley, *Rev. Mod. Phys.* **67**, 1995 (279).
5. F. Pederiva, S. A. Vitiello, K. A. Gernoth, S. Fantoni, and L. Reatto, *Phys. Rev.* B **53**, 1996 (15129).
6. K. A. Gernoth, *Ann. Phys. (NY)* **285**, 2000 (61).
7. K. A. Gernoth, submitted to *Ann. Phys. (NY)*.

8. D. Marx, S. Sengupta, and P. Nielaba, *J. Chem. Phys.* **99**, 1993 (6031).
9. D. Marx, P. Nielaba, and K. Binder, *Phys. Rev.* B **47**, 1993 (7788).
10. M. Pierce and E. Manousakis, *Phys. Rev.* B **59**, 1999 (3802).
11. S. Moroni, D. E. Galli, S. Fantoni, and L. Reatto, *Phys. Rev.* B **58**, 1998 (909).
12. B. S. Pudliner, V. R. Pandharipande, J. Carlson, S. C. Pieper, and R. B. Wiringa, *Phys. Rev.* C **56**, 1997 (1720).
13. R. Martoňák, W. Paul, and K. Binder, *J. Chem. Phys.* **106**, 1997 (8918).
14. R. Martoňák, W. Paul, and K. Binder, *Phys. Rev.* E **57**, 1998 (2425).
15. R. P. Feynman and A. R. Hibbs, *Quantum Mechanics and Path Integrals* (McGraw-Hill, New York, 1965).
16. J. D. Doll, R. D. Coalson, and D. L. Freeman, *Phys. Rev. Lett.* **55**, 1985 (1).
17. N. Metropolis, A. W. Rosenbluth, M. N. Rosenbluth, A. H. Teller, and E. Teller, *J. Chem. Phys.* **21** (1953), 1087.
18. M. H. Kalos and P. A. Whitlock, *Monte Carlo Methods*, Vol. 1, (Wiley, New York, 1986).
19. C. S. Barret and L. Meyer, *J. Chem. Phys.* **42**, 1965 (107).
20. C. S. Barret, L. Meyer, and J. Wasserman, *J. Chem. Phys.* **44**, 1966 (998).
21. *Rare Gas Solids*, Vols. 1 and 2, Eds. M. L. Klein and J. A. Venables (Academic Press, London, 1976/77).
22. L. Meyer, C. S. Barret, and P. Haasen, *J. Chem. Phys.* **40**, 1964 (2744).
23. C. S. Barret and L. Meyer, *J. Chem. Phys.* **41**, 1964 (1078).
24. O. Bostanjoglo and R. Kleinschmidt, *Z. Naturf.* A **21**, 1966 (2106).
25. G. Burns and A. M. Glazer, *Space Groups for Solid State Scientists* (Academic Press, Boston, 1990).
26. *International Tables for Crystallography*, Vol. A, Ed. T. Hahn (Kluwer, Dordrecht, 1996).
27. D. N. Timms, A. C. Evans, M. Boninsegni, D. M. Ceperley, J. Meyers, and R. O. Simmons, *J. Phys.: Condens. Matter* **8**, 1996 (6665).
28. N. W. Ashcroft and N. D. Mermin, *Solid State Physics* (Holt, Rinehart and Winston, New York, 1976).
29. S. Elliott, *The Physics and Chemistry of Solids* (John Wiley&Sons, Chichester, 1998).
30. S. W. Lovesey, *Theory of Neutron Scattering from Condensed Matter*, Vol. 1 (Clarendon Press, Oxford, 1984).
31. D. A. Peek and R. O. Simmons, *J. Chem. Phys.* **94**, 1991 (3169).
32. D. A. Peek, M. C. Schmidt, I. Fujita, and R. O. Simmons, *Phys. Rev.* B **45**, 1992 (9671); 9680.
33. E. Krotscheck, in *Microscopic Quantum Many-Body Theories and Their Applications*, Lecture Notes in Physics, Vol. 510, Eds. J. Navarro and A. Polls (Springer-Verlag, Berlin, 1998), p. 187.
34. V. Apaja, J. Halinen, and M. Saarela, *J. Low. Temp. Phys.* **113**, 1998 (909).
35. J. Halinen, V. Apaja, K. A. Gernoth, and M. Saarela, *J. Low. Temp. Phys.* **121**, 2000 (531).

36. K. A. Gernoth, in *Proceedings of the 165. We-Heraeus-Seminar, Theory of Spin Lattices and Lattice Gauge Models*, Lecture Notes in Physics, Vol. 494, Eds. J. W. Clark and M. L. Ristig (Springer-Verlag, Berlin, 1997), p. 84.
37. K. A. Gernoth, in *Condensed Matter Theories*, Vol. 12, Eds. J. W. Clark and P. V. Panat (Nova Science Publishers, Commack, NY, 1997), p. 331.
38. W.-K. Tung, *Group Theory in Physics* (World Scientific, Singapore, 1985).
39. J. F. Cornwell, *Group Theory in Physics*, Vol. I, (Academic Press, London, 1994).
40. J. C. Tolédano and P. Tolédano, *The Landau Theory of Phase Transitions* (World Scientific, Singapore, 1987).
41. H. T. Stokes and D. M. Hatch, *Isotropy Subgroups of the 230 Crystallographic Space Groups* (World Scientific, Singapore, 1988).

THE DESCRIPTION OF STRONGLY INTERACTING SYSTEMS BASED ON JASTROW CORRELATIONS AND CONFIGURATION INTERACTION

R. GUARDIOLA

Departamento de Física Atómica y Nuclear, Universidad de Valencia, Avda. Dr. Moliner 50, E-46100 Burjassot, Spain
E-mail: Rafael.Guardiola@uv.es

J. NAVARRO

Instituto de Física Corpuscular, (CSIC-Universidad de Valencia), Edificio Institutos de Paterna, Apdo. 22085, E-46071 Valencia, Spain
E-mail: navarro@ific.uv.es

In this work we describe the structure of a self-adapting family of trial functions appropriate to describe variationally the ground state properties of finite many-body strongly-interacting systems. The trial functions are based on the interplay between the familiar two-body Jastrow ansatz and a special coordinate-space configuration-interaction expansion. To illustrate the method we present recent calculations of drops of ^{4}He and ^{3}He atoms.

1 Configuration interaction in coordinate space

About ten years ago it was discovered that the description of the ground state of a finite boson system based on a special configuration interaction scheme (with one and two particle excitations) is equivalent to a coordinate-space scheme using additive pair correlations.[1] This configuration interaction description is based on a *reference state* and a complete set of single-particle orbitals. In configuration space one usually uses creation and annihilation operators, $a^\dagger_\nu$ and a_ν respectively, with commutation relations

$$[a^\dagger_\nu, a^\dagger_\mu] = 0, \quad [a_\nu, a_\mu] = 0, \quad [a_\nu, a^\dagger_\mu] = \delta_{\nu\mu}.$$

Bosonic systems are particularly simple, because the *reference* state for an N particle system is

$$|\Phi\rangle = \left[a^\dagger_0\right]^N |0\rangle, \tag{1}$$

where 0 labels the lowest energy single-particle state. Excitations are generated by promoting one particle to a higher energy orbital (1p-1h), or two particles (2p-2h), and so on. In general, the np-nh excitation operator has the simple structure $a^\dagger_\mu a^\dagger_\nu \cdots a^\dagger_\rho a^n_0$. The analysis carried out in Ref. 1 paid special attention to construct *translationally invariant* wave functions, both for the reference state and the np-nh excitations. To this end it was found practical to use as single-particle basis the set of eigenstates of a harmonic oscillator hamiltonian. In this case, the (unnormalized) coordinate representation of the reference state is

$$\langle \boldsymbol{r}_1 \boldsymbol{r}_2 \cdots \boldsymbol{r}_N | \Phi \rangle = \prod_{i=1}^{N} e^{-\frac{1}{2}\alpha^2 r_i^2} \equiv e^{-\frac{1}{2}\alpha^2 R^2} \times \prod_{i<j}^{N} e^{-\frac{1}{2N}\alpha^2 r_{ij}^2}, \tag{2}$$

where $\alpha = \sqrt{m\omega/\hbar}$ is the inverse length oscillator parameter and $\boldsymbol{R}$ the center-of-mass coordinate. The second row of Eq. (2) shows that even if the reference state is not translationally invariant, it defines in a unique way a translationally invariant part. In this manner, one may properly deal with intrinsic coordinates in the single-particle representation.

When dressing the reference state with np-nh excitations one must take care that the center of mass does not get excited. In the case of 1p-1h and 2p-2h it was found in Ref. 1 that the most general excitation, rotationally and translationally invariant, in configuration space is

$$|\Psi\rangle = \left\{ 1 + \sum_{n=1}^{\infty} \mathcal{S}_n \sum_{n_1 n_2 l_1} \langle n0000|n_1 l_1 n_2 l_2 0\rangle [a^\dagger_{n_1 l_1} \times a^\dagger_{n_2 l_1}]^0_0 \, a^2_{000} \right\} |\Phi\rangle \,, \tag{3}$$

where the cross in the product of $a^\dagger$ operators means Clebsch-Gordan coupling and the Brody-Moshinsky bracket ensures that the center of mass remains in the $0s$ state. The amplitudes $\mathcal{S}_n$ are the quantities to be determined variationally in the configuration interaction scheme.

In coordinate space the corresponding wave function is

$$\Psi = \sum_{n=0}^{\infty} \mathcal{S}_n \sum_{i<j}^{N} \left[\frac{2^n n!}{(2n+1)!!} \right]^{1/2} L_n^{1/2}(\frac{1}{2}\alpha^2 r_{ij}^2) \prod_{i=1}^{N} \phi_0(r_i), \tag{4}$$

where $L_n^{1/2}$ are the Laguerre polynomials of the harmonic oscillator and $\phi_0(r_i)$ is the coordinate representation of the $0s$ harmonic oscillator state. In this equation the summation index n has been extended from 0, to include the 1 appearing in Eq. (3).

Because of the completeness of the Laguerre polynomials, Eq. (4) may be reinterpreted as

$$\Psi(\boldsymbol{r}_1, \ldots, \boldsymbol{r}_N) = \sum_{i<j} f(r_{ij}) \prod_{i=1}^{N} \phi_0(r_i), \tag{5}$$

where $f(r_{ij})$ is a general function, explicitly translational and rotational invariant (the only restriction is that the wave function must be normalizable). This is the translationally invariant configuration interaction (TICI) scheme, at order 2, found in Ref. 1. It represents the most general 1p-1h plus 2p-2h configuration excitation, where all numeric amplitudes S_n are included in the pair correlation function $f(r)$.

At this point one may abandon the harmonic oscillator basis, and define a CI-2 ansatz in the general form

$$\Psi(\boldsymbol{r}_1, \ldots, \boldsymbol{r}_N) = \sum_{i<j} f(r_{ij}) \Phi_{\mathrm{REF}}(\boldsymbol{r}_1, \ldots, \boldsymbol{r}_N), \tag{6}$$

where now Φ_{REF} must be understood as a general (but translationally invariant) reference state. One may further generalize the CI-2 ansatz Eq. (6) to a CI-3,

$$\Psi(\boldsymbol{r}_1, \ldots, \boldsymbol{r}_N) = \sum_{i<j<k} f(r_{ij}, r_{ik}, r_{jk}) \Phi_{\mathrm{REF}}(\boldsymbol{r}_1, \ldots, \boldsymbol{r}_N), \tag{7}$$

where now $f(s,t,u)$ is symmetric in the three triangular coordinates s, t and u.

The CI-2 scheme has been used to determine nuclear properties for semirealistic interactions in configuration space,[1] in coordinate space after a gaussian expansion of the pair correlation[2] and a full variational solution after obtaining and solving the Euler-Lagrange equation for the pair correlation.[3] Comparison with diffusion Monte Carlo results[4] showed that CI-2 was a very good approximation to the ground state energy, but that higher order correlations were not negligible.

More interesting for our future purposes is the check that the calculations carried out with the gaussian expansion are indistinguishable from the Euler-Lagrange calculations, thus indicating the adequacy of the expansion in this non-orthogonal basis.

It should be mentioned that this way of carrying out the interaction of configurations is quite common in atomic physics. The equivalent to the CI-2 approximation is referred in quantum chemistry as including *singles and doubles* correlations. However, it is very different of the common way in nuclear physics, where currently the excited configurations fill the next major shell, in all possible np-nh forms. This procedure is justified when one uses effective interactions, but is not convenient when using a bare nucleon-nucleon interaction.

A final comment regards the case of fermionic systems. The reference state is not as simple as Eq. (1) and Pauli's exclusion principle requires to have several shells occupied. In consequence there may be several ways of exciting particles and, for example, several $f(r)$ functions in the CI-2 scheme. Just to clarify this point one may consider that the pair function exciting spin parallel particles should be different of the pair function for spin antiparallel. Even more, to have full singles and doubles excitations one should take into account the shells from which particles are promoted. A partial remedy to this limitation of the simple CI scheme is to assign an operatorial dependence to the excitation operator, such as spin dependence, or isospin dependence and so on. This generalization was carried out in Ref. 5 (spin/isospin dependence) and in Ref. 6 (tensor correlations).

2 The Jastrow plus Configuration Interaction scheme

The simple CI scheme is not able to deal properly with strongly interacting systems, specifically with interactions with a strong short-range repulsion. A natural way of extending its applicability is to combine the self-adapting character of the CI scheme with the ability of Jastrow correlations of controlling the short-range repulsion of many realistic interactions. This is the Jastrow plus configuration interaction (J-CI) scheme, already introduced in Ref. 6 and applied there to a very light system.

Our present view of the J-CI scheme follows the simple rule of *divide and conquer*. The variational trial function is constructed with three basic pieces. There is a *reference* part, denoted by $\Phi_{\rm REF}(\boldsymbol{R})$, where $\boldsymbol{R}$ represents the set of all constituent coordinates, with the basic role of incorporating the symmetry or antisymmetry of the system as well as to describe the basic properties (size, angular momentum quantum numbers, ...). There is a two-body truly short-range Jastrow correlation factor, basically related to the interaction of the constituents at short distances, denoted by $F_{\rm J}(\boldsymbol{R}) = \prod_{i<j}^{N} f_{\rm J}(r_{ij})$, and finally the CI part. Altogether, the wave

function reads

$$\Psi(\boldsymbol{R}) = F_{\rm CI}(\boldsymbol{R})F_{\rm J}(\boldsymbol{R})\Phi_{\rm REF}(\boldsymbol{R}). \tag{8}$$

In the case of three-particle CI correlations the factor $F_{\rm CI}$ may be written as in Eq. (7) in terms of a fully symmetric function $f_{\rm CI}(r_{ij}, r_{ik}, r_{jk})$ to be determined variationally. To this end, our previous experience with the expansion in Gaussians suggests to parametrize the CI3 correlation as

$$F_{CI}(\boldsymbol{R}) = \sum_{\nu} C_\nu G_\nu(\boldsymbol{R}), \tag{9}$$

where we have introduced a set of gaussian functions G_ν given by

$$G_\nu(\boldsymbol{R}) = \sum_{i<j<k} \mathcal{S}\{g_p(r_{ij})g_q(r_{ik})g_r(r_{jk}\}. \tag{10}$$

In this equation the index ν is equivalent to the triplet $\{p, q, r\}$ and $g_p(r) = \exp[-\beta_p r^2]$. Once a set of exponents β is chosen, there remains only to compute the linear amplitudes C_ν. Note that if one of the exponents equals zero, say $\beta_1 = 0$, the CI correlation includes the non-correlated term ($p = q = r = 1$), pair correlations ($p = q = 1$, but $r \neq 1$), and truly triplet correlations (all p, q and r greater than 1). The set of parameters required to fully specify the variational trial function contains those required to specify the reference state (for example, a harmonic oscillator parameter to fix the set of single-particle wave functions), the parameters related to the Jastrow correlation and the sets $\{C_\nu\}$ and $\{\beta_p\}$ related to the linear CI correlations. In practice, there are many fewer parameters than expected: the Jastrow two-body correlation may be obtained by solving the two-particle Schrödinger equation at short distances with an adequate healing condition. Regarding the β exponents our own experience indicates that their values are not much relevant, as far as they are limited to medium- and long-range correlations and that the normalizability of the trial function is not spoiled. There remain actually two kinds of parameters: the *non-linear* harmonic oscillator (or equivalent) parameter, which must be explicitly minimized and the *linear* C_ν amplitudes, which are obtained by solving a generalized eigenvalue problem

$$\mathcal{H}_{\mu\nu} C_\nu = E \mathcal{N}_{\mu\nu} C_\nu, \tag{11}$$

where the normalization ($\mathcal{N}$) and hamiltonian ($\mathcal{H}$) matrices are given by

$$\mathcal{N}_{\mu\nu} = \int d\boldsymbol{R} \Phi^*_{\rm REF}(\boldsymbol{R}) F^*_J(\boldsymbol{R}) G^*_\mu(\boldsymbol{R}) G_\nu(\boldsymbol{R}) \Phi_{\rm REF}(\boldsymbol{R}) F_J(\boldsymbol{R}),$$

and

$$\mathcal{H}_{\mu\nu} = \int d\boldsymbol{R} \Phi^*_{\rm REF}(\boldsymbol{R}) F^*_J(\boldsymbol{R}) G^*_\mu(\boldsymbol{R})\ H\ G_\nu(\boldsymbol{R}) \Phi_{\rm REF}(\boldsymbol{R}) F_J(\boldsymbol{R}).$$

With the exception of very light systems, the evaluation of the norm and hamiltonian matrices cannot be done analytically, because of the presence of Jastrow correlations. However, they may be computed by means of Monte Carlo methods. The algorithm consists in running a Metropolis random walk, having as transition probability function the positive definite quantity

$$W(\boldsymbol{R}) = |\Phi_{\rm REF}(\boldsymbol{R}) F_J(\boldsymbol{R})|^2, \tag{12}$$

i.e., generating random positions $\{\boldsymbol{R}_i\}$ corresponding to the probability distribution function $W(\boldsymbol{R})$. With the same random vectors one computes the norm matrix by appealing to the central limit theorem

$$\mathcal{N}_{\mu\nu} = \lim_{N\to\infty} \frac{1}{N} \sum_i G_\mu^*(\boldsymbol{R}_i) G_\nu(\boldsymbol{R}_i), \tag{13}$$

resulting in a *positive definite* norm matrix. An analogous procedure is followed for the evaluation of the hamiltonian matrix, this time sampling the quantity

$$\mathcal{H}_{\mu\nu} = \lim_{N\to\infty} \frac{1}{N} \sum_i G_\mu^*(\boldsymbol{R}_i) \frac{1}{G_\nu(\boldsymbol{R}_i) F_J(\boldsymbol{R}_i) \Phi_{\rm REF}(\boldsymbol{R}_i)}\ H\ G_\nu(\boldsymbol{R}_i) F_J(\boldsymbol{R}_i) \Phi_{\rm REF}(\boldsymbol{R}_i). \tag{14}$$

Usually the norm matrix is badly conditioned, having eigenvalues close to zero, which indicates the existence of poorly defined orthogonal basis vectors. It is then very convenient to carry out a singular value analysis of the norm matrix, removing all those eigenvectors which are obtained with large errors. Moreover, it is also recommended to re-run the random walk, once the linear amplitudes C_μ have been determined.

3 Drops made of helium atoms

Helium atoms may stick together to form a *drop*, i.e., a bound system. One may consider drops made of only ^{4}He atoms (a bosonic system), or made of ^{3}He atoms (a fermionic system) and finally drops with a mixture of the two species. The existence of these drops has been proved by means of molecular beam diffraction from a transmission grating.[8] With this procedure one has been able to detect and identify the bosonic dimer $^4He_2$ and trimer $^4He_3$[9] as well as small mixed systems.[10] This experimental device has even permitted the measurement of the dimer binding energy,[11] with a value around 1 mK.

The He–He interaction has been derived almost *ab initio*. For many years the *old* Aziz potential HFDHE2[12] was very extensively used to describe both bosonic and fermionic liquids. A further elaboration of the potential produced the *new* Aziz potential, HFD-B(HE),[13] which has been used in Quantum Monte Carlo calculations, giving values for the binding energy per particle of the bosonic[14] and the fermionic[15] liquid in very good agreement with the experimental values.

From the point of view of many-body physics, clusters of He are rather simple systems, because the interaction is simple. Particularly, in the fermion case there are no spin nor tensor forces, thus avoiding a non-trivial operatorial problem. However, the algebraic simplicity is widely compensated by the strength of the forces: actually, systems made of He atoms are real strongly interacting systems, with a very strong short-range repulsion an a rather long tail. With the exception of the very light drops, they are very dense, much more than nuclear matter or finite nuclei. In consequence, choosing as illustrative examples of our J-CI scheme the various drops of He corresponds to carrying out an astringent test.

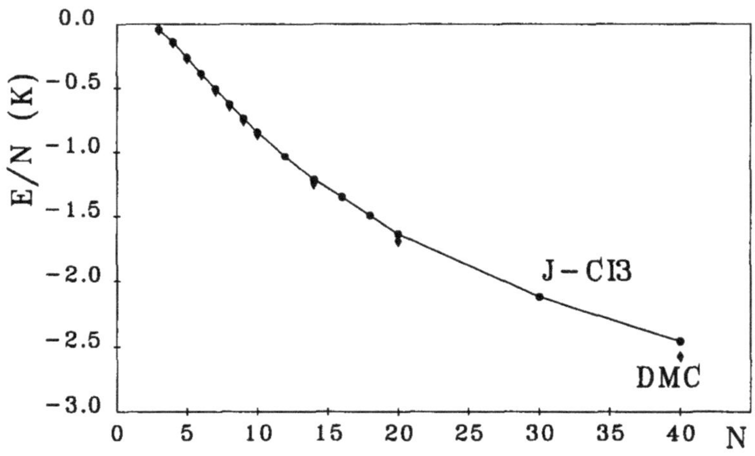

Figure 1. Ground state energies per particle, in K, as a function of the number of constituents of ^{4}He drops. Diamonds correspond to DMC calculations, and circles to J-CI variational calculations.

3.1 Drops of ^{4}He

The study of boson drops is very convenient, because as far as there exists exact Diffusion Monte Carlo (DMC) calculations for drops of various numbers of constituents one may test the goodness of the gaussian expansion as well as to check the statistical error of the Monte Carlo sampling. In other words, the comparison of our approach with these exact (within statistical precision) results serves to ascertain the performance and the limitation of our scheme.

Figure 1 compares our variational upper bounds[16] for the energy per particle with the values obtained with the DMC method by Lewerenz[17] ($N \leq 14$) and by Whaley.[18] All calculations consider the Aziz potential HFD-B(HE).[13] For small systems our results are indistinguishable from the DMC calculations, but the agreement is not as good in the case of heavy ($N = 40$) drops. Even if the difference between our upper bonds and the exact calculations is small (near a 2%) this failure reveals the main limitation of the J-CI method, namely the improper behavior when increasing the number of constituents. Using the usual many-body terminology, we recognize the lack of *size-extensivity* of the CI method.

The J-CI calculation was done with a gaussian reference wave function $\Phi_{\rm REF} = \prod_{i<j}^{N} \exp(-\alpha^2 r_{ij}^2/2N)$, a McMillan form for the Jastrow pair correlation, $f_{\rm J}(r) = \exp(-(b/r)^{\nu}/2)$ and a set of six Gaussians to describe the additive pair and triplet correlations. Specific details of our J-CI calculation may be found in Ref. 16. It should be mentioned that the Jastrow factor is independent of the number of particles, since its role is mainly to control the short-range repulsion of the potential, and that harmonic oscillator constant α is thus the only (non-linear) adjustable parameter.

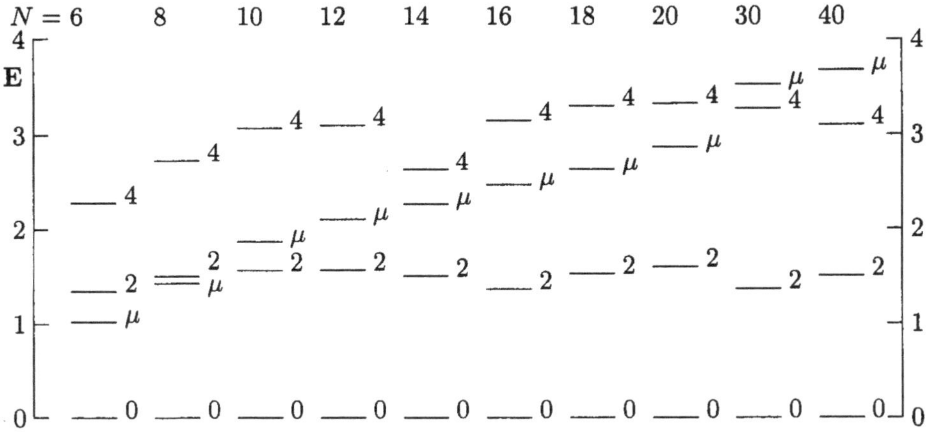

Figure 2. The vibrational excitation spectrum (in K) of ^{4}He drops. For each number of constituents the levels are labeled with their angular momentum. The line labeled with μ is the limit of bound states (the chemical potential).

3.2 Excited levels of ^{4}He drops

The excitation of bosonic drops may be studied in several ways. Chin and Krotscheck[19] studied the $L = 0$ and $L = 2$ excitations by means of the Feynman procedure[20] to deal with collective excitations of quantum liquids. Its method consists in functional minimization of an excitation trial function $F(\boldsymbol{R})$ which acts on the DMC determined exact ground state. Krisna and Whaley[21] had previously done a similar treatment, but using a specific Gram-Schmidt orthogonalized basis.

Our approach[22] is different, and limited to the lowest energy state of a given angular momentum L band. The ground state ($L = 0$) wave function is modified by multiplying with a symmetric operator

$$F_L(\boldsymbol{R}) = \sum_{i<j} r_{ij}^L Y_{L0}(\Omega_{ij}) \tag{15}$$

and the linear configuration-interaction amplitudes are re-obtained by solving the new generalized eigenvalue problem. The excitation energies are approximated by the difference of the ground state upper bound ($L = 0$) and the new ($L = 2, 4$) upper bounds. Figure 2 presents the excitation energies for drops of several number of atoms. It is quite appealing that the excitation energies are fairly independent of the number of atoms. The spectrum has some resemblance with the nuclear or molecular vibrational structure, but the troubles in obtaining higher $L = 0$ and $L = 2$ levels within the variational scheme does not permit a firm conclusion on the presumably vibrational nature of the excitation. The figure shows also the binding threshold (the chemical potential), and the levels above this threshold have no physical significance. We have studied also these levels by means of the *energy weighted sum rules* and our results are in a fairly good agreement with the previous calculations of Chin and Krotscheck.[19]

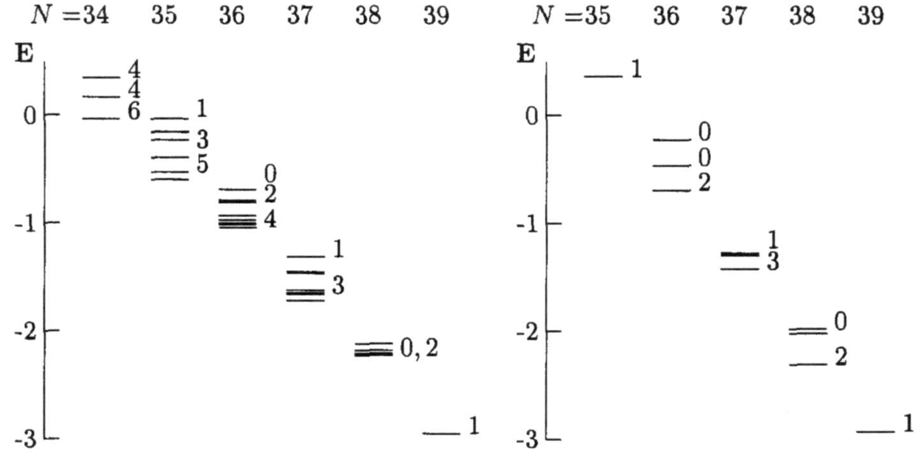

Figure 3. The level scheme (in K) of drops of ^{3}He from $N = 34$ to $N = 39$ constituents. The left figure represents some of the lowest levels related to the configuration $2p^6 1f^n$, for $n = 8$ to 14 (active f−shell), and the right figure to the configuration $2p^n 1f^{14}$ (active p−shell), with the $N = 20$ core $1s^2 1p^6 2s^2 1d^{10}$. Levels are grouped according to the value of the spin, and labeled with twice its value.

3.3 Fermionic drops

Apart from our own work, there is only one paper dealing with the microscopic description of drops of ^{3}He.[23] There have been, however, several papers using a density-functional-theory framework,[24–26] particularly the work by Barranco *et al*,[27] which have explored several relevant questions such as which is the minimum number of constituents to form a bound system, or the (L, S) ordering of the possible states for a given number of constituents.

Drops of ^{3}He differ from drops of ^{4}He in two important points: first, the antisymmetry related to the fermionic character of the constituents and second the mass difference. Otherwise, the basic interaction is the same for both kinds of drops. Pioneering microscopic studies by Pandharipande *et al* found that the system with 40 atoms was bound, but the system with 20 atoms was not bound. These are the *magic* numbers corresponding to a harmonic oscillator ordering of levels, the major shell corresponding to the configuration $1s^2 1p^6 2s^2 1d^1 0$, for $N = 20$ atoms, and the filling of $2p^6 1f^1 4$ for $N = 40$ atoms. In the analysis of Barranco and coworkers[27] in a non-local finite-range density functional theory, allowing for the $(2p,1f)$ configuration mixing, it was found that even a system with 29 atoms could give raise to a bound system. Moreover, they found that the lowest energy states were those with a maximum spin of the active shell(s), compatible with Pauli exclusion principle. Note that this does not mean that the fermionic fluid is *fully* polarized: actually, there is a spin-saturated *core* of 20 particles, but particles in the active shell prefer to align as much as possible their spins.

We have analyzed[28, 29] these fermionic systems within our J-CI approach. The variational wave function has a Slater-determinant part (in many cases a linear combination of Slater determinants) constructed with the harmonic oscillator single-particle states, the Jastrow correlation factor of the same structure as in the bosonic case, and the self-adjustable configuration interaction three-body correlation. In addition, the Slater determinant has been correlated with the so called ***backflow*** correction, which basically modulates the shape and position of the nodal surfaces. The study was carried out initially in a cartesian-like single-particle basis[28] and afterwards it was reformulated in the appropriate LS (Russell-Saunders) coupling scheme,[29] with good orbital and spin angular momentum quantum numbers.

Our microscopic calculations result in an important improvement with respect to the previous one.[23] Moreover, they partially confirm the results obtained in the density functional formalism, even if our calculations cannot have the full generality related to the $2p - 1f$ mixing. We have found an upper bound to the minimum number of constituents required to form a bound state, $N = 35$, significantly larger then the value of 29 obtained in.[27] We have also found a shell ordering, the system preferring to fill up first the $2p$ shell, and afterwards the $1f$shell. Moreover, we have found that the lowest energy terms are those with maximum spin, and that the energy of levels is almost L independent.

4 Final remarks

The combination of Jastrow correlations with our special form of configuration interaction has proven to be a general scheme to deal with strongly interacting systems. The main advantage of the proposed family of trial functions is their *self-adaptability*, avoiding the need of minimizing with respect to a large number of parameters. In the examples presented here there is just one single parameter, the harmonic oscillator parameter α which must be explicitly minimized. The linear CI amplitudes result from the solution of a simple generalized eigenvalue problem.

Within our scheme, the role of Jastrow correlations is limited to the very short-range part of the interaction. Actually, it may be found by solving the two-body problem at short distances with adequate healing conditions. Another significant piece, the reference wave function, must be tailored to account for the general properties of the system, such as size, angular momentum quantum numbers, statistics and so on. Finally, a great advantage results from the use of a self-adaptive linear correlation, including two- and three-body pieces. The method, however, is computationally costly, mainly because of the need of determining very accurately the normalization matrix. Here it is appropriate to mention that the use of a gaussian basis is not the most convenient one. We followed the inertia of our uncorrelated TICI scheme, where all calculations were carried out semi-numerically in the gaussian basis. However, as far as we must use a Monte Carlo algorithm, we are no longer tied to this basis. Preliminary calculations with functions with exponential tail shows an improvement in the convergence, specially in light systems.

Acknowledgments This work is supported by DGESIC (Spain) under contract Nb. PB97–1139.

References

1. R. F. Bishop, M. F. Flynn, M. C. Boscá, E. Buendía and R. Guardiola, *Phys. Rev.* C **42**, 1341 (1990).
2. R. F. Bishop, E. Buendía, M. F. Flynn and R. Guardiola, *J. Phys. G (Nucl. Part. Phys.)* **17**, 857 (1991)
3. R. F. Bishop, E. Buendía, M. F. Flynn and R. Guardiola, *J. Phys. G (Nucl. Part. Phys.)* **18**, 1157 (1992).
4. R. F. Bishop, E. Buendía, M. F. Flynn and R. Guardiola, *J. Phys. G (Nucl. Part. Phys.)* **18**, L21 (1992); *ibid.* **19**, 1163 (1993).
5. R. Guardiola, P. I. Moliner, J. Navarro, R. F. Bishop, A. Puente and Niels R. Walet, *Nucl. Phys.* A **609**, 218 (1996).
6. R. F. Bishop, R. Guardiola, I. Moliner, J. Navarro, M. Portesi, A. Puente and N. R. Walet, *Nucl. Phys.* A **643**, 243 (1998).
7. E. Feenberg and J. W. Clark, *Phys. Rev.* **113**, 388 (1959); E. Feenberg and C. W. Woo, *Phys. Rev.* **137**, 391 (1965); J. W. Clark and P. Westahus, *Phys. Rev.* **141**, 833 (1966); E. Krotscheck and J. W. Clark, *Nucl. Phys.* A **328**, 73 (1979).
8. W. Schoellkopf and J. P. Toennies, *Science* **256**, 1345 (1993).
9. W. Schoellkopf and J. P. Toennies, *J. Chem. Phys.* **104**, 1155 (1996).
10. W. Schoellkopf and J. P. Toennies, private communication.
11. R. E. Grissenti, W. Schöllkopf, J. P. Toennies, G. C. Hegerfeldt, T. Köhler and M. Stoll, *Phys. Rev. Lett.* **85**, 2284 (2000).
12. R. A. Aziz, V. P. S. Nain, J. S. Carley, W. L. Taylor and G. T. McConville, *J. Chem. Phys.* **70**, 4330 (1976)
13. R. A. Aziz, F. R. McCourt, and C. C. K. Wong, *Mol. Phys.* **61**, 1487 (1987).
14. J. Boronat and J. Casulleras, *Phys. Rev.* B **49**, 8920 (1994).
15. J. Casulleras and J. Boronat, *Phys. Rev. Lett.* **84**, 3121 (2000).
16. R. Guardiola, J. Navarro, and M. Portesi, *Phys. Rev.* B **60**, 6288 (1999).
17. M. Lewerenz, *J. Chem. Phys.* **106**, 4596 (1997);
18. K. B. Whaley, *Int. Rev. Phys. Chem.* **13**, 41 (1994).
19. S. A. Chin and E. Krotscheck, *Phys. Rev.* B **45**, 852 (1992).
20. R. P. Feynman, *Phys. Rev.* **94**, 262 (1954).
21. M. V. Rama Krisna and K. B. Whaley*J. Chem. Phys.* **93**, 6738 (1990).
22. R. Guardiola, J. Navarro and M. Portesi, Con. Matter Theories (2001), in press.
23. V. R. Pandharipande, S. C. Pieper and R. B. Wiringa, *Phys. Rev.* B **34**, 4571 (1986).
24. S. Stringari and J. Treiner, *J. Chem. Phys.* **87**, 5021 (1987).
25. S. Weisberger and P. G. Reinhard Z. Phys. D **23**, 275 (1992).
26. M. Barranco, D. M. Jezek, E. S. Hernández, J. Navarro and Ll. Serra, *Z. Phys.* D **28**, 257 (1993).
27. M. Barranco, J. Navarro and A. Poves, *Phys. Rev. Lett.* **78**, 4729 (1997).
28. R. Guardiola and J. Navarro, *Phys. Rev. Lett.* **84**, 1144 (2000).
29. R. Guardiola, *Phys. Rev.* B **62**, 3416 (2000).

THE MANY-BOSON SYSTEM IN ONE-DIMENSION: APPLICATION TO ^{4}HE

M. D. MILLER
Department of Physics, Washington State University, Pullman, WA 99164-2814, USA
E-mail: mdm@wsu.edu

E. KROTSCHECK
Institut für Theoretische Physik, Johannes Kepler Universität, A4040 Linz, Austria
E-mail: kro@grizzly.tphys.uni-linz.ac.at

For three model systems: hard rods, the Morse potential, and the Lennard-Jones 6-12 potential, we calculate equations of state and the low-lying excitations using non-perturbative variational theory. We examine the effect of dimerization on the many-body equation of state and argue that dimerization is manifested as the appearance of a many-body bound state. We show that in one dimension ^{4}He at absolute zero is just barely self-bound with a binding energy of 0.002 K at a density of 0.036 Å^{-1}. We calculate the Feynman excitation spectrum and we demonstrate the presence of strong anomalous dispersion in the phonon regime. Finally, we introduce ^{3}He impurities and calculate the zero concentration chemical potential, the ^{3}He-^{3}He effective interaction in the ^{4}He background and the energy of dimerization.

1 Introduction

In recent years, there has been increasing interest in the possibility of studying ^{4}He in a quasi one-dimensional environment. Initial experiments have been reported by Wada and coworkers[1] in a material they denoted as FSM-16. The substance consists of a honeycomb of hexagonally shaped tubes approximately 18 Å in diameter with an undisclosed length. The geometry of this system is thus characterized by *tubes* with one macroscopic spatial dimension and two transverse microscopic spatial dimensions. This type of system is termed *quasi* one-dimensional. As recently suggested by Cole and coworkers,[2] buckytubes may provide a useful matrix for experiments. In addition, substances with mesoporous substrates like K-L zeolite may also be viable.[3,4]

Some years ago, the properties of one-dimensional many-boson systems were the object of intense theoretical scrutiny.[5,6] Exact solutions were obtained for model systems such as the repulsive delta-function potential,[7–9] an attractive inverse-square model,[10] and a model of an anharmonic lattice.[11] As discussed by Sutherland,[12] the common thread connecting these various solutions is that exact or asymptotic solutions are given by Bethe's ansatz.

In this note we shall report on the application of the Jastrow-Feenberg variational method to calculate the static and dynamic properties of three model systems (hard rods, Morse potential, and Lennard-Jones potential) and ^{4}He (Aziz potential) in strictly one dimension. An exact theory with tubular geometry can be treated microscopically within the extension of the Jastrow-Feenberg variational method to inhomogeneous geometries.[13,14]

2 Jastrow-Feenberg Theory

The many-boson Hamiltonian in one dimension can be written as

$$H = -\frac{\hbar^2}{2m}\sum_{i=1}^{N}\frac{d^2}{dz_i^2} + \sum_{i<j} v(|z_i - z_j|). \tag{1}$$

The system consists of N bosons which uniformly fill a tube of length L. The particles in the tube are treated as a one-dimensional system with pair potential $v(|z_i - z_j|)$.

The ground-state wave function is written as a variational ansatz of the Jastrow-Feenberg form:

$$\Psi_0\left(z_1, ..., z_N\right) = \exp\left[\frac{1}{2}\left\{\sum_{i<j} u_2(z_i, z_j) + \sum_{i<j<k} u_3(z_i, z_j, z_k) + \dots\right\}\right]. \tag{2}$$

The most important component of the variational wave function is the two-body function $u_2(z_i, z_j)$, which describes both the short- and long-range correlations between pairs of particles.

An important aspect of the variational theory is the optimization of the correlations. The two-body and three-body functions are determined by the minimization of the energy-expectation value, formally written as

$$\frac{\delta}{\delta u_n}\left[\frac{\langle\Psi_0|H|\Psi_0\rangle}{\langle\Psi_0|\Psi_0\rangle}\right] = 0\,, \qquad n = 2, 3\,. \tag{3}$$

The additional information needed to solve these equations is the connection between the n-body functions and the physically observable distribution functions. This connection is provided by the hypernetted-chain (HNC) equations.[17] These equations are derived by diagrammatic analysis of the two-body distribution function $g(z)$ in terms of the two-body function. The analysis leads to the HNC relationships

$$g(z) = \exp\left[u_2(z) + N(z) + E(z)\right]. \tag{4}$$

The function $E(z)$ represents an infinite series of "elementary" diagrams which can be expressed as multi-dimensional integrals involving $g(z)$. The sum of nodal diagrams, $N(z)$, can be expressed conveniently in momentum space. Introducing the dimensionless Fourier transform

$$\tilde{f}(k) = \rho\int dz f(z)e^{ikz}\,, \tag{5}$$

where ρ denotes the linear density, and the static structure function

$$S(k) = 1 + \rho\int_{-\infty}^{\infty} dz e^{ikz}[g(z) - 1]\,, \tag{6}$$

the function $\tilde{N}(k)$ has the form

$$\tilde{N}(k) = \frac{[S(k) - 1]^2}{S(k)} = \frac{\tilde{X}^2(k)}{1 - \tilde{X}(k)}\,, \tag{7}$$

where we have also introduced, for further reference, the "direct correlation function" or the "non-nodal function"

$$\tilde{X}(k) = S(k) - 1 - \tilde{N}(k)\,. \tag{8}$$

The level of the HNC approximation is defined by the choice of $E(z)$; e.g., HNC/0 neglects the elementary diagrams altogether, note that triplet correlations can be implemented through a modification of the definition of "elementary diagrams". The combination of the HNC equations (4) and the Euler equations (3) are generally referred to as the hypernetted-chain Euler-Lagrange (HNC-EL) theory. With the quantities introduced above, the correlation energy can be written as

$$E = E_r + E_k + E_e + E_3 \tag{9}$$

with

$$\frac{E_r}{N} = \frac{\rho}{2}\int_{-\infty}^{\infty} dz\left[g(z)v(z) + \frac{\hbar^2}{m}\left|\frac{d}{dz}\sqrt{g(z)}\right|^2\right], \tag{10}$$

$$\frac{E_k}{N} = -\frac{1}{4}\int_{-\infty}^{\infty}\frac{dk}{2\pi\rho}t(k)(S(k)-1)\tilde{N}(k)\,, \tag{11}$$

$$\frac{E_e}{N} = -\frac{1}{4}\int_{-\infty}^{\infty}\frac{dk}{2\pi\rho}t(k)(S(k)-1)\tilde{E}(k)\,, \tag{12}$$

where $t(k) = \hbar^2 k^2/2m$ is the kinetic energy of a free particle, E_3 is the contribution from triplet correlations. E_3 can be expressed in terms of the three-body correlation function $u_3(z_1, z_2, z_3)$ and the three-body distribution function. Our working formulas[18–20] have turned out to be quite sufficient for all practical purposes.

The details of the HNC-EL scheme have been discussed in many contexts, for a comprehensive review of the method see Ref. 21. The Euler-Lagrange equation can be conveniently written in coordinate space for the radial distribution function,

$$-\frac{\hbar^2}{m}\frac{d^2}{dz^2}\sqrt{g(z)} + \left[v(z) + \Delta V_{\rm ele}(z) + w_{\rm I}(z)\right]\sqrt{g(z)} = 0\,, \tag{13}$$

where the "induced interaction" is

$$\begin{aligned}\tilde{w}_{\rm I}(k) &= -t(k)\left[S(k)-1\right] - \frac{1}{2}t(k)\left[\frac{1}{S^2(k)} - 1\right]\\ &= -t(k)\left[S(k)-1\right] - \tilde{V}_{\rm p-h}(k)\,,\end{aligned} \tag{14}$$

and $\Delta V_{\rm ele}(z)$ is a term that arises from triplet correlations and elementary diagrams,

$$\Delta V_{\rm ele}(z) = \frac{2}{N\rho}\frac{\delta(E_3 + E_e)}{\delta g(z)}\,. \tag{15}$$

The coordinate-space formulation of the Euler equation (13) is readily identified with the boson Bethe-Goldstone equation, which sums the dominant diagrams in the strong-coupling limit.

A momentum space formulation of the Euler equations equivalent to (13) can be given in terms of the structure factor $S(k)$,

$$S(k) = \left[1 + \frac{2}{t(k)}\tilde{V}_{\text{p-h}}(k)\right]^{-1/2} . \tag{16}$$

This equation is formally identical to the boson-RPA expression for the structure factor; the HNC-EL theory supplements the RPA with a microscopic theory of the particle-hole interaction

$$V_{\text{p-h}}(z) = g(z)\left[v(z) + \Delta V_{\text{ele}}(z)\right] + \frac{\hbar^2}{m}\left|\frac{d}{dz}\sqrt{g(z)}\right|^2 + \left[g(z) - 1\right] w_{\text{I}}(z) . \tag{17}$$

Thus the HNC-EL theory sums *both* important sets of diagrams self-consistently.

The pressure of the system is calculated by varying the energy per particle with respect to the density. It can be expressed entirely in terms of the ground-state structure functions,

$$\frac{P}{\rho} = \rho\frac{dE/N}{d\rho} . \tag{18}$$

The chemical potential of the system is defined as $\mu = E/N + P/\rho$, and the hydrodynamic compressibility K_T is obtained by differentiating the chemical potential with respect to density

$$mc_s^2 = \frac{1}{\rho K_T} = \rho\frac{d\mu}{d\rho} , \tag{19}$$

where c_s is the isothermal speed of sound. We note that in the $\lim_{k\to 0} S(k) \to \hbar k/2mc_s$ thus from Eq. (16) we can also identify $\tilde{V}_{\text{p-h}}(0^+) = mc_s^2$. The c_s^2 from $\tilde{V}_{\text{p-h}}(0^+)$ and from Eq. (19) are inconsistent except for an exact calculation. The use of an optimized theory with correct long-range correlations, i.e. $S(k) \propto \hbar k/2mc$ as $k \to 0$ also guarantees that none of the one-dimensional systems discussed here has a zero-temperature Bose-Einstein condensate.

3 Model Systems

Some years ago, the properties of one-dimensional many-boson systems were the object of intense theoretical scrutiny. Exact solutions were obtained for model systems such as the repulsive delta-function potential, an attractive inverse-square model, and a model of an anharmonic lattice.[6]

In recent work,[22] we reported that the variational energies agree to better than 4 percent with the energies of the exactly known hard rod system. Our analysis of the Morse system and the Lennard-Jones 6-12 potential indicated that, for these types of systems, the existence of a many-body bound state (a zero pressure, finite density system with negative energy) occurs only if there is also a dimer. We showed that in the Lennard-Jones approximation one-dimensional ^{4}He has *no* many-body bound state and, also, it has no dimer. All three model systems showed signs of a high density transition. In higher dimensions the onset of this liquid-solid phase

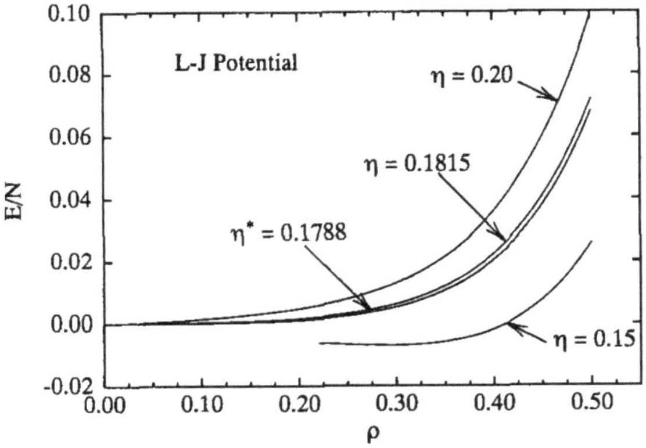

Figure 1. Equations of state for four representative Lennard-Jones systems. The system $\eta^* = 0.1788$ separates those Lennard-Jones systems with 2-body bound states from those with no 2-body bound states. The system with $\eta = 0.1815$ is a model for ^{4}He. In the Lennard-Jones approximation, ^{4}He has no one-dimensional dimer and no many-body bound state.

transition was signaled by an inability to obtain convergent solutions from the variational equations for a homogeneous ground-state beyond some maximum density. In one dimension such a transition can occur only at absolute zero temperature.

We consider a system of N mass m bosons on a line of length L interacting by means of a Lennard-Jones [LJ] pair potential, v_{LJ}. This function has the form

$$v_{LJ}(z) = 4\epsilon\left[(\sigma/z)^{12} - (\sigma/z)^{6}\right]. \tag{20}$$

In the following we shall measure energies in units of ϵ and lengths in units of σ. In the Hamiltonian, the only surviving parameter is the dimensionless coefficient of the kinetic energy, the quantum parameter η, where

$$\eta = \frac{\hbar^2}{m\epsilon\sigma^2}. \tag{21}$$

The properties of the many-body system depend only on η and the dimensionless density $\rho = N\sigma/L$.

In Fig. 1 we show the equations of state for four representative LJ systems. Following our experience with the Morse system we first determine the value of η^* which is defined as being the value of η below which LJ systems have 2-body bound states and above which they do not. For the LJ system this calculation needs to be done numerically and we find $\eta^* = 0.1788$ to four significant figures. We note that the values of the LJ parameters for ^{4}He are $\epsilon = 10.22$ K and $\sigma = 2.556$ Å. These yield $\eta = 0.1815$. Thus, in the LJ approximation, ^{4}He has no 2-body bound state in one-dimension. The energies as a function of density for $\eta = 0.1788,\ 0.1815,\ 0.20$ are shown in Fig. 1 and in agreement with the Morse system results *we find no many-body bound state in any LJ system for which $\eta > \eta^*$.*

Also in agreement with the results for both the hard rod and Morse systems, for each LJ system there is a maximum density above which we can find no variational solutions for a homogeneous ground-state. For those systems with $\eta < \eta^*$ we also

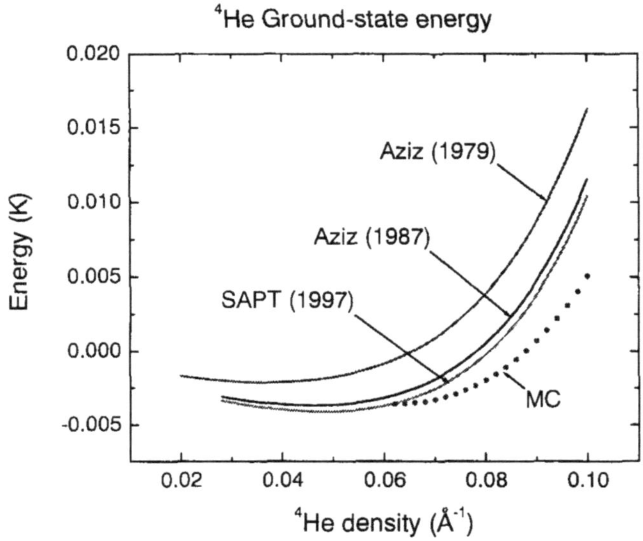

Figure 2. The equation of state for ^{4}He in one-dimension. The solid curves are the results with three different helium potentials as discussed in the text. The dotted line is the Monte Carlo results of Ref. 28.

find a minimum density below which we can find no solutions and as above for the Morse system we identify this occurrence with the vanishing of the speed of sound.

In Ref. 22 we also show a schematic phase diagram, η *vs.* density, for the one-dimensional LJ system at absolute zero. This phase diagram resembles the phase diagram in three dimensions strongly with gas, liquid and solid regions with a triple point which occurs in the region $\eta \approx 0.12$.

We note that a Lennard-Jones approximation for molecular hydrogen, H_2, yields $\eta = 0.076$.[24] Thus, the above results imply that one-dimensional molecular hydrogen will be a solid at $T = 0$ K.

4 ^{4}He

In recent work,[23] we have reported on calculations for the properties of the ground-state and low-lying excitations for ^{4}He in one spatial dimension. We found that at zero temperature, the ^{4}He will condense into a weakly bound one-dimensional liquid with a binding energy of approximately 0.002 K at a density of 0.036 Å^{-1}. The Feynman spectrum at zero-pressure is free particle-like and with increasing density smoothly develops the well-known phonon-roton shape.

We note that because of the large cancellation of kinetic and potential energies, these results are sensitive to the form of the helium potential function which is utilized. In Ref. 22, we used the older Aziz potential.[25] In Fig. 2, we compare the ground-state results for one-dimensional ^{4}He calculated with the older Aziz potential with that of two newer potentials: the new Aziz[26] and one determined from symmetry-adapted perturbation theory.[27] It is clear that the equation of state with the two newer potentials are in good agreement with one another and differ significantly from the equation of state with the older potential. The variational results for the 1987 Aziz potential show a binding energy of 0.0037 K at a density of 0.046 $Å^{-1}$. The Monte Carlo results of Gordillo, Boronat, and Casulleras[28] are shown in this figure as the dotted line. Their results show a binding energy of 0.036 ± 0.002 K at a density 0.062 ± 0.001 $Å^{-1}$. Thus our binding energies are in excellent agreement; however, our saturation density is too low.

As the density increases towards 0.19 $Å^{-1}$, the inverse of the static response function shows signs of an imminent transition in the region of $k = 1.5$ $Å^{-1}$. We have tentatively identified this density as the place where a phase transition is occurring.

We investigated the behavior of a ^{3}He atom in the ^{4}He and showed that the minimum energy state corresponds to the single ^{3}He atom completely separated from the line of ^{4}He. In a strictly one-dimensional system of course a ^{3}He atom cannot move past ^{4}He atoms to find this minimum energy configuration. However, it is probable that these results will remain valid for the quasi one-dimensional system where re-arrangements are possible.

Finally, we showed that a pair of ^{3}He atoms can form a very weakly bound dimer in the presence of the one-dimensional ^{4}He. The dimer binding energy increases rapidly with increasing ^{4}He linear density and thus its presence should be most apparent at the highest attainable ^{4}He densities.

5 Conclusion

We investigated the properties of one-dimensional ^{4}He and some model one-dimensional systems using variational techniques which were developed and honed for many-boson systems in two and three dimensions. We showed that the phase diagrams in one-dimension appeared similar to those in two and three dimensions. We showed[23] that strictly speaking none of these systems had Bose condensates. However, there does seem to be a high density solid-like phase. This phase is apparent in Monte Carlo studies of these systems.[28] The nature of this phase is still under investigation since true long-range order should not be present in these systems, even at absolute zero. Finally, in comparing the variational energies in the liquid phase to the results of Monte Carlo calculations we find good qualitative agreement although, at densities near the transition to the solid, the variational results are significantly higher than Monte Carlo. The possible sources of this discrepancy are under examination at this time.

Acknowledgements

The authors wish to congratulate Profs. John W. Clark, Alpo Kallio, Manfred L. Ristig, and Sergio Rosati on the occasion of their 65th birthdays. This work was supported by the Austrian Science Fund under grant No. P11098-PHY. The authors would like to thank John Wojdylo for many useful discussions.

References

1. H. Yano, S. Yoshizaki, S. Inagaki, Y. Fukushima, and N. Wada, *J. Low. Temp. Phys.* **110**, 573 (1998).
2. G. Stan and M. W. Cole, *Surf. Sci.* **395**, 20 (1998).
3. E. P. Bashkin, *Sov. Phys. JETP* **51(1)**, 181 (1980).
4. E. P. Bashkin and J. Wojdylo, *Phys. Rev.* B **62**, 6614 (2000).
5. E. H. Lieb and D. C. Mattis, *Mathematical Physics in One Dimension* (Academic Press, New York, 1965).
6. D. C. Mattis, *The Many-Body Problem; An Encyclopedia of Exactly Solved Models in One Dimension* (World Scientific, Singapore, 1993).
7. E. H. Lieb and W. Liniger, *Phys. Rev.* **130**, 1605 (1963).
8. E. H. Lieb, *Phys. Rev.* **130**, 1616 (1963).
9. C. N. Yang and C. P. Yang, *J. Math. Phys.* **10**, 1115 (1969).
10. B. Sutherland, *J. Math. Phys.* **12**, 246 (1971); *ibid.* **12**, 251 (1971); *Phys. Rev.* A **4**, 2019 (1971).
11. F. Calogero, *J. Math. Phys.* **12**, 419 (1971).
12. B. Sutherland, *Rocky Mountain J. Math.* **8**, 413 (1978).
13. E. Krotscheck, Q.-X. Qian, and W. Kohn, *Phys. Rev.* B **31**, 4245 (1985).
14. B. E. Clements, J. L. Epstein, E. Krotscheck, and M. Saarela, *Phys. Rev.* B **48**, 7450 (1993).
15. C. E. Campbell, in *Progress in Liquid Physics*, edited by C. A. Croxton (Wiley, London, 1977), Chap. 6, pp. 213-308.
16. J. W. Clark, in *Progress in Particle and Nuclear Physics*, edited by D. H. Wilkinson (Pergamon Press, Oxford, 1979), Vol. 2, pp. 89-199.
17. E. Feenberg, *Theory of Quantum Fluids* (Academic, New York, 1969).
18. C. E. Campbell, *Phys. Lett.* A **44**, 471 (1973).
19. C. C. Chang and C. E. Campbell, *Phys. Rev.* B **15**, 4238 (1977).
20. E. Krotscheck, *Phys. Rev.* B **33**, 3158 (1986).
21. E. Krotscheck and M. Saarela, *Phys. Rep.* **232**, 1 (1993).
22. E. Krotscheck, M. D. Miller, and J. Wojdylo, *Phys. Rev.* B **60**, 13028 (1999).
23. E. Krotscheck and M. D. Miller, *Phys. Rev.* B **60**, 13038 (1999).
24. M. D. Miller, L. H. Nosanow and L. J. Parish, *Phys. Rev.* B **15**, 214 (1977).
25. R. A. Aziz, V. P. S. Nain, J. S. Carley, W. L. Taylor, G. T. McConville, *J. Chem. Phys.* **70**, 4330 (1979).
26. R. A. Aziz, F. R. W. McCourt, C. C. K. Wong, *Mol. Phys.* **61**, 1487 (1987).
27. T. Korona, H. L. Williams, R. Bukowski, B. Jeziorski, K. Szalewicz, *J. Chem. Phys.* **106**, 5109 (1997).
28. M. C. Gordillo, J. Boronat and J. Casulleras, *Phys. Rev.* B **61**, R878 (2000).

THE GROUND STATE OF TRAPPED BOSONS BEYOND THE GROSS-PITAEVSKII APPROXIMATION

A. POLLS
Departament d'Estructura i Constituents de la Matèria, Universitat de Barcelona, Diagonal 647, Barcelona 08028, Spain
E-mail:artur@ecm.ub.es

We propose a modified Gross-Pitaevskii equation to describe a system of Bose hard spheres trapped in an isotropic harmonic potential. The new equation is based on a local density approximation which uses as local correlation energy, the energy functional derived from the low density expansion of the energy of the uniform hard-sphere gas.

1 Introduction

It is really a pleasure to participate in this conference in honour of John Clark, Alpo Kallio, Fred Ristig, and Sergio Rosati. My closest relations are with Sergio and John. I met Sergio in 1978 when my thesis advisor (Rafa Guardiola) sent me to Pisa to learn the secrets of the Hypernetted Chain theory (HNC). From the very beginning, I was captivated by the friendly atmosphere in Sergio's group and I started to learn a lot of things not only in physics but also from the Tuscany philosophy of life. One of the things that was impressing me, was the ability of Sergio to figure out if the numbers that I was presenting to him after many controls in my programs were right. Just by smelling the numbers he was able to discern the validity of the results.

I got to know John when I was visiting Washington University in 1987 to collaborate with him and with Wim Dickhoff. I enjoyed the help of John in scientific discussions and I was impressed by his patience and ability to improve my manuscripts written in a kind of "catalan-english" that however after being in his hands were going very smoothly through the referees.

In this talk I will discuss our recent results on the description of the ground state of dilute trapped bosons.[1] These systems can be described rather well in a mean field approximation. However, we can not avoid thinking as many-body physicists and immediately try to look for correlation effects.

The recent experimental realization of Bose-Einstein condensation (BEC) of magnetically trapped alkali atoms has generated a huge amount of experimental and theoretical activity. Presently, there are over twenty experimental groups around the world that can produce such atomic condensates almost in a routine way. The explosion in the number of papers is not less spectacular and more than one thousand articles on the subject have been published by now. The present status of the field has been recently reviewed by F. Dalfovo et al.[2]

The first atomic Bose condensate was achieved in 1995, in a laser cooled and magnetically trapped dilute gas of ^{87}Rb atoms.[3] The trapped gas was cooled further by taking advantage of evaporative cooling until the transition temperature was reached. For the number of trapped atoms in that experiment ($N \sim 10^3$ atoms), the transition temperature was around 100 nK. The magnetic trap is well described by a harmonic oscillator potential, usually with cylindrical symmetry. However, througout this paper we will consider a spherical potential well confining the atoms.

In order to have quantum effects, i.e., wave behaviour, we need a de Broglie wave length $\lambda = (2\pi\hbar^2/mT)^{1/2}$ of the order of the distance between the atoms ($\rho\lambda^3 \sim 1$). On the other hand, the system should be kept dilute, therefore the critical temperature will be extremely low, of the order of nanokelvins.

Up to now, the experimental conditions were such that the atomic gas was very dilute, i.e., the average distance between the atoms is much larger than the range of the interaction. As a consequence, the physics should be dominated by two-body collisions, generally well described in terms of the s-wave scattering length a. The case of a positive scattering length is equivalent to considering a very dilute system of hard spheres, whose diameter coincides with the scattering length itself. Everything happens as if at those densities the atoms would not distinguish the shape of the potential.

Typical scattering lengths are 53 Åfor ^{87}Rb and 28 Åfor ^{23}Na. On the other hand, the size of the trap is defined by the harmonic oscillator length $a_{HO} = (\hbar/m\omega)^{1/2}$ which is of the order of 10^4 Å. The corresponding distance between the energy levels associated with this potential well is around 4 nK. For those initial experiments, a common ^{87}Rb atom density in the trap was $\rho \sim 10^{12} - 10^{14}$ atoms/cm^3 giving an average inter-atom distance $d \sim \rho^{-1/3} \sim 10^4$ Å. Therefore, the effective atom size, defined by the scattering length is usually small compared to both the trap size and the inter-atom distance. The crucial parameter that defines the condition of diluteness is $x = \rho a^3$, which until very recently was kept rather small (i.e., $x \sim 10^{-5}$). Under these conditions, the Gross-Pitaevskii equation,[4] which assumes all the particles in the condensate, seems the logical tool to study those systems. Of course when correlations are taken into account, the occupation of the condensate can be depleted.

The situation is a little different in homogeneous liquid ^{4}He. In this case, BEC means a macroscopic occupation of the zero momentum state, which is measured by the condensate fraction, i.e., the fraction of the total number of particles in this state. However, there is only indirect evidence for this macroscopic occupation. Theoretical calculations and the analysis of inelastic neutron scattering data predict a condensate fraction of the order of 10 %.[7] This large depletion is an indication that ^{4}He liquid is a very correlated system.

There are two ways to bring x outside the regime of validity of the mean field description. One is by increasing the density and the other is by changing the effective size of the atoms. Recent experiments have explored both possibilities. On one side they have reached a very high number of atoms in the condensate, around 10^8, and on the other hand they have been able to change the scattering length of the atoms. This is the case of a recent experiment with ^{85}Rb in which by taking advantage of the presence of a Feshbach resonance, it is possible to vary the scattering length from negative to very high positive values. Under these conditions, effects beyond the mean field approximation should be observable, and the depletion of the condensate should be taken into account.[5,6]

We will start by discussing a homogeneous system of Bose hard spheres and use the results to determine the regime of validity of the Gross-Pitaevskii equation.

After that, we will discuss an extension of the GP equation, still in the framework of mean field theory that will allow to give a first estimation of the expected corrections to the GP results in these new scenarios. We will end up by calling for a fully microscopic many-body calculation in the framework of the correlated basis functions (CBF) which is one of the main subjects which has kept busy our honoured friends during an important part of their scientific career.

2 Uniform hard-sphere Bose gas

We consider a system of N spinless bosons with mass m described by the many-body Hamiltonian

$$H = -\frac{\hbar^2}{2m}\sum_i \nabla_i^2 + \sum_{i<j} V(r_{ij}). \tag{1}$$

To describe the uniform system one considers the thermodynamic limit by allowing $N \to \infty$ and $\Omega \to \infty$ keeping the density, $\rho = N/\Omega$, constant. The potential for hard-spheres is defined by $V(r) = \infty$ when $r < a$ and $V(r) = 0$ when $r > a$.

CBF theory provides a very efficient way to handle the correlations induced by the interactions between the particles (for a review, see Ref. 8). In the most simple version, one can take a Jastrow correlated wave function[9]

$$\Psi_T(1, ..., N) = \prod_{i<j} f(r_{ij}), \tag{2}$$

where the Jastrow correlation function, $f(r)$, depends only on the interparticle distance. Once the trial function is defined, the variational principle ensures that if we are capable to calculate the expectation value of the Hamiltonian,

$$E_{\rm CBF} = \frac{\langle \Psi_T \mid H \mid \Psi_T \rangle}{\langle \Psi_T \mid \Psi_T \rangle}, \tag{3}$$

then $E_{\rm CBF}$ will be an upper bound to the ground state energy. The correlation function, $f(r)$, is variationally determined by minimizing $E_{\rm CBF}$. Although conceptually it looks very simple, the evaluation of the expectation value is by no means an easy task and very sophisticated methods, with fundamental contributions of our honoured friends, have been devised during the last years.

$E_{\rm CBF}$ may be calculated either by Monte Carlo techniques or by evaluating the two-body distribution function $g(r)$, by means of cluster expansions. Finally, the energy per particle can be expressed as

$$e = \frac{1}{2}\rho \int d^3r\ g(r) \left[V(r) - \frac{\hbar^2}{2m}\nabla^2 \ln f(r) \right]. \tag{4}$$

In the particular case of hard spheres, the distribution function is strictly zero for $r < a$ and the previous expression reduces to the kinetic energy part

$$e = -\frac{1}{2}\rho \int d^3r\ g(r) \frac{\hbar^2}{2m}\nabla^2 \ln f(r). \tag{5}$$

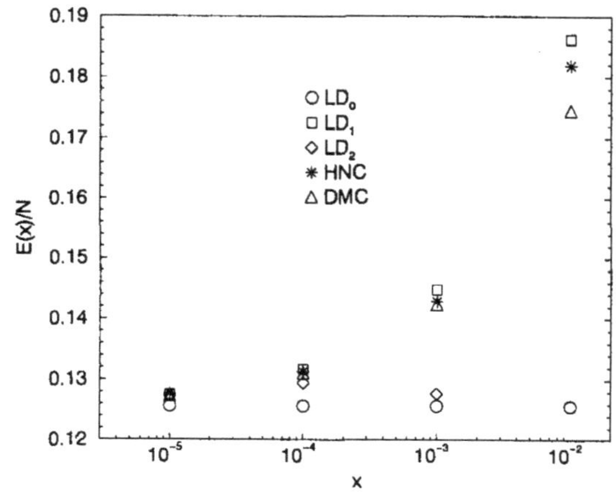

Figure 1. Energy per particle (in units of $\hbar^2/2ma^2$) for homogeneous hard spheres as a function of x. The symbols correspond to the low-density expansion results obtained by keeping only the first term (LD_0) or by adding the second (LD_1) and the third (LD_2) ones, and to the diffusion Monte Carlo (DMC) and HNC energies.

The distribution function $g(r)$ is evaluated by using HNC theory, that is an integral equation method which allows for massive summations of the cluster diagrams associated with $g(r)$.

The optimal choice for the Jastrow factor would be the one satisfying the Euler equation $\delta E_{\text{CBF}}/\delta f(r) = 0$. Otherwise, parametrized functional forms may be chosen whose parameters are found through the minimization process. We adopt here the correlation function minimizing the lowest-order energy of a homogeneous Bose gas with a healing condition at a distance d (taken as a variational parameter). For the hard-spheres case, $f(r < a) = 0$ and $f(r > a) = u(r)/r$, where $u(r)$ is the solution of the Schrödinger-like equation $-u'' = k^2 u$, $f(r)$ has the form[10]

$$f(r) = \frac{d \sin[K(r-a)]}{r \sin[K(d-a)]}, \tag{6}$$

where the healing conditions, $f(r \geq d) = 1$ and $f'(r \geq d) = 0$, are fixed by the relation: $\cot[K(d-a)] = (Kd)^{-1}$.

An alternative calculation, based on perturbation theory in the expansion parameter $x = \rho a^3$, leads to the following low-density expansion for the energy density:[11]

$$\frac{E}{\Omega} = \frac{2\pi\rho^2 a\hbar^2}{m}\left[1 + \frac{128}{15}\sqrt{\frac{\rho a^3}{\pi}} + 8\left(\frac{4}{3}\pi - \sqrt{3}\right)\rho a^3 \,\ln(\rho a^3) + O(\rho a^3)\right]. \tag{7}$$

Up to this order of the expansion, the details of the potential do not show up, and any potential with the same scattering length would give the same results. This universal behaviour has recently been checked by means of a diffusion Monte Carlo calculation (DMC).[12]

Fig. 1 shows the energy per particle in units of $\hbar^2/2ma^2$ for homogeneous hard spheres as a function of x. The energies have been multiplied by $10^{3(2,1)}$ at $x = 10^{-5(-4,-3)}$, respectively. The figure shows a comparison between the energy computed by retaining different expansion terms in Eq. (7). The LD_0 values correspond to the first term, whereas LD_1 and LD_2 are obtained by adding the second and third terms, respectively. The HNC results have been obtained disregarding the elementary diagrams (HNC/0) and using the correlation function of Eq. (6). The DMC results correspond to diffusion Monte Carlo calculations,[12] which can be considered as the exact results.

The agreement between the HNC/0 and the DMC results is excellent in the wide range of densities considered. However, we have checked that for higher densities the agreement is not so satisfactory and the contribution of elementary diagrams together with an optimization of the correlation function should be taken into account. The LD_0 results are only accurate at very low densities, while the LD_1 gives also a good representation of the exact DMC results. On the contary the addition of the logarithmic term spoils the agreement already at intermediate densities (the result for $x = 10^{-2}$ lies outside the frame of the figure and it has not been plotted).

Now we will describe the trapped bosons by performing a local density approximation (LDA). The local value of the parameter x in the trap will give an idea of the differences that we can expect by using the different energies reported in Fig. 1 as inputs to build the energy functional.

3 Trapped hard spheres

The energy functional associated with the Gross-Pitaevskii theory is simply obtained in the local-density approximation by keeping only the first term in the low density expansion (Eq. (7)):

$$E_{\rm GP}[\Psi] = \int dr \left[\frac{\hbar^2}{2m} |\nabla\Psi(r)|^2 + \frac{m}{2}\omega^2 r^2 |\Psi(r)|^2 + \frac{2\pi\hbar^2 a}{m} |\Psi(r)|^4 \right], \tag{8}$$

where the wave function $\Psi(r)$, in which all the atoms have condensated, is normalized to N. By performing a functional variation of $E_{\rm GP}[\Psi]$ one finds the Gross-Pitaevskii equation,

$$\left[-\frac{\hbar^2}{2m}\nabla^2 + \frac{m}{2}\omega^2 r^2 + \frac{4\pi\hbar^2 a}{m}|\Psi(r)|^2 \right] \Psi(r) = \mu\Psi(r), \tag{9}$$

where μ is the chemical potential. This equation has the form of a nonlinear stationary Schrödinger equation, and it has been solved for several types of traps using different numerical methods.

The next logical step, in the spirit of LDA, is to include in the energy functional the next terms of the correlation energy of the uniform system (Eq. (7)). After the discussion of the behaviour of the different terms (Fig. 1) it seems clear that it is reasonable to consider only the first correction term. However, before proceeding further it is convenient to simplify the notation by expressing lengths and energies in harmonic oscillator units. The spatial coordinates, the energy, and the wave function are rescaled as $\mathbf{r} = a_{\rm HO}\bar{\mathbf{r}}$, $E = \hbar\omega\bar{E}$, and $\Psi(r) = (N/a_{\rm HO}^3)^{1/2}\Psi_1(\bar{r})$, where $\Psi_1(\bar{r})$ is normalized to unity.

Using these new variables and taking into account the second term of the expansion, we obtain the modified Gross-Pitaevskii (MGP) energy functional for the energy per atom, $\bar{e}_{\rm MGP} = \bar{E}_{\rm MGP}/N$,

$$\bar{e}_{\rm MGP}[\Psi_1] = \int d\bar{\mathbf{r}} \left[\frac{1}{2}|\nabla_{\bar{r}}\Psi_1|^2 + \frac{1}{2}\bar{r}^2|\Psi_1|^2 + 2\pi\bar{a}N|\Psi_1|^4 + \frac{256\sqrt{\pi\bar{a}^5N^3}}{15}|\Psi_1|^5\right], \tag{10}$$

and the corresponding modified Gross-Pitaevskii equation,

$$\left[-\frac{1}{2}\nabla^2_{\bar{r}} + \frac{1}{2}\bar{r}^2 + 4\pi\bar{a}N|\Psi_1(\bar{r})|^2 + \sqrt{\pi\bar{a}^5N^3}\frac{128}{3}|\Psi_1(\bar{r})|^3\right]\Psi_1(\bar{r}) = \mu_1\Psi_1(\bar{r}), \tag{11}$$

where $\bar{a} = a/a_{\rm HO}$ and $\bar{\mu}$ is the chemical potential in harmonic oscillator units.

Without going beyond mean field theory we have explored also the alternative of using as local correlation energy the one provided by the HNC calculation for the uniform system. This option has the advantage that one is not limited to the use of hard spheres but in principle one can consider any type of potential for the two-body interaction. In this case, the local correlation energy $V_{\rm corr}^{LD}$, is given by

$$V_{\rm corr}^{LD} = \frac{1}{N}\int d\bar{\mathbf{r}}\rho_1(\bar{r})e_{\rm HNC}^{hom}(\rho_1). \tag{12}$$

where $e_{\rm HNC}^{hom}(\rho_1)$ is the HNC homogeneous gas energy per particle at density ρ_1. The minimization of the energy gives the HNC correlated Hartree equation ($\rm CH_{HNC}$),

$$\left[-\frac{1}{2}\nabla^2_{\bar{r}} + \frac{1}{2}\bar{r}^2 + \bar{e}_{\rm HNC}^{hom}(x_{\rm loc}) + x_{\rm loc}\frac{\partial\bar{e}_{\rm HNC}^{hom}(x_{\rm loc})}{\partial x_{\rm loc}}\right]\Psi_1(\bar{r}) = \mu_1\Psi_1(\bar{r}), \tag{13}$$

where we have also introduced the scaled unities and the local gas parameter, $x_{\rm loc}(\bar{r}) = \rho_1(\bar{r})a^3 = N\bar{a}^3|\Psi_1(\bar{r})|^2$.

These equations, GP, MGP, and $\rm CH_{HNC}$ have been solved by the steepest descent method for an isotropic harmonic oscillator trap, whose parameters will be described below.

There are several relationships between the different contributions to the total energy per atom or to the chemical potential that are useful in checking the numerical accuracy of the numerical procedure. By direct integration of the GP equation one finds the useful relation

$$\bar{\mu} = \bar{e}_{\rm kin} + \bar{e}_{\rm HO} + 2\bar{e}_{\rm int}^{(1)}, \tag{14}$$

where $\bar{e}_{\rm kin} = -1/2\int d^3\bar{r}\Psi_1\nabla^2\Psi_1$, $\bar{e}_{\rm HO} = 1/2\int d^3\bar{r}\Psi_1 r^2\Psi_1$, and $\bar{e}_{\rm int}^{(1)} = \int d^3\bar{r}\Psi_1(2\pi aN\Psi_1^2)\Psi_1$ are the different terms contributing to the total energy. Further relationships can be obtained by means of the virial theorem,

$$2\bar{e}_{\rm kin} - 2\bar{e}_{\rm HO} + 3\bar{e}_{\rm int}^{(1)} = 0. \tag{15}$$

Also important is to notice that the dimensionless parameter characterizing the effects of the interaction in the GP equation is given by $\bar{a}N$. This implies that one can get the same results with a proper rescaling of the variables N and $\bar{a}$. As can be seen by simple examination of the equations, this scaling property is lost in the MGP equation. Besides, the relation between the different contributions to the chemical potential changes to

$$\bar{\mu} = \bar{e}_{\text{kin}} + \bar{e}_{\text{HO}} + 2\bar{e}^{(1)}_{\text{int}} + \frac{5}{2}\bar{e}^{(2)}_{\text{int}}, \tag{16}$$

where

$$\bar{e}^{(2)}_{\text{int}} = \int d^3\bar{r}\ \Psi_1(\bar{r}) \left[\frac{256}{15}\sqrt{\pi\bar{a}^5N^3}\Psi_1(\bar{r})^3\right] \Psi_1(\bar{r}). \tag{17}$$

In this case the relation implied by the virial theorem is

$$2\bar{e}_{\text{kin}} - 2\bar{e}_{\text{HO}} + 3\bar{e}^{(1)}_{\text{int}} + \frac{9}{2}\bar{e}^{(2)}_{\text{int}} = 0. \tag{18}$$

A simple approach, valid for large N, is obtained by neglecting the kinetic energy term. This approach, loosely called the Thomas-Fermi (TF) approximation, allows, in the case of the GP equation, to derive simple analytical expressions, which are useful to make quick estimations of the different quantities.[13] In the TF approximation, $\bar{\mu} = 1/2(15\bar{a}N)^{2/5}$, while the energy is related to the chemical potential by $\bar{e} = 5\bar{\mu}/7$. The local value of $x = N\bar{a}^3\rho_1(0)$ at the center of the density distribution of the trapped bosons is given by $x_{\text{TF}}(0) = \left(15^2\bar{a}^{12}N^2\right)^{1/5}/(8\pi)$.

In order to see the effects of the new terms introduced in the GP equation we will present results for two situations. First, we consider ^{87}Rb in an isotropic trap characterized by an angular frequency $\omega/2\pi = 77.78$ Hz, and a scattering length $a = 52.9$ Å, which means $\bar{a} = 0.00433$. To explore situations were the corrections could be sizeable, we consider a quite large number of atoms, $N = 10^7$. For this case, $\bar{\mu}_{\text{TF}} = 105.68$, $\bar{\mu}_{\text{GP}} = 105.70$, $\bar{\mu}_{\text{MGP}} = 107.97$, and $\bar{\mu}_{\text{CH-HNC}} = 107.20$. In this situation, $x_{\text{TF}}(0) = 1.6 \times 10^{-4}$. Therefore, looking at Fig. 1 we expect corrections of the order of 1 or 2 %, and in fact this is the case. For the energy per particle, we have $e_{\text{TF}} = 75.49$, $\bar{e}_{\text{GP}} = 75.52$, $\bar{e}_{\text{MGP}} = 76.94$, and $\bar{e}_{\text{CH-HNC}} = 76.85$. As expected, for this large number of particles the TF and GP results are practically identical and there is also a very good agreement between the MGP and CH-HNC results.

The different contributions to the energy for the GP(MGP) equations are: $\bar{e}_{\text{kin}} = 0.0294(0.0292)$, $\bar{e}_{\text{HO}} = 45.306(46.57)$, $\bar{e}^{(1)}_{\text{int}} = 30.184(28.96)$ and $\bar{e}_{\text{int}} = 0(1.379)$. In both cases the virial theorem is well fulfilled. In order to respect the virial theorem, the contribution of $\bar{e}^{(2)}_{\text{int}}$ in the MGP case forces the rearrangement of the other terms. In the case of no interaction between the particles, $\bar{e}_{\text{kin}} = \bar{e}_{\text{HO}} = 3/4$, which correspond to the ground state of a three-dimensional isotropic harmonic oscillator.

By changing the number of trapped atoms in the range of experimental availability, the average values of x are such that the corrections to the GP equation are kept small, of the order of 2 % in the case of $N = 10^7$ atoms. However, the recent experiments, where the scattering length can largely be manipulated, open the door to explore higher values of x. In fact, in order to vary x, it is much more efficient to change the scattering length than the number of atoms. The available experimental results are for ^{85}Rb atoms, which have a Feshbach resonance allowing the modulation of the scattering length. The number of trapped atoms is $N \sim 10^4$ and the trap is anisotropic. However, in order to estimate the corrections induced by

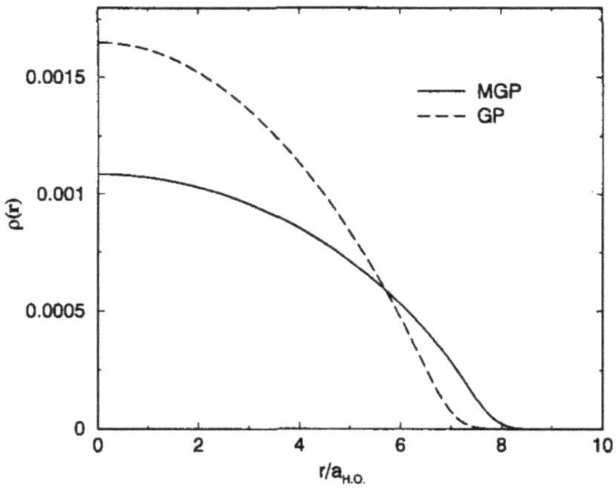

Figure 2. Density profiles for $N = 10^4$ in the case of ^{85}Rb atoms with $\bar{a} = 0.123$, in different approaches. Densities are normalized to unity and distances are in units of a_{HO}

the MGP equation, we will consider an isotropic trap characterized by a frequency $\omega/2\pi = 10$ Hz. This ω is clearly smaller than the frequencies used for ^{87}Rb and therefore the a_{HO} is larger in this case. We take $\bar{a} = 0.1228$, which is in the range considered by the experiments. It corresponds to $a = 8000a_0$, where a_0 is the Bohr radius of the hydrogen atom. In this case, $x_{TF}(0) = 0.03$, which is a little beyond the plotted points in Fig. 1. The energy per atom turns out to be $\bar{e}_{GP} = 18.25$ and $\bar{e}_{MGP} = 21.85$, while for the chemical potential we have $\bar{\mu}_{GP} = 25.48$ and $\bar{\mu}_{MGP} = 31.09$. The corrections are of the order of 20 %. Furthermore, as one can see in Fig. 2 the effects on the density profile are also important. The repulsive character of the additional terms in the MGP equation produces a wider density distribution and a depletion of the central value of the density in comparison to the GP results. As a consequence, the root-mean-square radius for the MGP distribution $\bar{R}_{rms} = 5.28$ is larger than the GP one $\bar{R}_{rms} = 4.67$. The HC-HNC results essentially agree with the MGP ones.

In conclusion, we find that the MGP equation induces corrections of 20% when the conditions of the recent experiments for ^{85}Rb are considered. As MGP is a mean field theory, i.e., tries to incorporate the correlations in the average mean field, it can not predict the depletion of the condensate. However we believe that the estimation of the energy, chemical potential, and density profile can still be accurate. Even though, at those densities it is legitimate to question also the use of a simplified interaction, in terms of hard spheres. In any case, it is clear that fully microscopic calculations which can take into account explicitly the depletion of the condensate are urgently required.[14]

Acknowledgments

This work was been made in collaboration with Adelchi Fabrocini and was supported by DGICYT (Spain) Grant No.PB98-1247, the program SGR98-11 from

Generalitat de Catalunya and the agreement CICYT (Spain)-INFN (Italy).

References

1. A. Fabrocini and A. Polls, *Phys. Rev.* A **60**, 2319 (1999).
2. F. Dalfovo *et al.*, *Rev. Mod. Phys.* **71**, 463 (1999).
3. M. H. Anderson et al., *Science* **269**, 198 (1995).
4. L. P. Pitaevskii, *Sov. Phys. JETP* **13**, 451 (1961); E.P. Gross, *Nuovo Cimento* **20**, 454 (1961).
5. S. L. Cornish et al., *cond-mat*/0004290
6. *Search and Discovery,* in *Phys. Today,* 17 (August 2000).
7. H. R. Glyde, *Excitations in Liquid and Solid Helium,* (Clarendon Press, Oxford, 1994).
8. S. Fantoni and A. Fabrocini, in *Microscopic Quantum Many Body Theories and Their Applications,* Eds. J. Navarro and A. Polls, Lecture Notes in Physics Vol. 150 (Springer-Verlag, Berlin, 1998), p. 119.
9. R. Jastrow, *Phys. Rev.* **98**, 1479 (1955).
10. V. R. Pandharipande and K. E. Schmidt, *Phys. Rev.* A **15**, 2486 (1977).
11. A. L. Fetter and J. D. Walecka, *Quantum Theory of Many-Particle Systems* (McGraw-Hill, New York, 1971).
12. S. Giorgini, J. Boronat and J. Casulleras, *Phys. Rev.* A **60**, 5129 (1999).
13. M. Edwards and K. Burnett, *Phys. Rev.* A **51**, 1382 (1995).
14. J. L. Dubois and H. R. Glyde, *cond-mat*/0008368.

PAIRING OF IMPURITIES IN QUANTUM FLUIDS

M. SAARELA
Physical Sciences/Theoretical Physics, University of Oulu, P.O. Box 3000, FIN-90014 Oulu, Finland
E-mail: Mikko.Saarela@oulu.fi

Dilute mixtures of ^{3}He impurities in liquid ^{4}He are studied in two and three dimensions at zero temperature. It is shown that ^{3}He atoms form loosely bound pairs, dimers, in two dimensions. That should be contrasted with the Cooper pairing predicted in three-dimensional mixture, which leads to the superfluidity of the ^{3}He component below $\approx 20\ \mu$K. The binding energy of the dimer ranges from milli- to micro-Kelvins with increasing ^{4}He density. By comparing the ^{3}He impurity chemical potential in ^{4}He with the one in pure ^{3}He we conclude that at finite pressures ^{3}He dimers form a mixture with ^{4}He with a maximum solubility of $\approx 3\%$. Increasing concentration screens the pairing interaction and the atomic mixture becomes stable at concentrations above 2-3%, depending on the pressure.

1 Introduction

In three dimensions pure ^{3}He and ^{4}He atomic systems form self bound superfluids at zero temperature. Two ^{4}He atoms in the free space form a bound state with binding energy E_b= 1.7 mK and the mixture of ^{3}He and ^{4}He atoms exists with maximal solubility less than 10% at zero temperature.[1] It has been predicted theoretically that the ^{3}He component in the mixture also becomes superfluid,[2,3] but this has not yet been verified experimentally. In two dimensions ^{4}He is a self-bound superfluid, but ^{3}He is a gas.[4] Yet, both ^{4}He and ^{3}He atoms form loosely bound states, dimers, with the binding energies of 40 mK and 20 μK, respectively.

In this work we highlight the main ingredients of a theory needed to analyze the bound state formation in quantum fluids.[5,6] As an application we present a careful theoretical analysis of the phase diagram of two-dimensional, low concentration ^{3}He-^{4}He mixtures.[7,8] We evaluate the effective ^{3}He-^{3}He interaction and give an estimate of the maximum solubility by comparing our calculated chemical potentials with recent Monte Carlo results for the pure two-dimensional ^{3}He gas.[4] We use the variational approach based on the Jastrow-Feenberg ansatz for the wave function *plus* correlated basis functions (CBF) perturbation theory to infinite order.[9] The method gives accurate results for helium fluids both in two and three dimensions.

2 Theory

In a microscopic theory for the ground-state properties of a ^{3}He-^{4}He mixture such as the energetics and structure one assumes knowledge of the empirical Hamiltonian

$$H = -\sum_{\alpha}\sum_{i=1}^{N_\alpha} \frac{\hbar^2}{2m_\alpha}\nabla_i^2 + \frac{1}{2}\sum_{\alpha,\beta}\sum_{i,j}^{N_\alpha,N_\beta}{}' V^{(\alpha\beta)}(|\mathbf{r}_i - \mathbf{r}_j|), \qquad (1)$$

that contains the local two-body interaction.[10] Our notation refers to particle species with greek subscripts α, β, ... and to individual particles with roman subscripts i, j, ... The prime on the summation symbol indicates that no two pairs (i, α), (j, β) can be the same. The number of particles of each species is N_α, and $N = N_3 + N_4$ is the total number of particles in the system. The ^{3}He concentration is defined as $x = N_3/N$ and the corresponding partial densities, $\rho_3 = x\rho$ and $\rho_4 = (1 - x)\rho$ are proportional to the total number density $\rho = N/\Omega$, where Ω is the volume occupied by the whole fluid.

2.1 Optimized variational wave functions

The Jastrow-Feenberg variational method and its extensions within the theory of correlated basis functions (CBF) provides the necessary tools for a precise ground state calculation.[5] A variational *ansatz* is made for the ground state wave function which incorporates both pair and triplet correlation functions of the form

$$\begin{aligned} \Psi_0(\{\boldsymbol{r}_i^{(\alpha)}\}) &= e^{\frac{1}{2}U(\{\boldsymbol{r}_i^{(\alpha)}\})}\Phi_0(\{\boldsymbol{r}_i^{(3)}\}), \\ U(\{\boldsymbol{r}_i^{(\alpha)}\}) &= \frac{1}{2!}\sum_{\alpha\beta}\sum_{i,j}^{N_\alpha,N_\beta}{}' u^{(\alpha\beta)}(\boldsymbol{r}_i,\boldsymbol{r}_j) + \frac{1}{3!}\sum_{\alpha\beta\gamma}\sum_{i,j,k}^{N_\alpha,N_\beta,N_\gamma}{}' u^{(\alpha\beta\gamma)}(\boldsymbol{r}_i,\boldsymbol{r}_j,\boldsymbol{r}_k)\,. \end{aligned} \quad (2)$$

Here $\Phi_0(\{\boldsymbol{r}_i^{(3)}\})$ is a Slater determinant of plane waves ensuring the antisymmetry of the fermion component of the wave function. The pair and triplet correlation functions $u^{(\alpha\beta)}(\boldsymbol{r}_i, \boldsymbol{r}_j)$ and $u^{(\alpha\beta\gamma)}(\boldsymbol{r}_i, \boldsymbol{r}_j, \boldsymbol{r}_k)$ are optimized by minimizing the total energy E_0,[6,11]

$$\frac{\delta E_0}{\delta u^{(\alpha\beta)}(\boldsymbol{r}_i,\boldsymbol{r}_j)} = 0\,, \qquad \frac{\delta E_0}{\delta u^{(\alpha\beta\gamma)}(\boldsymbol{r}_i,\boldsymbol{r}_j,\boldsymbol{r}_k)} = 0\,. \quad (3)$$

The structure of the mixture is determined by the pair distribution functions $g^{(\alpha\beta)}(\boldsymbol{r}_i, \boldsymbol{r}_j)$ and their Fourier transforms, the static structure functions

$$S^{(\alpha\beta)}(k) = \delta_{\alpha\beta} + \sqrt{\rho_\alpha\rho_\beta}\int d^2r\left[g^{(\alpha\beta)}(r) - 1\right]e^{i\boldsymbol{r}\cdot\boldsymbol{k}}\,. \quad (4)$$

Relationships between the correlation functions $u^{(\alpha\beta)}(\boldsymbol{r}_i, \boldsymbol{r}_j)$ and the distribution functions $g^{(\alpha\beta)}(\boldsymbol{r}_i, \boldsymbol{r}_j)$ are provided by the (Fermi) hypernetted-chain ((F)HNC) hierarchy of integral equations.

2.2 Dilute mixtures

We are interested in ^{3}He-^{4}He mixtures with a dilute ^{3}He component because this situation is closest to what can be achieved experimentally. For low concentrations of the Fermion component it is adequate to use the single loop approximation, which sums self-consistently all chain and parallel-connected diagrams, but omits propagator corrections. We introduce a 2×2 matrix notation, for example for the interacting and non-interacting static structure functions $\boldsymbol{S}(k)$ and $\boldsymbol{S}_F(k)$ we have respectively

$$\boldsymbol{S}(k) \equiv \left(S^{(\alpha\beta)}(k)\right) \quad \text{and} \quad \boldsymbol{S}_F(k) \equiv \begin{pmatrix} S_F(k) & 0 \\ 0 & 1 \end{pmatrix}, \quad (5)$$

where $S_F(k)$ is the static structure function of the non-interacting Fermi component. The single loop approximation connects the "dressed" correlation functions $\tilde{\mathbf{\Gamma}}(k) = \left(\tilde{\Gamma}^{(\alpha\beta)}(k)\right)$ with the structure function matrix

$$\boldsymbol{S}(k) = \boldsymbol{S}_F(k) + \boldsymbol{S}_F(k)\tilde{\mathbf{\Gamma}}(k)\boldsymbol{S}_F(k)\,. \tag{6}$$

Without going into further details of the derivations, we assert that the coupled Euler equations (3) for the $u^{(\alpha\beta)}(r)$ are equivalent[6] to

$$\left[\boldsymbol{S}^{-1}\boldsymbol{H}_1\boldsymbol{S}^{-1}\right](k) - \left[\boldsymbol{S}_{\boldsymbol{F}}^{-1}\boldsymbol{H}_1\boldsymbol{S}_{\boldsymbol{F}}^{-1}\right](k) = 2\tilde{\boldsymbol{V}}_{\text{p-h}}(k)\,. \tag{7}$$

The matrix $\boldsymbol{H}_1(k)$ contains only free particle kinetic energies on the diagonal and $\tilde{\boldsymbol{V}}_{\text{p-h}}(k)$ is the Fourier transform of the so-called particle-hole interaction or the *static,* effective interaction, which is defined in coordinate space as

$$\begin{aligned} V_{\text{p-h}}^{(\alpha\beta)}(r) &= \left[1+\Gamma^{(\alpha\beta)}(r)\right]\left[V^{(\alpha\beta)}(r)+\Delta V^{(\alpha\beta)}(r)\right] \\ &+ \left[\frac{\hbar^2}{2m_\alpha}+\frac{\hbar^2}{2m_\beta}\right]\left|\nabla\sqrt{1+\Gamma^{(\alpha\beta)}(r)}\right|^2 + \Gamma^{(\alpha\beta)}(r)w_{\text{I}}^{(\alpha\beta)}(r)\,. \end{aligned} \tag{8}$$

The quantities $\Delta V^{(\alpha\beta)}(r)$ are corrections due to elementary diagrams and triplet correlations; they must be calculated individually and are, in this sense, an external input to the theory. The "induced interaction" matrix is

$$\tilde{\boldsymbol{w}}_{\text{I}}(k) = -\tilde{\boldsymbol{V}}_{\text{p-h}}(k) - \frac{1}{2}\left[\boldsymbol{S}_{\boldsymbol{F}}^{-1}\boldsymbol{H}_1\tilde{\mathbf{\Gamma}} + \tilde{\mathbf{\Gamma}}\boldsymbol{H}_1\boldsymbol{S}_{\boldsymbol{F}}^{-1}\right](k)\,. \tag{9}$$

For any given choice of elementary diagrams and triplet correlations, the equations (7), (8), and (9) form a closed set of equations that can be solved by iteration until convergence is reached. Given the resulting distribution and structure functions, the variational total energy E_0 can then be calculated.[6]

2.3 Two impurity limit

In the two-impurity limit Eq. (7) can be written in the form of a Schrödinger equation,[12]

$$\left[-\frac{\hbar^2}{2m_3}\nabla^2 + V_{\text{scat}}^{\alpha\beta}(r) - E_b\right]\phi(\boldsymbol{r}) = 0\,, \tag{10}$$

where $V_{\text{scat}}^{(33)}(r) = V^{(33)}(r) + w_{\text{I}}^{(33)}(r) + \Delta V^{(33)}(r)$ is a local effective interaction, with $w_{\text{I}}^{(33)}$ the Fourier transform of the induced interaction

$$\tilde{w}_{\text{I}}^{(33)}(k) = -\frac{[S^{(34)}(k)]^2}{2S^{(44)}(k)}\left[\frac{\hbar^2k^2}{m_3} + \frac{\hbar^2k^2}{2m_4S^{(44)}(k)}\right]\,, \tag{11}$$

which is determined after the pure ^{4}He and one ^{3}He impurity structure functions $S^{(44)}(k)$ and $S^{(34)}(k)$ are calculated. Only the zero energy solution $E_b = 0$ is consistent with the mixture equations and for $E_b < 0$ the mixture is unstable against dimer formation.

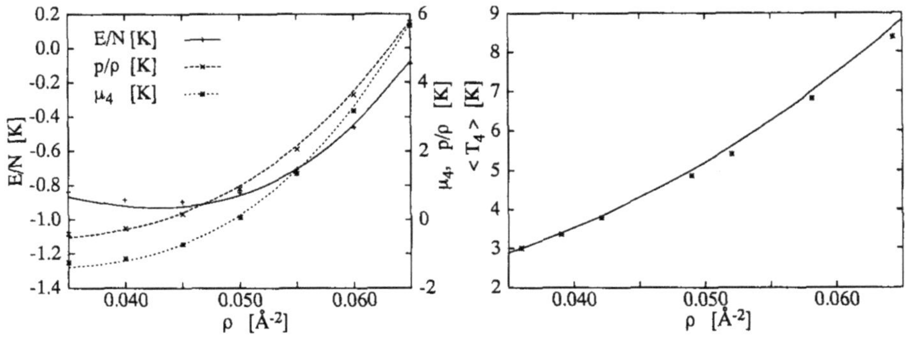

Figure 1. The figure on the left shows the HNC-EL results for the energy per particle E/N, (solid line, left-hand scale), the ratio of pressure and density p/ρ (short-dashed line, right-hand scale) and the chemical potential μ_4 (long-dashed line, right-hand scale). Also shown are DMC data for the energy[13] (+-symbols), the pressure per density (crosses) and the chemical potential (stars). All energies are given in degrees Kelvin. The figure on the right shows our result for the kinetic energy of ^{4}He (solid line) in comparison with the Monte Carlo results[13] (markers).

3 Pairing in dilute ^{3}He-^{4}He mixtures

Pairing of ^{3}He impurities in ^{4}He is a very delicate phenomenon and requires accurate knowledge of the ground state and single-impurity energetics. We begin the discussion by studying the accuracy of our method in two dimensions. Direct experimental measurements of the ground-state properties do not exist and we will make comparison with Monte Carlo simulation data[13] for ^{4}He. Fig. 1 shows the two-dimensional equation of state of ^{4}He calculated with the Aziz-II potential.[10] The agreement with Monte Carlo data is quite satisfactory. The same holds for the derived quantities pressure per density p/ρ, chemical potential μ_4, and the kinetic energy per particle $\langle T_4 \rangle$ which are also shown in Fig. 1.

By comparing the chemical potentials of ^{3}He impurities in the mixture with the chemical potential in the pure ^{3}He gas we can study the existence of the mixture. In three dimensions the gain in energy at zero pressure by putting the ^{3}He impurity into the mixture is about 0.3 K.[14] In two dimensions no simulation data are available at this time for a direct comparison, but we can calculate a rigorous upper bound for the impurity chemical potential using the "average correlation approximation (ACA)",[15,16]

$$\mu_3^I \leq \left(\frac{m_4}{m_3} - 1 \right) \langle T_4 \rangle + \mu_4 \equiv \mu_3^{ACA} , \tag{12}$$

where both $\langle T_4 \rangle$ and μ_4 can be obtained from simulations.[13]

The comparison of our results with the ACA result is shown in Fig. 2 (left figure). The gain in energy due to relaxing the ACA is about 0.2 - 0.3 K in the whole density range. As expected Monte Carlo results for the ACA are slightly below our approximate result and thus we can conclude that our full calculation gives a reliable upper bound for μ_3^I. Also shown in the figure is the ratio of pressure and density and we find that μ_3^I is positive at zero pressure.

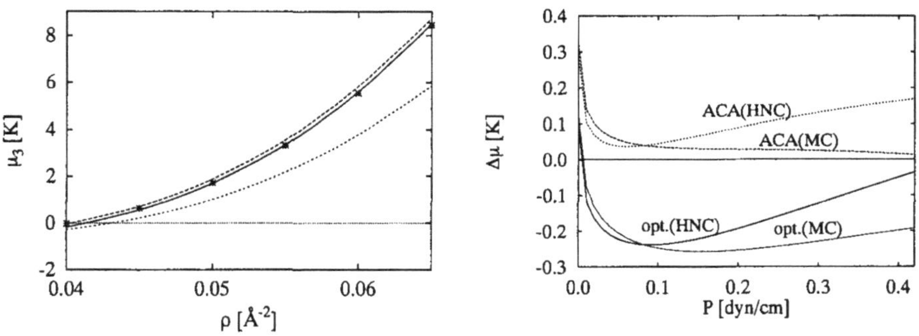

Figure 2. The figure on the left shows the ^{3}He impurity chemical potential as a function of density. The result of our full calculation (solid line) is compared with our ACA result (long-dashed line). The stars give the ACA result of the Monte Carlo simulations.[13] Also plotted is the ratio pressure/density (short dashed line). In the figure on the right the difference $\Delta\mu$ between the ^{3}He impurity chemical potentials and the pure ^{3}He chemical potential is shown as a function of pressure. The solid line shows the result of our full calculation and the short dashed line is our ACA estimate (12), the long dashed line is the ACA estimate from Monte Carlo simulations and the dotted line is our estimate for $\Delta\mu$ obtained by supplementing the ACA from Monte Carlo results by our enhancement of the binding due to relaxing the ACA.

Recent Monte Carlo simulations exist for the two-dimensional equation of state of the pure ^{3}He,[4] from which we can calculate the chemical potential $\mu_3^{\rm pure}(P)$. In Fig. 2 (right figure) we show the difference,

$$\Delta\mu(P) = \mu_3^I(P) - \mu_3^{\rm pure}(P)\,, \tag{13}$$

as a function of pressure for different approximations. A positive $\Delta\mu$ indicates phase separation, whereas a negative $\Delta\mu$ indicates mixing. The ACA using Monte Carlo yields a rigorous upper limit for the impurity chemical potential which is less than 0.1 K above $\mu_3^{\rm pure}(P)$ for pressure $P > 0.01$ dyn/cm. Our result with ACA gives a slightly higher upper limit. The situation changes when the impurity-background correlations are fully optimized. Then the chemical potential is lowered by 0.2-0.3 K. The maximum difference in the chemical potentials is $\approx$ -0.25 K at $P = 0.1$ dyn/cm.; we can therefore conclude that the mixture is stable at pressures $P > 0.002$ dyn/cm. At zero pressure, the ^{3}He impurity chemical potential is still positive, $\mu_3^I = +0.13K$, whereas the chemical potential of the pure ^{3}He gas approaches zero proportional to $\sqrt{P}$. This means that no stable mixture can exist at zero pressure. The most realistic estimate is obtained by supplementing the ACA from Monte Carlo data with our *correction* to the chemical potential due to relaxing the ACA. As shown in Fig. 2 it lowers the chemical potential of the impurity even further.

3.1 Two impurities

The bare interaction is sufficient to bind ^{3}He dimers in two dimensions by 20 μK. However, the longest range part of the effective interaction $V_{\rm scat}^{33}(r)$ comes from the phonon exchange and induces a strong density dependence of the dimer binding

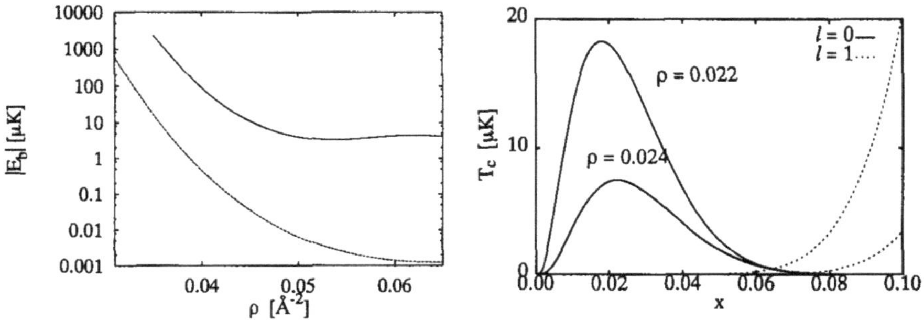

Figure 3. In the left figure is the binding energy of a dimer of ^{3}He atoms in two-dimensional ^{4}He on a logarithmic scale as a function of density. The solid line is the result of the full calculation, the dashed line shows the result from the HNC approximation. In the right figure is the critical temperature of the superfluid phase transition in three dimensions as a function of concentration in s- and p-channels for two different densities marked in the figure.

shown in Fig. 3. At zero pressure the binding energy of the dimer is 26μK. Because the binding is very weak, the wave function decays very slowly. The evaluation of the root-mean-square (rms) radius of the dimer is dominated by the tail of the wave function $\phi(r) \sim K_0(\sqrt{2m_3|E_b|/\hbar^2}\, r)$ and that gives a simple approximation for the rms-radius

$$\sqrt{< r^2 >} \approx \sqrt{\frac{\hbar^2}{3m_3|E_b|}}, \tag{14}$$

which is accurate within 3% for the whole density range. Its value increases from 70 Å at the density 0.035 Å^{-2} to 1000 Å at 0.065 Å^{-2}.

In three dimensions no bound dimer states are formed, yet the ^{3}He particles can form Cooper pairs. We have estimated the critical temperature in the weak-coupling approximation,[3]

$$k_B T_c \approx E_F \exp\left[\frac{\pi^2\, \hbar^2}{m^* k_F\; T_\ell(2E_F)}\right] \tag{15}$$

by calculating the phase shifts and T_ℓ-matrix for different concentrations giving the Fermi energy E_F. The results are shown in Fig. 3 (right figure). The best possibility of finding the superfluid phase transition is at about 2% concentration below 20 μK.

3.2 Finite-concentration mixtures

As the concentration of impurities is increased the Fermi statistics screens the interaction, dimers dissolve and an atomic mixture can be formed. A full microscopic theory is required for calculating the properties of such a mixture. The most interesting result is the concentration dependence of the second sound speed. The mixture exhibits an instability where the second sound becomes soft when the concentration is lowered below 2-3% as shown in Fig. 4 for three different densities for

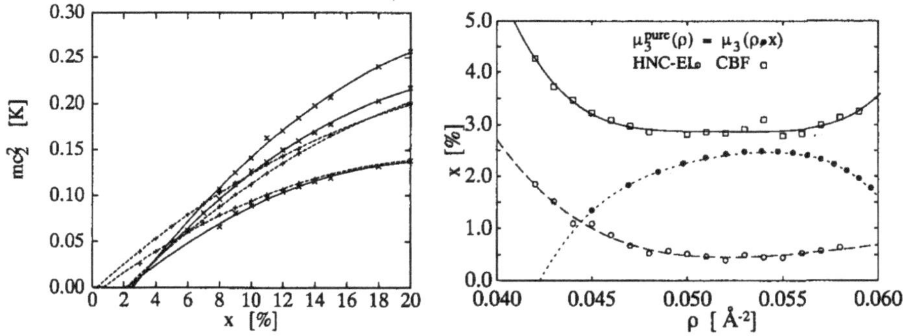

Figure 4. The speed of second sound of the two-dimensional mixture is shown for the densities $\rho = 0.045$, 0.055, and 0.065 Å^{-2}, as a function of concentration x for both the (F)HNC-EL calculation (+-symbols) and the CBF corrected calculation (crosses). Also shown are the fits to these data that determine the estimates for the lowest concentration that is stable against concentration fluctuations (dashed and solid lines, respectively) (left figure). The figure shows the extrapolated critical concentration where the atomic mixture becomes unstable against *infinitesimal* concentration fluctuations. The squares and circles show the actually extrapolated values from the CBF and FHNC-EL calculation, respectively; the long-dashed and the solid line give a smooth interpolation of these results. The short-dashed line marked with filled dots gives the upper bound of the concentration below which the mixture is globally stable (right figure).

both the (F)HNC-EL calculation and the CBF corrected calculations. It is very difficult to get close to the actual phase transition because this instability is driven by the appearance of the ^{3}He-^{3}He dimer.

Finally the phase diagram of the two-dimensional ^{3}He-^{4}He mixture with critical concentrations is shown in the right figure of Fig. 4. The markers of the upward bending interpolating curves are concentrations where the speed of the second sound becomes zero. The upper curve is the full result including the CBF corrections. Below that curve the atomic mixture is unstable against the dimer formation. The downward bending curve gives the critical concentrations where the chemical potential in the mixture is equal to the one in the pure ^{3}He. Above that line the mixture is globally unstable and can exist only in a meta-stable state. Below that line the stable mixture can exist, but only in the dimerized state.

4 Summary

At low concentrations in two dimensions ^{3}He component in ^{4}He forms dimers. This dimer phase is stable below 2-3 % concentrations. The two-dimensional ^{3}He-^{4}He atomic mixture can exist in a super-saturated state above 3% concentrations, because concentrations above that are unstable against phase separation.

In three dimensions the ^{3}He atoms in the ^{3}He-^{4}He mixture do not form bound dimers instead they form Cooper pairs leading to the superfluid phase transition of the ^{3}He component. We predict that the highest transition temperature achievable experimentally is for a s-wave superfluid phase transition of the 2% mixture at about 20 μK.

Acknowledgments

The work was supported, in part, by the Academy of Finland under project 163358. I like to thank E. Krotscheck, K. Schörkhuber and J. Paaso of collaboration on this project.

References

1. C. Ebner and D. O. Edwards, *Phys. Rep.* **2**, 77 (1971).
2. J. Bardeen, G. Baym, and D. Pines, *Phys. Rev.* **156**, 207 (1967).
3. E. Krotscheck, J. Paaso, M. Saarela, K. Schörkhuber, and R. Zillich, *Phys. Rev.* B **58**, 12282 (1998).
4. J. Boronat, private communication, 1999.
5. E. Feenberg, *Theory of Quantum Fluids* (Academic Press, New York, 1969).
6. E. Krotscheck and M. Saarela, *Phys. Rep.* **232**, 1 (1993).
7. E. Krotscheck, J. Paaso, M. Saarela, and K. Schörkhuber, *Phys. Rev. Lett.* **85**, 2344 (2000).
8. E. Krotscheck, J. Paaso, M. Saarela, and K. Schörkhuber, *Phys. Rev.* B , submitted (2000).
9. J. W. Clark, in *Progress in Particle and Nuclear Physics*, volume 2, Ed. D. H. Wilkinson (Pergamon Press Ltd., Oxford, 1979), p. 89.
10. R. A. Aziz, F. R. W. McCourt, and C. C. K. Wong, *Mol. Phys.* **61**, 1487 (1987).
11. C. E. Campbell and E. Feenberg, *Phys. Rev.* **188**, 396 (1969).
12. J. C. Owen, *Phys. Rev. Lett.* **47**, 586 (1981).
13. S. Giorgini, J. Boronat, and J. Casulleras, *Phys. Rev.* B **54**, 6099 (1996).
14. R. de Bruyn Ouboter and C. N. Yang, *Physica* B **144**, 127 (1986).
15. G. Baym, *Phys. Rev. Lett.* **17**, 952 (1966).
16. J. Boronat, A. Fabrocini, and A. Polls, *Phys. Rev.* B **39**, 2700 (1989).

Strongly Correlated Electrons

SPIN POLARIZATIONS OF QUANTUM HALL STATES

TAPASH CHAKRABORTY
Institute of Mathematical Sciences, Taramani, Madras 600 113, India

KARRI NIEMELÄ AND PEKKA PIETILÄINEN
Theoretical Physics, University of Oulu, Linnanmaa, FIN-90570 Oulu, Finland

We present our results on temperature dependence of spin polarizations at $\nu = 1$ in the lowest as well as in the next higher Landau level that compare well with recent experimental results. At $\nu = 3$ the behavior of the spin polarization is not much influenced by higher Landau levels, except by having a much smaller magnitude. In sharp contrast we predict that for filling fraction $\nu = \frac{8}{3}$, unlike the case of $\nu = \frac{2}{3}$, the system remains fully spin polarized even at vanishingly small Zeeman energies.

1 Introduction

The quantum Hall effects (QHE) have provided us interesting systems to explore the properties of interacting electrons in two dimensions.[1] It has long been known[2] that the electron spin degree of freedom plays an important role in these systems. For the Landau level filling factor $\nu = 1$ ($\nu = n_e/n_s$, where n_e is the electron number and $n_s = AeB/\hbar c = A/2\pi\ell_0^2$ is the Landau level degeneracy, A is the area of the system and ℓ_0 is magnetic length) the ground state is fully polarized with total spin $S = n_e/2$,[3] even without the Zeeman energy. A fully spin-polarized state is also expected for $\nu = \frac{1}{3}$, while filling factors $\nu = 2/m$, where m is an odd integer[1] exhibit unpolarized ground states. Recently, a new method to investigate spin polarizations in these systems was introduced by Barrett *et al.*[4] (see also Ref. 5) in their work on spin excitations around $\nu = 1$ and also temperature dependence of spin polarizations at $\nu = 1$. This method has been used by several groups to explore spin polarization of various other filling factors. The basic idea in these experiments is to exploit the magnetic interaction between ^{71}Ga nuclei and conduction electrons in GaAs quantum wells by using nuclear magnetic resonance (NMR) spectroscopy. Direct information about electron spin polarization of quantum Hall states is obtained via measurement of the Knight shift of ^{71}Ga NMR signal due to conduction electrons in the GaAs well. For a fully polarized ground state, as is the case for $\nu = 1$ and $\nu = \frac{1}{3}$, experimental results indicate that spin polarization saturates to its maximum value at very low temperatures and drops rapidly as the temperature is raised (Fig. 1, see also Ref. 4, 5). At large T, spin polarization is expected to decay as T^{-16-8} and was found experimentally to behave that way.[4,5]

More recently, Song *et al.*[10] reported on an experimental set up similar to that of Barret *et al.*[4] in order to explore $\nu = 1$ and $\nu = 3$. Interestingly, temperature dependence of spin polarization at $\nu = 3$ revealed a different behavior as compared to that at $\nu = 1$. More specifically, their results indicate that even at the lowest temperature studied, the spin polarization at $\nu = 3$ does not show any indication of saturation and with increasing temperature it drops sharply down to zero (Fig. 1(b)).

In this paper, we present our results on spin-polarization versus temperature for $\nu = 1$ in the lowest Landau level as well as in the next Landau level. A comparison of our results with the experimental results of Ref. 10 is also made. We find that at low temperatures, the behavior of spin polarization at $\nu = 3$ is similar to that at $\nu = 1$ but of much smaller magnitude.

The results agree reasonably well with available experimental data at $\nu = 3$. However, discrepancies between the theoretical results and the experimental data remain at higher temperatures. We also present theoretical results for $\nu = \frac{2}{3}$ in the next higher Landau level. At $\nu = \frac{2}{3}$, convincing evidence exists about the spin polarization in the lowest Landau level,[1,11,13] but there are no experimental data available as yet for spin polarizations in the next higher Landau level, i.e., for $\nu = \frac{8}{3}$.

2 Model

To explore the temperature dependence of spin polarizations of QHE states in the lowest and second lowest Landau levels we apply the exact diagonalization method in a periodic rectangular geometry.[1] This numerical method has been widely used in the quantum Hall effect literature[1] and is known to be very accurate in determining the ground state and low-lying excitations in the system. Since even at the lowest experimental magnetic field the Landau level separation $\hbar\omega_c$ is still an order of magnitude greater than typical energies due to the Coulomb interaction, electrons in the lowest Landau level can be treated as inert. In the calculations that follow we can therefore consider the lowest Landau level to be a uniform background causing merely a constant shift to interaction energies. The higher Landau levels then enter the system Hamiltonian via a modified interaction potential.[12] More specifically, for a finite number of active electrons N_e in a rectangular cell and choosing the Landau-gauge vector potential, the Hamiltonian in the $n = 0, 1$ Landau levels is (ignoring the kinetic energy and single-particle terms in the potential energy which are constants[1]),

$$\mathcal{H} = \sum_{j_1, j_2, j_3, j_4} \mathcal{A}_{nj_1, nj_2, nj_3, nj_4} a^\dagger_{nj_1} a^\dagger_{nj_2} a_{nj_3} a_{nj_4}.$$

Here the matrix element is defined via

$$\mathcal{A}_{nj_1, nj_2, nj_3, nj_4} = \delta'_{j_1+j_2, j_3+j_4} \mathcal{F}_n(j_1 - j_4, j_2 - j_3), \tag{1}$$

$$\begin{aligned}\mathcal{F}_n(j_a, j_b) &= \frac{1}{2ab} \sum_{\mathbf{q}} {\sum_{k_1}}' \sum_{k_2} \delta_{q_x, 2\pi k_1/a} \delta_{q_y, 2\pi k_2/b} \delta'_{j_a k_2} \\ &\quad \times \frac{2\pi e^2}{\epsilon q} \left[\frac{8 + 9(q/b') + 3(q/b')^2}{8(1 + q/b')^3} \right] \\ &\quad \times \mathcal{B}_n(q) \exp\left(-\frac{1}{2} q^2 \ell_0^2 - 2\pi i k_1 j_b / n_s \right),\end{aligned} \tag{2}$$

$$\mathcal{B}_n(q) = \begin{cases} 1 & \text{for } n = 0 \\ (1 - \frac{1}{2} q^2 \ell_0^2)^2 & \text{for } n = 1 \end{cases}. \tag{3}$$

$$n_e = \begin{cases} N_e & \text{for } n = 0 \\ \frac{\nu}{\nu-2} N_e & \text{for } n = 1 \end{cases} , \tag{4}$$

and a and b are the two sides of the rectangular cell that contains the electrons. The Fang-Howard variational parameter b' is associated with the finite-thickness correction,[1] ϵ is the background dielectric constant, and the results are presented in terms of the dimensionless thickness parameter $\beta = (b'\ell_0)^{-1}$. The Kronecker δ with prime means that the equation is defined mod n_s, and the summation over q excludes $q_x = q_y = 0$.

We then calculate the temperature dependence of spin polarization for different filling factors from,[6-8]

$$\langle S_z(T)\rangle \equiv \frac{1}{Z}\sum \mathrm{e}^{-\varepsilon_j/kT}\langle j|S_z|j\rangle$$

where $Z = \sum_j \mathrm{e}^{-\varepsilon_j/kT}$ is the canonical partition function and the summation is over all states including all possible polarizations. Here ε_j is the energy of the state $|j\rangle$ with Zeeman coupling included. They are evaluated for finite-size systems in a periodic rectangular geometry.[1] Our earlier theoretical results indicated that at small values of the Zeeman energy, temperature dependence of spin polarization is non-monotonic for filling factors $\nu = 2/m$, $m > 1$ being an odd integer. In particular, for $\nu = \frac{2}{3}$ and $\nu = \frac{2}{5}$, we found that spin polarization initially *increases* with temperatures, reaching a peak at $T \sim 0.01$ K when it falls as $1/T$ with increasing temperature. Appearence of the peak was associated with spin transitions at these filling factors and was found to be in good agreement with the experimental observation.[11] For $\nu = 1$ and $\nu = 1/3$, our results are also in excellent agreement[7] with the earlier available experimental results.[4,5]

3 Results and discussion

Here we present our numerical results for spin polarization versus temperature for filling fractions $\nu = 1$, $\nu = 3$, $\nu = \frac{2}{3}$, and $\nu = \frac{8}{3}$. Our results for $\langle S_z(T)\rangle/\max\langle S_z(T)\rangle$ versus T for an eight-electron system in a periodic rectangular geometry at $\nu = 1$ are presented in Fig. 1 (a) where we also present the experimental data of Ref. 10 for comparison.

Here the temperature is expressed in units of $e^2/\epsilon\ell_0$ and the conversion factor to K is $e^2/\epsilon\ell_0[K] = 51.67(B[\,\mathrm{T}])^{\frac{1}{2}}$, appropriate for systems studied experimentally. In our calculations, we fix the parameters tothe experimental values: the Landé g-factor is 0.44 and the magnetic field is $B = 9.4$ T. The curves that are close to the experimental data (and presented here) are for $\beta = 2 - 4$. As we discussed above, at low temperatures there is a rapid drop in spin polarization and for high temperatures, spin polarizations decay as $1/T$. Our results are in good agreement with those experimental features. They were also in good qualitative agreement with the earlier experimental results at this filling factor.[7] These results are presented with the intention of comparing them with the temperature dependence of spin polarization at $\nu = 3$. The results in the latter case are shown in Fig. 1 (b) (again for an eight-electron system in a periodic rectangular geometry). In drawing this figure, we have taken the following facts from the experimental results of

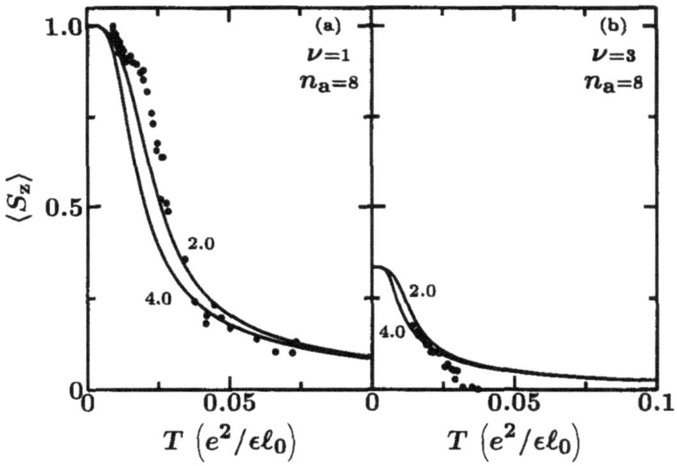

Figure 1. Temperature dependence of spin polarization at (a) $\nu = 1$ and (b) $\nu = 3$ for two different values of finite-thickness parameter β. Experimental data points are from Ref. 10.

Ref. 10 into consideration: (i) that the maximum $\langle S_z \rangle$ is in fact, 1/3 and not 1 as in $\nu = 1$, (ii) the experimental scale at $\nu = 3$ of Ref. 10 is the same as that at $\nu = 1$, and (iii) spin polarization at $\nu = 3$ is drawn in Fig. 1 (b) in the same scale as for $\nu = 1$. All the parameters except the magnetic field are kept the same as in the case of $\nu = 1$. Just as in the experiment, we fix the magnetic field for $\nu = 3$ at the much lower value of $B = 4.4$ T. The filled Landau levels, however, are still found to be inert at this low field and does not influence our chosen Hamiltonian. As seen in Fig. 1 (b), numerical values of spin polarization are much smaller here than those for $\nu = 1$. Our theoretical results for $\beta = 2 - 4$ agree reasonably well with the experimental results of Ref. 10 except in the high temperature regime where the experimental data drop down to zero. Our results, in contrast, have the usual $1/T$ tail. We should point out however, that due to discreteness of the energy spectrum for finite number of electrons the terms with S_z and $-S_z$ in the polarization cancel each other at high temperatures like $1/T$ and we will always end up with $1/T$ decay of $\langle S_z(T) \rangle$ versus T.[6] Therefore, we cannot predict with certainty how a macroscopic system would behave at high T. However, given the fluctuations in data points for $\nu = 1$ and $\nu = 3$ and the fact that the last few data points for $\nu = 3$ are extremely small, it is not clear if one expects saturation of points with $1/T$ behavior or the spin polarization actually vanishes. Clearly, experimental data at high temperatures do not show any sign of saturation and in order to settle the question of actual vanishing of $\langle S_z(T) \rangle$ it would be helpful to have more data in the high temperature regime. Saturation is also not visible in the low-temperature region of the experimental data. In order to clarify many of these outstanding issues, it is rather important to have more experimental probes of temperature dependence at this filling factor.

In sharp contrast to the filling fraction $\nu = 3$, the influence of higher Landau levels is found to be quite significant for $\nu = \frac{2}{3}$. As we have demonstrated earlier,[6]

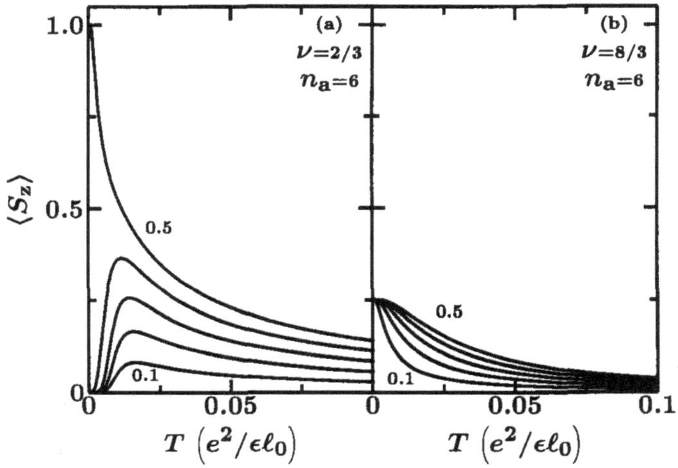

Figure 2. Temperature dependence of spin polarization at (a) $\nu = \frac{2}{3}$ and (b) $\nu = \frac{8}{3}$ for $\beta = 0$ and different values of Landé g-factor ($g = 0.1 - 0.5$).

at low Zeeman energies the system at this filling factor is spin unpolarized and with increasing Zeeman energies, the system undergoes a phase transition to a fully spin polarized state. Similar result is also expected for $\nu = \frac{2}{5}$. This is now well established through a variety of experiments.[11,13,15–18] Our results for $\langle S_z(T)\rangle$ versus T at $\nu = \frac{2}{3}$ and $\nu = \frac{8}{3}$ are shown in Fig. 2 where we present results for a six-electron system and a magnetic field value of 4.4 tesla.

We present the results for $\beta = 0$, but the results are found to be insensitive to finite-thickness correction. We also consider several different values of Landé g-factor in the range 0.1-0.5. Interestingly, the results indicate that the total spin S of the active electrons, unlike in the lowest Landau level, is at its maximum value $S = N_e/2$ even without Zeeman coupling. Even an infinitesimal Zeeman coupling will orient the spins in the active system resulting the polarization to be 1/4 contradicting the simple minded composite fermion model which predicts fractions of the form $2 + 2/m$, m odd, to be unpolarized. This somewhat surprising behaviour can be thought to be due to the more repulsive effective interaction forcing the electrons, according to Hund's rule, to settle down in the maximum spin state more effectively than for electrons on the lowest Landau level. In Ref. 9 we considered also the case of a very small Zeeman energy ($g = 0.02$), but the results still indicate full spin polarization of the active system. At this low Zeeman energy, spin polarization drops rather rapidly from its maximum value as the temperature is increased. In this context, we should mention that the idea of an extremely small Zeeman energy is not that far fetched: in recent experiments, a significant reduction in Zeeman energy has been achieved by application of a large hydrostatic pressure on the heterostructure.[17,19,20] It is even possible to have situations close to zero Zeeman energy.[21] With the help of all the different techniques available in the literature to study spin polarization, it should be possible to explore $\langle S_z(T)\rangle$ for $\nu = \frac{8}{3}$.

4 Summary

In closing, we have studied spin polarization as a function of temperature for $\nu = 1$ and $\nu = \frac{2}{3}$ in the higher Landau level. Our results indicate that for $\nu = 3$ the calculated temperature dependence of spin polarization is not much influenced by the higher Landau level (except being much lower in magnitude). Available experimental results are incomplete at low and high temperature regions where no saturation of data points have been observed. Our results at $\nu = \frac{8}{3}$ reveal that the system is always fully spin polarized even at very small Zeeman energies. This is in contrast to the behavior at $\nu = \frac{2}{3}$ which, at low Zeeman energies has a spin unpolarized state[1] that is well supported by various experimental investigations. More experimental data points at $\nu = 3$ in the low and high-temperature regime would be very helpful. Experimental probe of $\nu = \frac{8}{3}$ with NMR and optical spectroscopy should be able to explore the spin states predicted in the present work.

References

1. T. Chakraborty and P. Pietiläinen, *The Quantum Hall Effects* (Springer, New York, 1995).
2. B. I. Halperin, *Helv. Phys. Acta* **56**, 75 (1983); T. Chakraborty and F. C. Zhang, *Phys. Rev.* B **29**, 7032 (1984).
3. F. C. Zhang and T. Chakraborty, *Phys. Rev.* B **64**, 7076 (1986).
4. S. E. Barrett *et al.*, *Phys. Rev. Lett.* **74**, 5112 (1995); P. Khandelwal *et al.*, *Phys. Rev. Lett.* **81**, 673 (1998).
5. M. J. Manfra, B. B. Goldberg, L. N. Pfeiffer, and K. N. West, *Phys. Rev.* B **54**, R17327 (1996); *Physica* E **1**, 28 (1997).
6. T. Chakraborty and P. Pietiläinen, *Phys. Rev. Lett.* **76**, 4018 (1996).
7. T. Chakraborty, P. Pietiläinen, and R. Shankar, *Europhys. Lett.* **38**, 141 (1997).
8. T. Chakraborty, K. Niemelä, and P. Pietiläinen, *Phys. Rev.* B **58**, 9890 (1998).
9. T. Chakraborty and P. Pietiläinen, *Phys. Rev. Lett.* **83**, 5559 (1999).
10. Y.-Q. Song, B. M. Goodson, K. Maranowski, and A. C. Gossard, *Phys. Rev. Lett.* **82**, 2768 (1999).
11. I. V. Kukushkin, K. v. Klitzing, and K. Eberl, *Phys. Rev.* B **55**, 10607 (1997).
12. T. Chakraborty and P. Pietiläinen, *Phys. Rev.* B **38**, 10097 (1988).
13. I. V. Kukushkin, K.v. Klitzing, and K. Eberl, *Phys. Rev. Lett.* **82**, 3665 (1999).
14. K. Niemelä, P. Pietiläinen, and T. Chakraborty, to be published (1999).
15. L. W. Engel *et al.*, *Phys. Rev.* B **45**, 3418 (1992).
16. S. Kronmüller *et al.*, *Phys. Rev. Lett.* **81**, 2526 (1998).
17. W. Kang *et al.*, *Phys. Rev.* B **56**, 12776 (1997); H. Cho *et al.*, *Phys. Rev. Lett.* **81**, 2522 (1998).
18. S. I. Dorozhkin *et al.*, *Phys. Rev.* B **55**, 4089 (1997).
19. N. G. Morawicz *et al.*, *Semicond. Sci. Technol.* **8**, 333 (1993).
20. S. Holmes *et al.*, *Semicond. Sci. Technol.* **9**, 1549 (1994).
21. D. R. Leadley *et al.*, *Phys. Rev. Lett.* **79**, 4246 (1997).

ELECTRONIC MOLECULES IN CONDENSED MATTER

FEODOR V. KUSMARTSEV
School of MAP, Loughborough University, Loughborough LE11 3TU, UK
E-mail: F.Kusmartsev@lboro.ac.uk

Molecules may be created not only from atoms but also from electrons. If for molecules consisting of atoms the glue which is binding atoms into molecules is the electron cloud around these atoms then for electronic molecules the glue which binds these electrons into an electronic molecule is the phonon cloud. But atomic molecules are drastically different from electronic molecules. Their shapes and sizes are not well defined although they vary within some limits. We show that electronic molecules may arise in solids with rather narrow bands. In many situations, for example in metal and magnetic oxides materials, these molecules have a linear shape and therefore we call them "electron strings". They arise primarily in oxide materials with not very wide bands, like HTSC, due to the electron-phonon and electron-electron interactions although antiferromagnetic correlations are also very important. We estimate the string length and the number of particles self-trapped into a single string, taking into account the typical parameters of cuprates, such as La_2CuO_4.

1 Introduction

It was Landau who first in 1933 suggested that an electron in an ionic solid may be surrounded by a polarization, phonon cloud in a similar way as atomic nuclei are surrounded by an electronic cloud. The associated quasi-particle was called a polaron. Since that time polarons have been studied very intensively. The first consistent theory was built by Pekar, and described the polarons in ionic solids. The polarons and their properties are now well-studied, both theoretically and experimentally. So it is a well established phenomenon. If the polaron may be viewed as an electronic atom, where electrons play the role of the nuclei and phonons replace the electronic orbitals, then a bipolaron is indeed an electronic molecule consisting of two such electronic atoms. The bipolarons have been studied less intensively than polarons, but some facts are known about their existence in some oxide materials, like, for example, $BaBiPbO_3$. Recently we have proposed that in materials with narrow bands there may arise many-particle electronic molecules having a linear shape, which are called strings. The linear shape is a result of the competition between the long-range electron-electron Coulomb repulsion and the short range effective attraction associated with the electron- phonon interaction. In some cases one also has to take into account the electron kinetic energy.

There is a growing body of experimental evidence[1–4] and there also exist theoretical arguments[5,6] indicating the existence of complex inhomogeneous mesoscopic structures created from holes and spins in high-temperature superconductors (HTSC) and other oxides. Such inhomogeneous structures ascribed to *a stripe phase* have been discussed in HTSC by Bianconi, Thurston, Tranquada, *et al*[12–17] and in theoretical papers.[7–11] In more recent papers[2,3,16,18] the one dimensional character of charge and spin fluctuations and the importance of lattice distortions in the creation of these mesoscopic stripe structures have been clearly demonstrated. This stripe structure has also been observed in other oxides, manganites[4] and there the lattice effects have been seen. These experiments show that the stripe phenomenon is generic and there the lattice deformations arise.

Recently we have proposed that in materials with narrow bands there may arise a long cigar-shaped object consisting of many self-trapping particles called a *string*. The string phenomenon is a many-particle generalization of a conventional self-trapping effect which normally leads to the formation of fluctuons, deformons, polarons and other single and two-particle objects studied very intensively in the past (see, for example, Ref. 19 and references therein). The electron strings may be created both by electron-phonon and electron-electron interactions. In the limit of very narrow bands studied in Refs. 5, 6, 18 the electron hopping vanishes and strings are insulating defects.

In the present work we have taken into account the finite bandwidth and treat the kinetic energy of electrons on equal footing with their potential energy. The increase of the bandwidth (when the contribution of electron kinetic energy increases) leads to a new type of strings - *conducting strings*. Since the creation of strings is related to atomic displacements they may be detected by all sorts of experiments, although primarily in scanning tunneling microscopy (STM) experiments. Photoemission, X-ray and neutron scattering experiments may also be very useful although the mesoscopic size of the strings may create an obstacle. The strong lattice fluctuations, which may be associated with conducting or insulating strings, have been detected by MeV helium ion channeling, an ultrafast real-space probe of atomic displacements.[2]

In general such strings may correspond to either a ground state or a metastable state. However in a doped antiferromagnet the criterion for the formation (of both conducting and insulating) strings is much better satisfied. Either type of strings may arise in the ground state.

The stripe phase observed in HTSC[1,3,12–17] may correspond to a liquid crystal consisting of highly conducting or insulating electron strings which coexists with itinerant free current carriers.[18] Depending on the distance between these strings the liquid crystal may be in either of two states: insulating or conducting. The conducting state may arise either due to percolation through the strings or due to the appearance free itinerant current carriers.[18]

2 Hamiltonian

We consider the general Hamiltonian of spinless fermions interacting via a strong long-range electron-electron repulsion and with different types of phonons on a d-dimensional hypercubic lattice

$$H = \left[-t \sum_{<i,j>} a_i^\dagger a_j + \sum_{q,i} \omega(q) n_i u_i(q) b_q + \sum_q \omega(q) b_q^\dagger b_q + hc\right] \sum_{i<j} V(i-j) n_i n_j, \quad (1)$$

where t is the electron hopping-integral, the operator $a_i^\dagger(a_i)$ creates (destroys) a fermion at a lattice site i , n_i is the occupation number operator $a_i^+ a_i$ and the operator $b_q^\dagger(b_q)$ is an operator of the creation (destruction) of a phonon. The summations in Eq. (1) extend over the lattice sites i and – as indicated by $< i, j >$ –

over the associated nearest-neighbour sites j. The matrix element of the electron-phonon interaction is equal to

$$u_n(q) = \frac{\gamma(q)\exp(iqn)}{\sqrt{2N}}. \quad (2)$$

The function $\gamma(q)$ and the phonon dispersion relation $\omega(q)$ are different for different types of electron-phonon interactions. For example, for optical longitudinal phonons (Pekar-Frölich interaction[20]) in the continuum limit the product $\gamma^2(q)\omega(q) = 4\pi e^2/(\epsilon^* q^2)$ with $1/\epsilon^* = 1/\epsilon_\infty - 1/\epsilon_0$. In general for the short-range electron phonon interaction $\gamma^2(q)\omega(q) = \text{const} = c$. The long-range part of the Coulomb interaction has the conventional type,[5]

$$V(i-j) = \frac{e^2}{\bar{\epsilon} \mid i-j \mid}, \quad (3)$$

where it is plausible to assume that for spinless fermions the effective dielectric constant $\bar{\epsilon}$ may be taken as $\bar{\epsilon} = \epsilon_\infty$. Such a Hamiltonian, we believe, may correctly describe the physics. Since it contains both a strong electron-electron repulsion and an electron-phonon interaction, it may lead to the formation of highly conducting electron strings.

The Hartree-Fock many-body wave function of the M self-trapped particles $\Psi(1,2,...,M)$, which we employ to calculate an expectation value of the Hamiltonian, Eq. (1) has the form of a Slater determinant[5,6]

$$\Psi(1,2,...,M) = \frac{1}{\sqrt{M!}} \det \| \psi_i(k_j) \|, \quad (4)$$

consisting of single particle wave functions

$$\psi_{m_x}(k_j) = \begin{cases} \frac{1}{\sqrt{N}} \exp(ik_j m_x) & \text{if} \quad 1 \leq \mathrm{m_x} \leq \mathrm{N} \\ \equiv 0, & \text{otherwise} \end{cases}. \quad (5)$$

Each of these wave functions describes the electron(hole) trapped by N neighboring sites (string potential well) with equal probability, $1/N$. If the string is oriented in the x direction and is located on the sites $m_x = 1,...,N$ the particle quasi-momentum k_j is determined by boundary conditions at the ends of the string. For simplicity we use periodic boundary conditions.

3 Jahn-Teller and optical phonons

With the use of this many body wave function we have estimated an expectation value of the Hamiltonian H, Eq. (1) taking into account both Jahn-Teller and Pekar-Frölich or longitudinal optical phonons. Using this many-body wave function, first, we calculated the one-body and the pair correlation functions and then with the use of the adiabatic approximation we have excluded slow (classical) phonon variables to get an expression for the adiabatic potential E_S including the Coulomb and exchange energies.[5,6] The calculated expression of the total energy E_S per particle has the form

$$\frac{E_S}{M} = 2dt - 2t\frac{\sin \pi n}{\pi n}\left(1 - \frac{n}{M}\right) - \frac{cn}{2} - \epsilon_c n \left[(1 - \frac{\epsilon_\infty}{\epsilon_0}) \log N - \log M\right], \quad (6)$$

where d is a dimension of the hypercubic lattice, the value n is the electron(hole) doping inside the string, $n = M/N$, and the value $\epsilon_c = e^2/(a\epsilon_\infty)$ with a as an interatomic distance. In the Hamiltonian (1) c is the coupling with Jahn-Teller phonons. The first two terms in the r.h.s. of Eq. (6) are associated with the electron kinetic energy, while the next terms are associated with the energies of electron-phonon and electron-electron interactions, respectively (see, for comparison, Ref. 5). Equation (6) represents a variational estimate of the total energy of M particles self-trapped into a string of length N valid for a wide range of values of c/t since it was obtained on the basis of an exact solution[5] found in the limit of very strong coupling $c/t \gg 1$. Therefore, in the framework of this variational approach we may get a reliable estimate of the number of particles, the length and the energy of an electron string valid for a wide range of the parameters of the Hamiltonian such as a coupling constant c, the bandwidth t and the characteristic Coulomb energy ϵ_c. Here the values M and n are variational parameters. The optimal number of particles trapped into the string of fixed length N is determined by a minimization of E_S/M with respect to M. If the ionicity of the solid is weak (i.e., $\epsilon_0 \sim \epsilon_\infty$) it is approximately given by

$$M = \frac{2t \sin \pi n}{\epsilon_c \pi n}. \tag{7}$$

After substitution of this expression into Eq. (6) for E_S we get the dependence $E_S = E_S(n)$ on the doping of the string $n = M/N$. Depending on the relation between the values of t, c and ϵ_c there may exist one or two types of solutions which correspond to two different types of strings: when $n = 1$ we obtain an *an insulating* string,[5] and when $n < 1$ we get a *a conducting* string. When $c \sim t > \epsilon_c$ the conducting string may be in its ground state. Then the number of particles trapped into the string is described by Eq. (7) and the value of the string doping must be determined numerically by minimization. When the coupling constant c is very large ($c \gg t$ and $c > \epsilon_c$) Eq. (7) is not applicable, since the associated solution describing a conducting string disappears while the solution associated with the marginal extremum $n = 1$ and describing insulating strings still exists. Thus, in extreme strong coupling limit the string is always insulating with $N = M$ and the number of particles in such a string is

$$M = \exp\left(\frac{c}{2\epsilon_c} - 1\right). \tag{8}$$

On the other hand the total energy $E_S(M)$, at large values of M decreases strongly with M and increases with N. In the case where the double occupation of the sites is prohibited, the number M can not be larger than N. Then the minimum energy E_S corresponds to $N = M$.

For a single hole in an antiferromagnet there is an increase in the exchange energy equal to $2dJ$, where J is an exchange constant. For M separated holes this energy increase is equal to $2dMJ$. On the other hand for M holes trapped in a string such an increase in exchange energy is equal to $J(2dM - M + 1)$. Therefore, the absolute minimum of the total energy of the string in a doped antiferromagnet is $E_{S-min} = [2dt - \epsilon_c + (2d-1)J]\, M + J$, where the value M is defined by Eq. (8). The comparison of this expression with the total energy of M separated self-trapped particles indicates that the strings may have a lower energy if the inequality $2\epsilon_c \leq c \leq 2\epsilon_c + 2(2d+1)J$ holds. Thus, the exchange interaction between anti- ferromagnetic spins significantly improves the physical conditions required for string formation in doped antiferromagnets. Therefore, at low temperatures, if this condition holds then M separated particles will condense into a string configuration. Such string solutions found originally for deformed type of strings[5] arise also in the case when the electron(hole) is interacting with other Jahn-Teller phonons, like those discussed in Ref. 21, or in general for any type of short-range electron-phonon interaction as, for example, with Holstein optical phonons. The number of particles in the string is defined by Eqs. (7,8) while the string length depends on the type of the string and for conducting strings must be estimated by a numerical minimization of $E_S(n)$ with respect to n. For each type of phonons which have a short-range interaction with electrons(holes) the coupling constant in Eqs. (7,8) must be defined, respectively, while the main Eqs. (6-8,) remain the same (for more details, see Ref. 22).

The case when a single electron or hole is interacting with polar phonons, i.e. with longitudinal optical phonons with frequency ω_0 (and with the constant of the electron-phonon interaction $\gamma^2(q) = 4\pi e^2/(q^2\epsilon^*\hbar\omega_0))$[20] is also relevant and important to most oxides having a considerable amount of ionic bonding. If $c = 0$, a minimization of the expression (6) with respect to M and n gives an estimate for the length of the string N and the number of particles M trapped into the string. For the value of M we get the analytic expression

$$M = \frac{2\sin(\pi n)}{(E_c - E_p)\pi n}. \tag{9}$$

In the limit of low density $n \ll 1$ the values of M and N or n may be presented by the analytic formulae,

$$M \approx \frac{2\epsilon_0 at}{e^2}, \tag{10}$$

and

$$n = \frac{M}{N} \approx (\frac{2\epsilon_0 at}{e^2})^{\beta} \exp(1.6\beta - .4), \tag{11}$$

where $\beta = \frac{\epsilon_\infty}{\epsilon_\infty - \epsilon_0}$.

4 Antiadiabatic Approach

The similar expression for the length of the string valid even beyond the adiabatic limit may be obtained[6] with the use of Lang-Firsov unitary transformation, which transforms the Hamiltonian H into the form

$$\tilde{H} = \exp(S)H\exp(-S) = -\sum_{<i,j>} \tilde{t}_{ij} a_i^\dagger a_j \quad - E_p \sum_i n_i + \\ + \sum_q \omega(q) b_q^\dagger b_q + \sum_{i<j} v_{ij} n_i n_j, \tag{12}$$

where $S = \sum_{q,i} n_i[w_i(q)b_q - h.c.]$, the hopping integral $\tilde{t}_{ij} = t\exp\left(\sum_q[w_i(q) - w_j(q)]b_q - h.c.\right)$, the polaron shift $E_p = \sum_q \omega(q)\gamma^2(q)/(2N)$, and the effective inter-particles interaction is

$$v_{ij} = \frac{1}{\epsilon_0 \mid i-j \mid} - \frac{1}{N}\sum_q \omega(q)\gamma^2(q)\cos(q \mid i-j \mid). \tag{13}$$

Then, with the use of this expression the total energy of the string when $M = N$ and when $t \to 0$, is equal to

$$E_{tot} = \epsilon_c M \log M + .6\epsilon_c M - cM/2. \tag{14}$$

The minimization of this expression with respect to the value M gives the equation (8) for the number M.

The same result may be obtained for the dispersionless optical (Holstein) phonons. However for Jahn-Teller phonons (in contrast to acoustical phonons) there arises a weak dispersion $\omega_q = \omega_0 + \delta\omega(\cos q_x + \cos q_y + \cos q_z)$ (with a q-independent matrix elements $\gamma(q) = \gamma$), and a weak attraction between particles on next-neighboring sites will be generated. This will give an extra contribution into the total energy,

$$E_{tot} = \epsilon_c M \log M + .6\epsilon_c M - ME_p. \tag{15}$$

The minimization of this expression with respect to the value M gives the length of the string $N = M$ as

$$M = \exp\left(\frac{E_p\epsilon_\infty a}{e^2} - 1.1\right). \tag{16}$$

However if we take into account both Jahn-Teller and longitudinal optical phonons (see, Eq. (6)), then the calculated expression of the total energy takes the form

$$E_{tot} = -\epsilon_c M\left[(1 - \frac{\epsilon_\infty}{\epsilon_0})\log N - \log M\right] - ME_p. \tag{17}$$

Then the expression for the number of particles trapped into the string is modified to the form

$$M = \exp\left(\frac{E_p\epsilon_0 a}{e^2} - 1.1\right).. \tag{18}$$

From the comparison of these two expressions, Eqs. (16) and (18), for the number of particles trapped into the string one sees that in ionic solids the conditions for the formation of electronic molecules are strongly improved due to a strong screening polaron effect associated with the polarisibility of ionic solids. The electronic molecules in ionic solids are very long.

From these two approaches (adiabatic and antiadiabatic) we obtain an universal expression for the number M of trapped particles strings, the logarithm of which is proportional to the electron-phonon interaction, c, or to the polaron shift E_p and is inversely proportional to the intersite Coulomb repulsion between holes, ϵ_c, or in an ionic solid, to $V = e^2/(\epsilon_0 a)$.

5 Strings and Stripes in HTSC

To estimate the length of the string we choose typical parameters of HTSC (for example, for La_2CuO_4, $\epsilon_0 = 30$, $\epsilon_\infty = 5$, the interatomic distance $a = 3.8$ Å and the hopping integral $t = .5$ eV). Then by a minimization of the total energy we obtain that 7 holes will be trapped into a string with a length equal to $N = 40$ interatomic distances or ~ 150 Å. Note, that this estimate gives only approximate parameters for the string. Many strings having nearly equal lengths are associated with nearly the same energy. This property of the strings leads to the phenomenon of string multistability, a glassy character of string creation. The string length and the number of the trapped electrons are not well defined.

The conclusion is that in polar oxide materials, like HTSC there may arise electronic strings which are linear multi-particle "electronic molecules". At low temperatures the electron strings may be ordered in CuO planes creating a nematic liquid crystal. The striped phase in HTSC observed in numerous experiments,[12,14,16,17] may correspond to such a liquid crystal of conducting strings, which coexists with itinerant current carriers.[18] With the doping of an antiferromagnet La_2CuO_4 the only change is in the distance between the strings while the structure of the strings (like the string doping n or the length N) is not changed. The metallic stripe phase arises due to a correlated percolation over these strings when the density of such strings is larger than the percolation threshold. For square lattices the percolation threshold is well known and is equal to $x_c \sim .5$.[24] Using this value and our estimate for the string doping in La_2CuO_4 as $n = 7/40$ we readily obtain the hole doping $\delta = nx_c \approx .09$ of the antiferromagnet La_2CuO_4 at which the metallic stripe phase may arise. However, if the strings are insulating than the metallic phase may arise when there appear, on doping, itinerant current carriers.[18] The spin-spin and hole-spin correlations will of course slightly modify these results. It seems that our conclusion about the important contribution of the phonons into the origin of the stripe phase is confirmed in recent experiments which show a large influence of isotopic effects on the critical temperature of the stripe ordering[3] and strong lattice fluctuations in YBCO,[2] which may be associated with the dynamics of the strings. With the isotope changes[3] the structure of individual strings is changed (for example, the strings become shorter), and therefore the critical temperature of the stripe ordering must change.

In summary, we find that in oxides HTSC there may arise electron strings – linear electronic molecules whose length and number of electrons are not well defined. A single electronic molecule has a cigar shape with the length of the order of 2-20 lattice spacings and consisting of 2-20 holes. For other oxides, due to the universality and multistability of the string creation, these string parameters will not be changed as much. It is also very natural that such "polymeric" electron molecules may form a liquid crystal which may be associated with the stripe phase of HTSC.[18]

Acknowledgments

I am very grateful to A. Bianconi, A. Bishop, S. Kivelson, D. Edwards, G. Gehring, V. Emery, E.I. Rashba, Danya Khomskii, H.S. Dhillon, and other participants of the workshop on strongly correlated electrons in Isaac Newton Institute (Cambridge) for illuminating discussions. The work has been supported by the Isaac Newton Institute, University of Cambridge.

References

1. H. A. Mook, Pengcheng Dai, F. Dogan, R. D. Hunt, *Nature* **404**, 729 (2000).
2. R. P. Sharma *et al*, *Nature* **404**, 736 (2000).
3. A. Lanzara, G. Zhao, N. L. Saini, A. Bianconi, K. Conder, H. Keller and K. A. Müller, *J. Phys. C (Cond. Matt.)* **11**, L541 (1999).
4. M. Uehara *et al*, Cheong, *Nature* **399**, 560 (1999) ; S. Mori, S. Chen *et al*, *Nature* **392**, 473 (1999).
5. F. V. Kusmartsev, *J. de Physique IV* **9**, Pr10-321 (1999).
6. F. V. Kusmartsev, *Phys. Rev. Lett.* **84**, 530, 5026 (2000).
7. E. L. Nagaev, *Sov. Phys. JETP* **16**, 558 (1972); V. A. Kaschin and E. L. Nagaev, *Zh. Eksp. Teor. Fiz.* **66**, 2105 (1974).
8. J. R. Zaanen and O. Gunnarson, *Phys. Rev.* B **40**, 7391 (1989).
9. U. Löw *et al*, *Phys. Rev. Lett.* **72**, 1918 (1994).
10. V. J. Emery, and S. A. Kivelson, *Nature* **374**, 434 (1995).
11. V. J. Emery *et al*, PhyB **163**, 306, (1990); *Phys. Rev. Lett.* **64**, 475 (1990).
12. A. Bianconi, *Phys. Rev.* B **54**, 12018 (1996); M. Zimmermann *et al*, *Europhys. Lett.* **41**, 629 (1998).
13. T. R. Thursten *et al*, *Phys. Rev.* B **40**, 4585 (1989).
14. J. M. Tranquada, *Nature* **375**, 561 (1995).
15. A. Bianconi *et al*, *Phys. Rev. Lett.* **76**, 3412 (1996); and see references therein.
16. H. A. Mook *et al*, *Nature* **395**, 58 (1998).
17. N. L. Saini, J. Avila, A. Bianconi *et al*, *Phys. Rev. Lett.* **79**, 3467 (1997)
18. F. V. Kusmartsev, D. Di Castro, G. Bianconi and A. Bianconi, *Phys. Lett.* A **275**, 118 (2000).
19. E. I. Rashba, in: *Excitons*, Eds. E. I. Rashba and M. D. Sturge (North-Holland, Amsterdam, 1982), p. 543.
20. S. I. Pekar, *Untersuchungen über die Elektronentheorie Kristalle* (Akademie Verlag, Berlin, 1954).
21. L. P. Gorkov and A. B. Sokol, *Pisma Zh. Eksp. Teor. Fiz.* **46**, 333 (1987).
22. F. V. Kusmartsev, in preparation.
23. V. J. Emery, S. A. Kivelson and J. M. Tranquada, *Proc. NAS USA* **96**, 8814 (1999).
24. B. I. Shklovskii and A. L. Efros, *Electronic Properties of Doped Semiconductors* (Springer-Verlag, New York, 1984).

THE METAL-INSULATOR TRANSITION IN 2D AND NEW PHASES OF QUANTUM LOCALISATION

D. NEILSON AND J. S. THAKUR
School of Physics, The University of New South Wales, Sydney 2052 Australia
E-mail: d.neilson@unsw.edu.au

We describe a theory for the 2D electron system with defects which takes into account both electron-electron and electron-defect interactions. The approach is based on known properties of electrons in the low density limit. Our results provide a basis for understanding some of the properties of the metal insulator transition. The zero temperature phase diagram of localised and delocalised states as a function of electron density, defect concentration, and magnetic field are in good agreement with experiment.[1-3] We identify two distinct localisation transitions as a function of decreasing density, a discontinuous transition from liquid to solid at low densities, and a continuous localisation from solid to liquid at higher densities, $r_s \lesssim 8$. We also find that a magnetic field parallel to the 2D plane tends to suppress the conducting phase.

1 Introduction

In conventional semiconductors the r_s value for the electron densities is so small (typically $r_s \lesssim 1$) that the electron-electron interaction strength is weak compared to the Fermi energy. This restriction is mainly due to the large effective Bohr radius, $40 \lesssim a_B^\star/a_B \lesssim 200$. In a 2D layer without any electron-electron interactions there can be no conducting state in two-dimensions in the presence of any defects.[4] The observation of a metal-insulator transition[1] in 2D electron layers in n-Si MOSFETs thus generated considerable interest. The transition has now been observed also in p-type and n-type GaAs-AlGaAs devices, and in p-type SiGe. This makes it unlikely that it is some material dependent effect. The nature of the insulating and conducting states remains unclear, and there is even some doubt that the observed properties persist in the zero temperature limit. The transition is only observed at low electron densities where the electron-electron interaction energies are much larger than the Fermi energy so it is likely the transition is driven by these interactions.

At low densities the exchange-correlation holes which surround each electron are strong and they are spatially extensive on the scale of the spacing between electrons. The density profile in the exchange-correlation hole has been determined using diffusion quantum Monte-Carlo numerical simulations (DQMC).[5,6] Wigner localisation is caused by the strong density exclusion around each electron, leading to localisation of the density fluctuations. In the pure system it occurs at extremely low electron densities,[5] $n \lesssim 3 \times 10^8\text{cm}^{-2}$ in n-type GaAs. The mechanism for Wigner localisation is quite different from that for Anderson localisation by defects. Wigner crystallisation has close parallels with close-packing solidification of hard spheres.

For electron densities $r \gtrsim 8$ there is a region of complete density exclusion in the exchange-correlation hole out to a distance r_h from the centre.[5,6] The electrons with their exchange-correlation hole resemble particles surrounded by a density exclusion region as if they had repulsive hard cores. As the electron density is decreased the ratio r_h/r_0 expands so that an increasing fraction of the total available area is excluded from occupancy by neighbouring electrons. At low electron densities in the presence of a small concentration of defects, localisation should continue to be driven by the strong electron-electron interactions. Disorder destroys long-range-order, so an electron solid would be a frozen electron glass with liquid-like short-range order. The defects only facilitate the localisation, acting as pinning centres. We find that weak substrate disorder can cause a phase transition from electron liquid to a glass phase. This is quite similar to a classical glass of hard discs near the close-packing limit. Observations of hysteresis would provide direct evidence for a metastable electron glassy phase.

For strong electron-electron interactions, the density fluctuations $\rho(\boldsymbol{q},t) = \sum_{\boldsymbol{k}<k_F} a^{\dagger}_{\boldsymbol{k}+\boldsymbol{q}}(t)a_{\boldsymbol{k}}(t)$ which cause Wigner localisation remain the appropriate basis set in leading order, even with weak disorder. This restriction averages out phase information of processes where the particles and holes propagate independently,[7] thus excluding Anderson localisation which should not be important at these low densities. We incorporate the scattering of defects in a way which preserves, within the density basis we are using, the Ward identities and particle conservation.

2 Theoretical method

The Hamiltonian for the 2D electron system with defects is

$$H = \sum_k \epsilon_k a_k^{\dagger} a_k + \frac{1}{2}\sum_q V(q)\rho(q)\rho(-q) + \sum_q U(q)\rho(-q), \tag{1}$$

where $\epsilon = \hbar^2 k^2/2m^{\star}$ is the single-particle kinetic energy, $V(q)$ is the Coulomb interaction between electrons, and $U(q) = W(q)\sum_i \exp(-\,\mathrm{i}\boldsymbol{q}.\boldsymbol{r}_i)$ is the defect potential for impurities randomly distributed at points $\boldsymbol{r}_i$.

We use a memory function formalism with mode-coupling theory, using the restricted basis of the density fluctuations $\rho(\boldsymbol{q},t)$. Details of the quantum generalisation of the mode-coupling formalism[8,9] are given in Ref. 10. The density correlation function $\langle\rho(\boldsymbol{r},t)\rho(0,0)\rangle$ gives the time dependence of the decay of the density fluctuations. In the liquid phase $\langle\rho(\boldsymbol{r},t)\rho(0,0)\rangle$ goes to zero when $t\to\infty$ because of the propagation of the diffusive mode. In contrast, in the glass phase the diffusion constant is zero, so if we approach the glass transition the decay of density fluctuations becomes very slow and the local structure and the local density fluctuations do not vanish when $t\to\infty$. Thus $\lim_{t\to\infty}\langle\rho(\boldsymbol{r},t)\rho(0,0)\rangle$ is non-zero. A narrow quasi-elastic peak builds up in the dynamic structure factor $S(\boldsymbol{q},\omega)$ which develops a singularity at zero frequency.

We define the Kubo relaxation functions,

$$\Phi(q,t) \equiv \lim_{t\to\infty}\Big(N(q,t)\Big|N(q,0)\Big), \tag{2}$$

where the normalised density dynamical variables $N(q,t) \equiv \rho(q,t)/\sqrt{\chi(q)}$. The order parameters $f(q)$ for the transition are the long-time limit of the Kubo relaxation functions, $\lim_{t\to\infty} \Phi(q,t)$, or in Laplace space,

$$\Phi(q,z) = \left(N(q) \left| \frac{1}{\mathcal{L} - z} \right| N(q) \right) , \tag{3}$$

$$f(q) = -\lim_{z\to 0} z\Phi(q,z) \begin{cases} > 0 & \text{for glass} \\ = 0 & \text{for liquid} \end{cases} , \tag{4}$$

where $\mathcal{L}$ is the Liouvillian.

The Kubo relaxation function is expressed in terms of force-force relaxation functions. We start by using the exact identity

$$P_N \frac{1}{\mathcal{L} - z} P_N = \frac{1}{P_N(\mathcal{L} - z)P_N - P_N\mathcal{L}Q_NQ_N\mathcal{L}Q_N - z^{-1}Q_N\mathcal{L}P_N} P_N \tag{5}$$

(where the projection operators,

$$P_N(q) = |N(q))\,(N(q)| , \qquad Q_N(q) = 1 - P_N(q), \tag{6}$$

project dynamical variables into and out of the subspace spanned by the density fluctuations $\rho(q)$, respectively), together with the relation $P_N\mathcal{L}P_N = 0$ (which comes from the equation of continuity $\mathcal{L}\rho(q) = -qJ(q)$), to express $\Phi(q,z)$ in Eq. (3) in terms of the current relaxation function $K(q,z)$,

$$\Phi(q,z) = -\frac{1}{z + \Omega(q)K(q,z)} , \tag{7}$$

where

$$\begin{aligned} K(q,z) &= \frac{1}{q^2} \left(Q_N\mathcal{L}N(q) \left| \frac{1}{Q_N\mathcal{L}Q_N - z} \right| Q_N\mathcal{L}N(q) \right) \\ &= \left(J(q) \left| \frac{1}{Q\mathcal{L}Q - z} \right| J(q) \right) . \end{aligned} \tag{8}$$

The last result follows from the equation of continuity. The spectral density function $\Omega(q)$ is given by,

$$\Omega(q) = (\mathcal{L}N(q)\ (\mathcal{L}N(q)) = q^2/(m\chi(q)) \tag{9}$$

where $\chi(\boldsymbol{q})$ is the electron static susceptibility.

Now introducing projection operators P_J and Q_J into and out of the current fluctuation subspace, respectively,

$$\begin{aligned} P_J &= |\mathcal{L}N(q)) \frac{1}{\Omega(q)} (\mathcal{L}N(q)| = q^2\, |J(q))\,(J(q)| , \\ Q_J &= 1 - P_J , \end{aligned} \tag{10}$$

we express $K(q,z)$ in terms of the force-force correlation function $M(q,z)$,

$$K(q,z) = -\frac{\Omega(q)}{z + M(q,z)}, \tag{11}$$

where

$$M(q,z) = \frac{m}{q^2} \left(\mathcal{L}^2N(q)Q_N \left| \frac{1}{\tilde{\mathcal{L}} - z} \right| Q_N\mathcal{L}^2N(q) \right) , \tag{12}$$

and $\tilde{\mathcal{L}} = Q_J \mathcal{L} Q_J$.

Using mode-coupling theory the force-force relaxation functions can be approximated by linear and bi-linear products of the relaxation functions. Starting from the equation of motion we have

$$Q_N \mathcal{L}^2 \rho(q) = Q_N \Big[H, [H, \rho(q)]\Big] \tag{13}$$

$$\begin{aligned} &= Q_N \Bigg\{ \frac{1}{2m} \sum_{\boldsymbol{k}} [q^2 + 2\boldsymbol{q} \cdot \boldsymbol{k}] a^\dagger_{q+k} a_k \\ &+ \frac{1}{m} \sum_{\boldsymbol{k}} V(k)(\boldsymbol{q} \cdot \boldsymbol{k}) \rho(k) \rho(\boldsymbol{q} - \boldsymbol{k}) \\ &+ \frac{1}{m} \sum_{\boldsymbol{k}} W(k)(\boldsymbol{q} \cdot \boldsymbol{k}) \rho(\boldsymbol{q} - \boldsymbol{k}) \Bigg\} , \end{aligned} \tag{14}$$

so we obtain,

$$\begin{aligned} M(\boldsymbol{q}, z) = \frac{1}{mq^2} \Bigg(&\Bigg[\sum_{\boldsymbol{k}} V(\boldsymbol{k})(\boldsymbol{q} \cdot \boldsymbol{k}) \rho(\boldsymbol{k}) \rho(\boldsymbol{q} - \boldsymbol{k}) + \sum_{k} W(\boldsymbol{k})(\boldsymbol{q} \cdot \boldsymbol{k}) \rho(\boldsymbol{q} - \boldsymbol{k}) \Bigg] \\ &\times Q_N \left| \frac{1}{\tilde{\mathcal{L}} - z} \right| Q_N \\ &\times \Bigg[\sum_{\boldsymbol{k}} V(\boldsymbol{k})(\boldsymbol{q} \cdot \boldsymbol{k}) \rho(\boldsymbol{k}) \rho(\boldsymbol{q} - \boldsymbol{k}) + \sum_{k} W(\boldsymbol{k})(\boldsymbol{q} \cdot \boldsymbol{k}) \rho(\boldsymbol{q} - \boldsymbol{k}) \Bigg] \Bigg). \end{aligned} \tag{15}$$

Provided the scattering from defects is not strong, we may neglect the cross terms in Eq. (15) and write

$$M(\boldsymbol{q}, z) = M_{ee}(\boldsymbol{q}, z) + M_{de}(\boldsymbol{q}, z), \tag{16}$$

where

$$\begin{aligned} M_{ee}(\boldsymbol{q}, z) = \frac{1}{mq^2} &\sum_{\boldsymbol{k}\boldsymbol{k}'} V(\boldsymbol{k}') V(\boldsymbol{k}) (\boldsymbol{q} \cdot \boldsymbol{k}') (\boldsymbol{q} \cdot \boldsymbol{k}) \\ &\times \left(\rho(\boldsymbol{k}) \rho(\boldsymbol{q} - \boldsymbol{k}) Q_N \left| \frac{1}{\tilde{\mathcal{L}} - z} \right| Q_N \rho(\boldsymbol{k}') \rho(\boldsymbol{q} - \boldsymbol{k}') \right), \end{aligned} \tag{17}$$

$$\begin{aligned} M_{de}(\boldsymbol{q}, z) = \frac{1}{m\boldsymbol{q}^2} \Bigg(&\sum_{\boldsymbol{k}} W(\boldsymbol{k})(\boldsymbol{q} \cdot \boldsymbol{k}) \rho(\boldsymbol{q} - \boldsymbol{k}) Q_N \left| \frac{1}{\tilde{\mathcal{L}} - z} \right| Q_N \\ &\times \sum_{\boldsymbol{k}'} W(\boldsymbol{k}')(\boldsymbol{q} \cdot \boldsymbol{k}') \rho(\boldsymbol{q} - \boldsymbol{k}') \Bigg). \end{aligned} \tag{18}$$

$M_{ee}(\boldsymbol{q}, z)$ is the contribution to $M(\boldsymbol{q}, z)$ from electron-electron scattering. It is a four-point density relaxation function. Classically the relaxation function and correlation function are equivalent, and since we usually expand the four-point density correlation function in lowest order as the product of two-point density correlation functions, the same can be done for the four-point relaxation functions. However, these steps cannot be directly followed in the quantum treatment since the quantum relation between the Fourier transforms of the correlation function and the relaxation function for any variables A and B is not one-to-one,

$$\left\langle A\middle|B\right\rangle_{\omega} = \frac{1-\mathrm{e}^{-\beta\hbar\omega}}{\hbar\omega}\left(A\middle|B\right)_{\omega}, \tag{19}$$

where $z = \omega + \mathrm{i}\delta$. In the limit $\omega \to 0$ the two functions in Eq. (19) do become identical, and in this special case we are justified in factorising the quantum four-point relaxation function as a superposition of two two-point relaxation functions. Restricting ω to be zero means that we can only identify the transition itself and that we can obtain no information about the low lying excited states for quantum systems.

With this factorisation we then obtain,

$$\begin{aligned} M_{ee}(\boldsymbol{q},z) = \frac{1}{mq^2}\sum_{\boldsymbol{kk}'} & V(\boldsymbol{k}')V(\boldsymbol{k})(\boldsymbol{q}\cdot\boldsymbol{k}')(\boldsymbol{q}\cdot\boldsymbol{k}) \\ & \times\Big[\langle\rho(\boldsymbol{k},t)\rho(\boldsymbol{k}',0)\rangle\langle\rho(\boldsymbol{q}-\boldsymbol{k})\rho(\boldsymbol{q}-\boldsymbol{k}')\rangle \\ & +\langle\rho(\boldsymbol{k},t)\rho(\boldsymbol{q}-\boldsymbol{k}',0)\rangle\langle\rho(\boldsymbol{q}-\boldsymbol{k})\rho(\boldsymbol{k}')\rangle\Big]. \end{aligned} \tag{20}$$

Returning to Eq. (18), the term $M_{de}(\boldsymbol{q},z)$ is the contribution to $M(\boldsymbol{q},z)$ from electrons scattering off defects. $M_{de}(\boldsymbol{q},z)$ is already a two-point relaxation function in the density. We write

$$\begin{aligned} M_{de}(\boldsymbol{q},z) &= \frac{1}{mq^2}\sum_{\boldsymbol{kk}'}\Big\langle(W(\boldsymbol{k})(\boldsymbol{q}\cdot\boldsymbol{k})W(\boldsymbol{k}')(\boldsymbol{q}\cdot\boldsymbol{k}')\Big\rangle \\ &\qquad \Big(\rho(\boldsymbol{q}-\boldsymbol{k})Q_N\Big|\frac{1}{\tilde{\mathcal{L}}-z}\Big|Q_N\rho(\boldsymbol{q}-\boldsymbol{k}')\Big) \end{aligned} \tag{21}$$

$$= \sum_{\boldsymbol{k}}\left[W^2(\boldsymbol{k})(\boldsymbol{q}\cdot\boldsymbol{k})^2\right]\times\left[\chi(|\boldsymbol{q}-\boldsymbol{k}|)\Phi(|\boldsymbol{q}-\boldsymbol{k}|,z)\right]. \tag{22}$$

The glass state is identified with infinite relaxation time and we define,

$$M(q) \equiv \lim_{t\to\infty} M(q,t) = -\lim_{z\to 0} zM(q,z) \tag{23}$$

to obtain for our final expressions,

$$M(q) = M_{ee}(q) + M_{de}(q), \tag{24}$$

$$\begin{aligned} M_{ee}(\boldsymbol{q}) = \frac{1}{2mq^2}\sum_{q'} & \Big[V(q')(\boldsymbol{q}\cdot\boldsymbol{q}') + V(\boldsymbol{q}-\boldsymbol{q}')(\boldsymbol{q}\cdot(\boldsymbol{q}-\boldsymbol{q}'))\Big]^2 \\ & \times\chi(q')\chi(|\boldsymbol{q}-\boldsymbol{q}'|)f(q')f(|\boldsymbol{q}-\boldsymbol{q}'|), \end{aligned} \tag{25}$$

$$M_{de}(q) = \frac{n_i}{mq^2}\sum_{q'}\left(W^2(q')(\boldsymbol{q}\cdot\boldsymbol{q}')^2\right)\chi(|\boldsymbol{q}-\boldsymbol{q}'|)f(|\boldsymbol{q}-\boldsymbol{q}'|), \tag{26}$$

where n_i is the impurity density.

The factors $\chi(\boldsymbol{q})$ in Eqs. (25) and (26) feed details of the quantum exchange-correlation hole into the vertex part of the memory function. The static structure factor $S(q)$ from the DQMC numerical simulation data[5,6] determines a static local field factor $G(\boldsymbol{q})$ in the expression,

$$S(\boldsymbol{q}) = \int_0^\infty \mathrm{d}\omega \, \frac{\chi_0(\boldsymbol{q}, \mathrm{i}\omega)}{1 + V(\boldsymbol{q})]1 - G(\boldsymbol{q})]\chi_0(\boldsymbol{q}, \mathrm{i}\omega)} \,, \tag{27}$$

and from $G(q)$ we obtain $\chi(\boldsymbol{q})$,

$$\chi(\boldsymbol{q}) = \frac{\chi_0(\boldsymbol{q})}{1 + V(\boldsymbol{q})[1 - G(\boldsymbol{q})]\chi_0(\boldsymbol{q})} \,. \tag{28}$$

Using Eqs. (4), (7) and (11), we write the order parameter in terms of $M(q)$,

$$f(q) = \lim_{z\to\infty} -z\Phi(q,z) = \frac{1}{1 + \frac{\Omega(q)}{M(q)}} . \tag{29}$$

Eqs. (24) to (26) and (29) form a closed set of non-linear equations for the order parameter $f(q)$ and $M(q)$. They are solved iteratively to self-consistency.

When correlations and disorder are weak $M(\boldsymbol{q})$ is small and the $f(\boldsymbol{q})$ identically vanish. This indicates the density fluctuations $\rho(\mathbf{q}, t)$ decay to zero for $t \to \infty$, and so the phase is a liquid. $M_{ee}(\boldsymbol{q})$ is increased by lowering the electron density. $M_{de}(\boldsymbol{q})$ is increased by increasing the level of disorder. When $M(\boldsymbol{q})$ reaches a critical value, the $f(\boldsymbol{q})$ discontinuously jump to non-zero values. This is the transition boundary to the glass phase. Thus for a fixed electron density, if the disorder level is sufficiently small there is only the trivial solution $f(\boldsymbol{q}) = 0$, implying a delocalised state. If the disorder is increased, then at the critical level of disorder the $f(\boldsymbol{q})$ suddenly and discontinuously jumps to a non-zero value. This signals that the system cannot relax into its ground state, since the relaxation time for density fluctuations goes to infinity. We identify this as an insulating phase that is a frozen electron glass.[11] The discontinuous behaviour implies a discontinuous metal-insulator transition to a localised state. For $r_s > 9$ this transition from liquid to solid occurs when the carrier density is decreased. If the level of disorder is decreased, then the transition occurs at a smaller electron density (that is, for stronger electron correlations). At very low carrier densities the system is always insulating.

If we now increase the electron density at a fixed level of disorder, then by $r_s \lesssim 8$ the $f(\boldsymbol{q})$ starts to continuously increase in value from zero. This implies a continuous localisation. We identify this as a separate insulating phase that is a single-particle localised state in which the electrons are independently localised. We thus conclude that there are two distinct transitions out of the conducting state, a discontinuous transition when the electron density is decreased, and a continuous one when the electron density is increased. The zero temperature phase diagram we obtain[11] is in good quantitative agreement with reported experimental phase diagrams.

To further investigate the nature of the conducting state, we extended studies of the stability of a 2D superconducting state to include the effects of defect scattering[12] using a response function which includes both electron-electron correlations and scattering of disorder. We found disorder does reduce the attraction between the electrons generated by the strong correlations, and that at sufficiently high levels of disorder the superconductivity is suppressed altogether. Superconductivity persists to levels of disorder where the plasmon is damped out. The transition temperature to the superconducting state is more sensitive to disorder at higher electron densities, where the correlations are weaker.

We find a suppression of the conducting phase by an external magnetic field parallel to the 2D electron plane. A field in this direction can only couple to the electron spins, partially aligning them. The aligned spins have additional exchange which has the effect of increasing the radius of the density exclusion region in the exchange-correlation hole. The increase is $\sim 15\%$ at $r_s = 10$.[6] The expansion of the exchange-correlation hole increases the probability for forming the coherent insulating state so that spin polarisation favours the localised insulator[13] and suppresses the conducting state. A magnetic field of less than 1 T is sufficient to stabilise the fully spin polarised state and suppress the conducting state.[13] This mechanism leads to good agreement with the experimental magnetic phase diagram reported in Ref. 3.

Acknowledgments

This work is supported by an Australian Research Council Grant.

References

1. S. V. Kravchenko, G. V. Kravchenko, and J. E. Furneaux, *Phys. Rev.* B **50**, 8039 (1994); S. V. Kravchenko, D. Simonian, M. P. Sarachik, Whitney Mason, and J. E. Furneaux, *Phys. Rev. Lett.* **77**, 4938 (1996).
2. D. Simonian, S. V. Kravchenko, M. P. Sarachik and V. M. Pudalov, *Phys. Rev. Lett.* **79**, 2304 (1997); M. Pudalov, G. Brunthaler, A. Prinz and G. Bauer, *Sov. Phys. JETP* **65**, 932 (1997).
3. A. R. Hamilton, M. Y. Simmons, M. Pepper, E. H. Linfield, P. D. Rose and D. A. Ritchie, *Phys. Rev. Lett.* **82**, 1542 (1999).
4. E. Abrahams, P. W. Anderson, D. C. Licciardello, and T. V. Ramachandran, *Phys. Rev. Lett.* **42**, 673 (1979).
5. B. Tanatar and D. M. Ceperley, *Phys. Rev.* B **39**, 5005 (1989).
6. F. Rapisarda and G. Senatore, *Aust. J. Phys.* **49**, 161 (1996).
7. D. Neilson, L. Świerkowski, A. Sjölander and J. Szymański, *Phys. Rev.* B **44**, 6291 (1991).
8. U. Bengtzelius, W. Götze and A. Sjölander, *J. Phys. C (Cond. Matt.)* **17**, 5915 (1984).
9. J. S. Thakur and J. Bosse, *Phys. Rev. Lett.* **59**, 998 (1987); *Phys. Rev.* A **43**, 4378 (1991).
10. J. S. Thakur and D. Neilson, *Phys. Rev.* B **54**, 7674 (1996).
11. J. S. Thakur and D. Neilson, *Phys. Rev.* B **59**, R5280 (1999).
12. J. S. Thakur and D. Neilson, *Phys. Rev.* B **57**, 1801 (1998).
13. J. S. Thakur and D. Neilson, *J. Phys. C (Cond. Matt.)* **12**, 4483 (2000).

Related Subjects

INFORMATION REPRESENTATION IN THE MULTI-LAYER PERCEPTRON

M. J. BARBER
Institut für Theoretische Physik, Universität zu Köln, 50937 Köln, Germany
E-mail: mjb@thp.uni-koeln.de

A typical criticism of neural networks is that they are "black boxes" whose function is unclear. Although neural networks may accurately model the statistical properties of a data set, the means by which they do so is usually unknown. We adapt current notions of population coding to more traditional artificial neural networks and develop means to interpret the synaptic weights of multi-layer perceptrons. These considerations give rise to neural network architectures with simpler functional relationships.

1 Introduction

The human brain is composed of about 100 billion neurons (Fig. 1). Each neuron forms synapses connecting it to other neurons. Typically, a neuron makes several thousand synapses to other cells. Information is transmitted between neurons as a chemical signal at these synapses. The chemical cascade that carries the signal is generated through a nonlinear process and can be very complex, so artificial neural networks frequently are composed of neuron-like units that are highly simplified models of real neurons.[3] These model neurons are connected to one another by weighted "synapses". The chemical cascade and resulting changes in the electric potential of the soma typically are replaced by a nonlinear "activation function" that determines the activation state of the neuron.

Methods for decoding information from the firing rates of populations of neurons were pioneered by Georgopoulos and collaborators.[2,5] They showed that a "population vector" derived from the firing rates of a population of cortical neurons (Fig. 2) can be used to predict the intended arm movements of monkeys This vector estimate of direction, $\mathbf{V}_{\mathrm{est}}$, is obtained from the neural firing rates a_i by

$$\mathbf{V}_{\mathrm{est}} = \sum_{i=1}^{N} a_i \mathbf{C}_i, \tag{1}$$

where the preferred direction vector $\mathbf{C}_i$ indicates the direction at which neuron i has its maximal firing response. The preferred direction vectors in general form a non-orthogonal, highly overcomplete set. The population vector approach has been refined and extended by several authors; in particular, Salinas and Abbott provide an excellent discussion of several such refinements, as well as introducing their own.[4] It is interesting to note that, despite the nonlinear process by which neuronal activation states are determined, the encoded signal is linearly decoded.

A complementary encoding process can be deduced by examining neural firing profiles. An appropriate rule to encode a signal $\mathbf{V}$ into firing rates for use with the above population vector is

$$a_i = [\mathbf{C}_i \cdot \mathbf{V}]_+ , \tag{2}$$

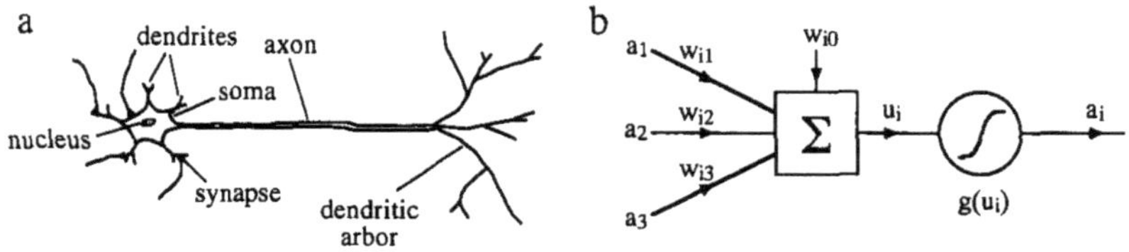

Figure 1. Neurons are basic elements of the nervous system. (a) Neurons process information by transmitting signals to one another through synaptic connections in the axon's dendritic arbor. The transmitted signals are generated through a complex, nonlinear process. (b) For model neurons, the signal generation process is often greatly simplified into a nonlinear thresholding function of a weighted sum of inputs to the neuron.

where the brackets indicate rectification of the form

$$[x]_+ = \begin{cases} 0 \text{ if } x < 0 \\ x \text{ if } x \geq 0 \end{cases}. \tag{3}$$

The resulting piecewise-linear profiles are biologically realistic (see for example figure 4 in Ref. 1).

The encoding and decoding procedure defined by Eqs. (1) and (2) can be rewritten in a form more appropriate for general vector spaces. We encode a vector $\mathbf{x}$ into neural activities as a (typically nonlinear) function g_x of an inner product of $\mathbf{x}$ with encoding vectors $\hat{\mathbf{x}}_i$, so that

$$a_i = g_x \left(\langle \hat{\mathbf{x}}_i, \mathbf{x} \rangle \right) \tag{4}$$

and recover a pattern vector from a set of firing rates with

$$\mathbf{x} = \sum_i a_i \mathbf{x}_i, \tag{5}$$

using corresponding decoding vectors $\mathbf{x}_i$. We generally expect that the sets of encoders $\{\hat{\mathbf{x}}_i\}$ and decoders $\{\mathbf{x}_i\}$ will be overcomplete representations for the vector spaces that can be encoded and decoded.

Encoding and decoding rules of this form allow a direct method to generate neural networks that encapsulate desired functions. Consider a second population of neurons representing a vector $\mathbf{y}$ through the rules

$$b_j = g_y \left(\langle \hat{\mathbf{y}}_j, \mathbf{y} \rangle \right), \tag{6}$$

$$\mathbf{y} = \sum_j b_j \mathbf{y}_j, \tag{7}$$

where $\{\hat{\mathbf{y}}_j\}$ and $\{\mathbf{y}_j\}$ are the sets of encoders and decoders for y, respectively. Further, take the two vectors $\mathbf{x}$ and $\mathbf{y}$ to be related through

$$\mathbf{y} = \mathbf{A}\mathbf{x}. \tag{8}$$

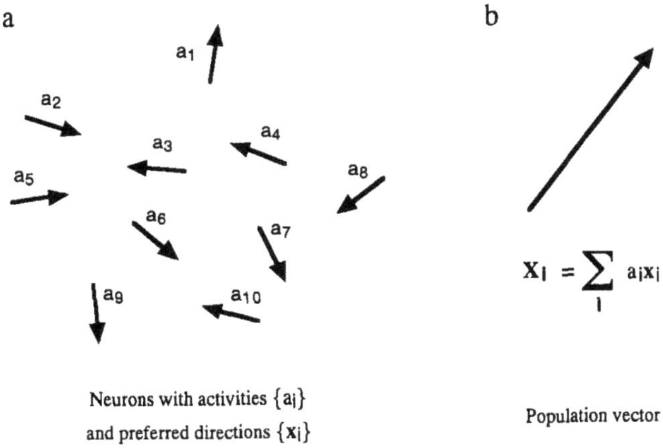

Figure 2. The arm movements of monkeys can be predicted using a "population vector" derived from the firing rates of a population of cortical neurons. (a) "Preferred direction" vectors are determined from the maximal response of the neurons. The firing rates of each neuron are measured for a given motion. (b) The population vector is a linear combination of the preferred direction vectors, with the firing rates as the components.

By combining Eqs. (5), (6), and (8), we can construct a two-layer neural network with inputs $\{a_i\}$ and outputs $\{b_j\}$ determined by

$$b_j = g_y\left(\sum_i \omega_{ji} a_i\right), \tag{9}$$

where the weights ω_{ji} are determined by

$$\omega_{ji} = \langle \hat{\mathbf{y}}_j, \mathbf{A}\mathbf{x}_i \rangle . \tag{10}$$

In accord with the general form of the representations we have defined, the output activities $\{b_j\}$ are determined through a nonlinear update rule, but may be linearly decoded using Eq. (7).

2 Population Codes and Perceptrons

An interesting parallel exists between neural systems of the form defined in Eqs. (4) and (5) and a popular version of the multi-layer perceptron (MLP). In the MLP architecture, the neurons are organized into distinct layers, most commonly with three layers.[3] An input pattern $\mathbf{x}^\mu$ is encoded into the activities of the layer of input neurons, $\{a_i^\mu\}$. These activities drive the activities $\{b_j^\mu\}$ of the layer of "hidden" neurons, which in turn drive the activities $\{c_k^\mu\}$ of the layer of output neurons. These output neurons are matched to an output pattern $\mathbf{z}^\mu$; presumably, the hidden neurons correspond to some implicit "hidden pattern" $\mathbf{y}^\mu$ as well.

The strength and nature of the connections between the different layers is measured by "synaptic" weights, so that

$$b_j^\mu = g\left(\sum_i w_{ji} a_i^\mu\right) \tag{11}$$

and

$$c_k^\mu = f\left(\sum_j v_{kj} b_j^\mu\right). \tag{12}$$

The functions f and g are typically monotone increasing, and serve to increase the range of input-output mappings that the neural network can represent. Typical choices for g are the logistic function and the hyperbolic tangent (we will utilize the hyperbolic tangent in this work). The same function is often chosen for f, but the identity function is also not uncommon, especially for regression (as opposed to classification) problems. The choice of the identity function is interesting; it suggests that the network learns a nonlinear mapping from the input to the hidden units, but that the resulting hidden-unit representation is optimized for linear decoding. This is directly analogous to the neural network defined by Eqs. (9) and (10) above, with the output units serving to decode the information represented in the hidden layer.

To strengthen the connection between population coding and the multi-layer perceptron, we introduce population codes for the different pattern vectors in the MLP. The forms of the population codes for $\mathbf{x}^\mu$ and $\mathbf{y}^\mu$ are as given in Eqs. (4) through (7). For the linear output units representing the output pattern $\mathbf{z}^\mu$, we take

$$c_j^\mu = \langle \hat{\mathbf{z}}_k, \mathbf{z}^\mu \rangle, \tag{13}$$

$$\mathbf{z}^\mu = \sum_k c_k \mathbf{z}_k \tag{14}$$

as the coding rules.

We assume the relationships between $\mathbf{x}^\mu$, $\mathbf{y}^\mu$, and $\mathbf{z}^\mu$ to have the forms

$$\mathbf{y}^\mu = \mathbf{A}\mathbf{x}^\mu, \tag{15}$$

$$\mathbf{z}^\mu = \mathbf{B}\mathbf{y}^\mu. \tag{16}$$

Through an identical approach to that used above in constructing a neural network (Eqs. (9) and (10)), we obtain

$$b_j^\mu = g_y\left(\sum_i \omega_{ji} a_i^\mu\right), \tag{17}$$

with weights

$$\omega_{ji} = \langle \hat{\mathbf{y}}_j, \mathbf{A}\mathbf{x}_i \rangle. \tag{18}$$

Similarly, the output units obey

$$c_k = \sum_j \nu_{kj} b_j, \tag{19}$$

with weights

$$\nu_{kj} = \langle \hat{\mathbf{z}}_k, \mathbf{B}\mathbf{y}_j \rangle . \tag{20}$$

This reproduces the usual MLP architecture, with the weights ω_{ji} and ν_{kj} paralleling the usual MLP weights w_{ji} and v_{kj}, respectively. In this work, we will focus on the relationship, given in Eqs. (17) and (18), between the input and hidden layers; similar reasoning may be applied to the relationship between the hidden and output layers.

The mapping from the input units to the hidden units can be examined in more detail. First, rewrite Eq. (18) in the equivalent matrix form of

$$\mathbf{b} = g\left(\hat{\mathbf{Y}}^{\mathrm{H}}\mathbf{A}\mathbf{X}^{\mathrm{H}}\mathbf{a}\right). \tag{21}$$

Here, $\hat{\mathbf{Y}}$ and $\mathbf{X}$ are constructed using encoding and decoding vectors as columns of the matrices, so that

$$\mathbf{X} = [\mathbf{x}_1|\mathbf{x}_2|\mathbf{x}_3|\cdots|\mathbf{x}_m], \tag{22}$$

$$\hat{\mathbf{Y}} = [\hat{\mathbf{y}}_1|\hat{\mathbf{y}}_2|\hat{\mathbf{y}}_3|\cdots|\hat{\mathbf{y}}_n], \tag{23}$$

where m and n indicate the number of input and hidden units, respectively.

Next, we replace $\mathbf{A}$ with its singular value decomposition (SVD), so that

$$\mathbf{A} = \mathbf{U}\mathbf{S}\mathbf{V}^{\mathrm{H}}, \tag{24}$$

with unitary matrices $\mathbf{U}$ and $\mathbf{V}$, and diagonal matrix $\mathbf{S}$. We further reduce these matrices, producing an "economy size" version of the SVD of $\mathbf{A}$. We exclude any zeros from the diagonal of $\mathbf{S}$, so that $\mathbf{S}$ is a non-singular square matrix with the singular values σ_1, $\sigma_2, \ldots$, σ_r on the diagonal. We eliminate the corresponding columns of $\mathbf{U}$ and $\mathbf{V}$ as well.

Substituting this new representation of $\mathbf{A}$ into Eq. (21), we have

$$\mathbf{b} = g\left(\hat{\mathbf{Y}}^{\mathrm{H}}\mathbf{U}\mathbf{S}\mathbf{V}^{\mathrm{H}}\mathbf{X}^{\mathrm{H}}\mathbf{a}\right). \tag{25}$$

The sets of orthonormal vectors that make up the columns of $\mathbf{U}$ and $\mathbf{V}$ are seen here to define bases for the vector spaces spanned by $\{\hat{\mathbf{y}}_j\}$ and $\{\mathbf{x}_i\}$, respectively. The dimensionality of these spaces is thus equal to the rank r of $\mathbf{A}$, regardless of the number of input and hidden units. Greater numbers of units may still be useful in terms of performance in the face of noisy data.

The decomposed weight matrix can be used to generate a new neural network architecture (Fig. 3). Define

$$\mathbf{p}^{\mu} = \mathbf{X}^{\mathrm{H}}\mathbf{a}^{\mu}, \tag{26}$$

$$\mathbf{q}^{\mu} = \mathbf{V}^{\mathrm{H}}\mathbf{p}^{\mu}, \tag{27}$$

$$\mathbf{r}^{\mu} = \mathbf{S}\mathbf{q}^{\mu}, \tag{28}$$

$$\mathbf{s}^{\mu} = \mathbf{U}\mathbf{r}^{\mu}, \tag{29}$$

with

$$\mathbf{b}^{\mu} = g\left(\hat{\mathbf{Y}}^{\mathrm{H}}\mathbf{s}^{\mu}\right), \tag{30}$$

$$\mathbf{c}^{\mu} = \mathbf{W}\mathbf{b}^{\mu}. \tag{31}$$

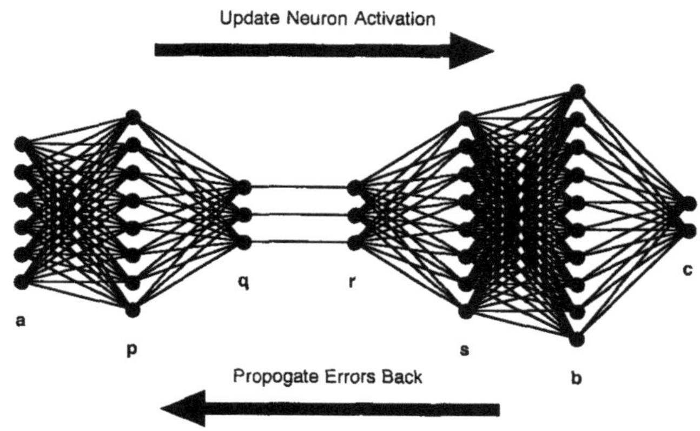

Figure 3. Introducing population coding into a three-layer perceptron produces another perceptron. The states of activation of the network units are updated in a purely feed-forward sense, but the usual backpropagation learning rule is modified slightly because of the orthonormalization constraint applied to two layers.

The weight matrix **W** defining the mapping between hidden and output units may in principle be decomposed in the same fashion as we have done with the input-to-hidden weights.

3 Learning

We will find all weights using a variant of the standard backpropagation learning rule. The learning objective is to minimize the difference between the network output $\mathbf{c}^\mu$ in response to an input ξ^μ and a target pattern ζ^μ. We must additionally constrain the optimization process so that the columns of **U**, treated as vectors, form an orthonormal set, with a similar constraint on the columns of **V**. To do this, we define a cost functional of the form

$$E = \alpha \sum_\mu \|\zeta - \mathbf{c}^\mu\|^2 + \beta \sum_{i,j} \left(\langle \mathbf{u}_i, \mathbf{u}_j \rangle - \delta_{ij}\right)^2 + \gamma \sum_{i,j} \left(\langle \mathbf{v}_i, \mathbf{v}_j \rangle - \delta_{ij}\right)^2 . \tag{32}$$

The constants α, β, and γ establish the relative importance of the three terms; they are all taken equal to unity in this work. The vectors $\mathbf{u}_i$ and $\mathbf{v}_j$ are the ith and jth columns of **U** and **V**, respectively. We utilize the gradient descent prescription

$$\Delta w_{ij} = -\eta \frac{\partial E}{\partial w_{ij}} \tag{33}$$

to minimize E; this gives rise to a modified version of backpropagation.

As a test of the neural network, we model data about the coordination number of borates that have chemical formulas of the form $A_x B_y O_z$. The neural network used has 10, 10, 3, 3, 10, 10, and 1 units in the layers, proceeding from the input layer to the output layer. The data (115 samples) is divided into a training set (75% of the data) and a test set (25% of the data). The training patterns are presented

a

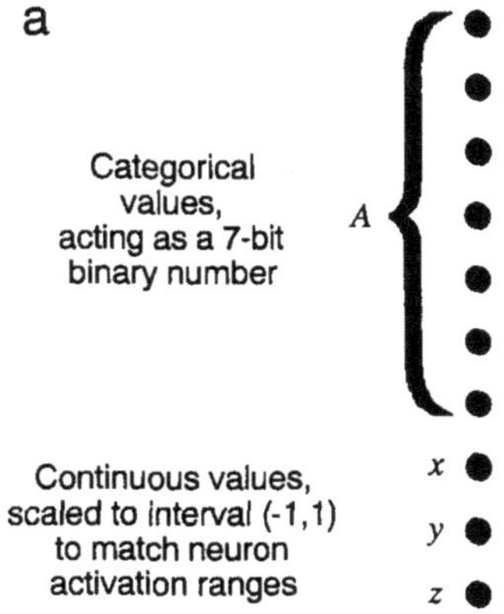

b

	Training set	Test set
Initial weights	2.8386	2.9402
Trained weights	0.0253	0.1070

c

	Training set	Test set
Initial weights	0.5814	0.6931
Trained weights	0.0484	0.1286

Figure 4. Learning in the modified neural network. (a) Data about borate crystals is processed into a suitable form for use as network inputs. The single output unit is a continuous variable, with the coordination numbers scaled into the interval $(-1, 1)$. (b) The modified neural network is able to learn a mapping between the encoded inputs and the coordination number of the borate crystal, as shown by the improvement in the cost function for the training and test sets. (c) Learning performance in a traditional three-layer feed-forward network is similar to the decomposed network.

to the network 1000 times in random order, with the gradient descent update rule (Eq. (33)) used to adjust the weights.

The decomposed network developed here performs comparably to a traditional three-layer feed-forward network (Fig. 4). Additionally, the structure of the neural network provides a direct means to gain some understanding of the means through which the network functions. For instance, the columns of **XV** show direct correlations in the inputs that are relevant to the neural network function. The diagonal elements of **S** show the relative importance of these correlations (Fig. 5).

4 Conclusions

The neural network architecture that we have developed in this work performs similarly to the standard multi-layer perceptron. The decomposition considered leads to extra layers in the new network, but these layers are linear and thus capable of performing the same computation as the direct input- to hidden-unit connections. As well, the performance of the decomposed network with respect to learning the statistical properties of a data set is empirically similar to performance of the standard network.

However, clearer meaning is attached to the synaptic weights in the decomposition process. Individual elements of the weight matrix are often simpler to understand, and can be used to identify correlations in the data sets that are relevant to the learned mappings. Additionally, the rank of the weight matrix is explicitly constrained by the matrix **S**, which opens the possibility of directly incorporating the rank into the learning rules to optimize generalization ability of the neural network.

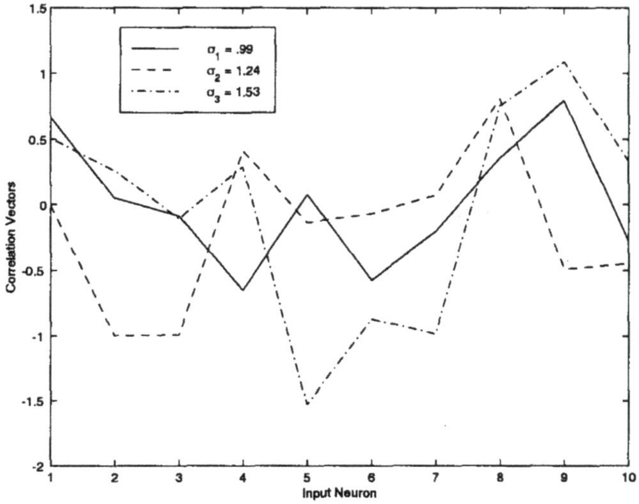

Figure 5. The columns of **XV** show direct correlations in the inputs that are relevant to the neural network function. Diagonal elements of **S** show the relative importance of these correlations.

The decomposition developed in this work provides a partial explanation to how neural networks capture the statistical properties of data sets. From this work, it has become clear to us that it is in fact the neural network architecture, and not the learning rules, that explains the network function. In future work, we will examine the correlations between hidden layer neurons in greater detail, to better understand the nature of the network architecture and hopefully to provide another part of the explanation for the network function.

Acknowledgments

This work was supported in part by the Graduiertenkolleg Azentrische Kristalle, GK549 of the DFG.

References

1. A. F. Fuchs, C. A. Schudder, and C. R. S. Kaneko, *J. Neurophys.* **60**, 1874 (1988).
2. A. P. Georgopoulos, A. B. Schwartz, and R. E. Kettner, *Science* **233**, 1416 (1986).
3. J. Hertz, A. Krogh, and R. G. Palmer, *Introduction to the Theory of Neural Computation* (Addison-Wesley, Reading, MA, 1991).
4. E. Salinas and L. F. Abbott, *J. Comp. Neurosci.* **1**, 89 (1994).
5. A. B. Schwartz *J. Neurophys.* **70**, 28 (1993).

CLASSICAL AND QUANTUM LYAPUNOV EXPONENTS IN THE PHASE-SPACE TOMOGRAPHIC APPROACH

R. VILELA MENDES

Grupo de Física-Matemática, Complexo Interdisciplinar, Univ. de Lisboa, Av. Gama Pinto, 2, 1699 Lisboa Codex Portugal

E-mail: vilela@cii.fc.ul.pt

Lyapunov exponents are constructed using the tomography map in classical and quantum phase space. The marginal distributions being well defined probabilities, the correspondence between classical and quantum notions is clear. Providing an unambiguous extension of the notion of Lyapunov exponent to quantum mechanics, the method is also computationally efficient to obtain analytical results for the Lyapunov exponent, both classical and quantum. Examples are worked out and some insight is obtained on the taming effect of quantum mechanics on classical chaos.

1 Introduction

For many-body systems, non-integrability of the dynamics is the generic condition. This is true both for classical and quantum systems. Therefore, to go beyond crude approximations and to characterize the complex behavior of many-body systems, tools had to be developed to characterize the occurrence of chaos, ergodic behavior, break of ergodicity, measures of entropy, etc. In classical mechanics these theoretical tools have reached a reasonable degree of maturity. By contrast, in quantum mechanics, even the notion of quantum chaos is still a controversial subject.

Classical chaotic motion is characterized by the existence of positive Lyapunov exponents or positive Kolmogoroff-Sinai entropy. These quantities being based on the properties of classical trajectories in phase space, it is not obvious what the corresponding quantities in quantum mechanics should be. To construct quantum mechanical functionals, with the same physical meaning as the classical quantities which characterize classical chaos, a phase-space formulation for quantum mechanics should be used, rather than the usual Hilbert space formulation. The difficulty here lies in the fact that quantum phase-space is a non-commutative manifold with the usual pointwise product of functions being replaced by the $*$-product. One possible solution would be to use the tools of non-commutative geometry for this formulation. Another approach, however, is to look for (commutative) quantities which have the same formal structure both in classical and in quantum mechanics.

The Wigner function,[1] which some authors have attempted to use for this purpose, is not the appropriate choice because, unlike the classical probability distributions, it is not positive definite. There is however a set of phase-space quantities that have the same mathematical nature in both classical and quantum mechanics. This is the set of marginal distributions of the symplectic tomography formulation,[2–9] which are, in both cases, well-defined probability distributions. In the symplectic tomography formulation of classical and quantum mechanics the dynamics is defined by a set of marginal probability distributions. The difference between classical and quantum mechanics comes only from the modification of the equations of motion.

Once the appropriate phase-space quantities are identified and classical Lyapunov exponents are formulated in terms of probability distributions, the transition to quantum mechanics is rather straightforward. Despite its apparent complex form, the marginal distribution formulation is a computationally effective way to obtain Lyapunov exponents, both classical and quantum. Some examples have been explicitly computed, namely kicked systems on the line, on the 2-torus and on the circle. In particular a characterization is obtained for the origin of the taming effect of quantum mechanics on classical chaos in the standard map. More details may be found in Ref. 10.

2 Symplectic tomography of classical and quantum states

2.1 Classical mechanics

States in classical statistical mechanics are described by a function $\rho(q, p)$, the probability distribution function in $2n$-dimensional phase space ($q \in R^n, p \in R^n$), with properties

$$\rho(q, p) \geq 0\,, \qquad \int \rho(q, p)\, d^n p = P(q)\,, \qquad \int \rho(q, p)\, d^n q = \widetilde{P}(p)\,.$$

$P(q)$ and $\widetilde{P}(p)$ are the probability distributions for position and momentum (the *marginals* of ρ).

The inverse Fourier transform of the characteristic function $\langle e^{ik\bullet X}\rangle$ for any vector observable $X(q, p)$

$$w(Y) = \frac{1}{(2\pi)^n} \int \langle e^{ik\bullet X}\rangle e^{-ik\bullet Y}\, d^n k \tag{1}$$

is a real normalized non-negative function.

Consider the following classical observable[6]

$$X(q, p) = \mu \circledast q + \nu \circledast p\,, \tag{2}$$

where $\circledast$ denotes the componentwise product of vectors

$$(\mu \circledast q)_i = \mu_i q_i\,,$$

and μ and ν are vector-valued real parameters.

The vector variable $X(q, p)$ may be interpreted as a coordinate of the system, when measured in a rotated and scaled reference frame in the classical phase space. For the coordinate (2) in the transformed reference frame, we obtain from Eq. (1) the distribution function (the tomography map)

$$w(X, \mu, \nu) = \frac{1}{(2\pi)^n} \int e^{-ik\bullet(X - \mu\circledast q - \nu\circledast p)}\, \rho(q, p)\, d^n q\, d^n p\, d^n k\,. \tag{3}$$

This function is homogeneous

$$w(\lambda X, \lambda\mu, \lambda\nu) = |\lambda|^{-n} w(X, \mu, \nu)\,, \tag{4}$$

and Eq. (3) has an inverse

$$\rho(q, p) = \frac{1}{(4\pi^2)^n} \int w(X, \mu, \nu) \exp[i(X - \mu \circledast q - \nu \circledast p) \bullet \mathbf{1}]\, d^n X\, d^n \mu\, d^n \nu\,. \quad (5)$$

Since the map

$$\rho(q, p) \Longrightarrow w(X, \mu, \nu)$$

is invertible, the information contained in the distribution function $\rho(q, p)$ is equivalent to the information contained in the marginal distributions $w(X, \mu, \nu)$.

The Boltzman evolution equation for the classical distribution function for a particle with mass $m = 1$ and potential $V(q)$,

$$\frac{\partial \rho(q, p, t)}{\partial t} + p \bullet \nabla_q \rho(q, p, t) - \nabla_q V(q) \bullet \nabla_p \rho(q, p, t) = 0\,, \quad (6)$$

can be rewritten in terms of the marginal distribution $w(X, \mu, \nu, t)$

$$\frac{\partial w}{\partial t} - \mu \bullet \nabla_\nu w - \nabla_x V\left(-\nabla_X^{-1} \circledast \nabla_\mu\right) \bullet (\nu \circledast \nabla_X w) = 0\,. \quad (7)$$

By $\Pi_{\mathrm{cl}}(X, \mu, \nu, X', \mu', \nu', t_2, t_1)$ we denote the classical propagator that connects two marginal distributions at different times t_0 and t $(t > t_0)$

$$w(X, \mu, \nu, t) = \int \Pi_{\mathrm{cl}}(X, \mu, \nu, X', \mu', \nu', t, t_0)\, w(X', \mu', \nu', t_0)\, d^n X'\, d^n \mu'\, d^n \nu'. \quad (8)$$

The propagator satisfies the equation

$$\frac{\partial \Pi_{\mathrm{cl}}}{\partial t_2} - \mu \bullet \nabla_\nu \Pi_{\mathrm{cl}} - \nabla_x V\left(-\nabla_X^{-1} \circledast \nabla_\mu\right) \bullet (\nu \circledast \nabla_X \Pi_{\mathrm{cl}}) = 0\,, \quad (9)$$

with boundary condition

$$\lim_{t_2 \to t_1} \Pi_{\mathrm{cl}}(X, \mu, \nu, X', \mu', \nu', t_2, t_1) = \delta^n(X - X')\, \delta^n(\mu - \mu')\, \delta^n(\nu - \nu')\,. \quad (10)$$

2.2 *Quantum mechanics*

For quantum mechanics the construction is similar and the mathematical nature of the quantities that are constructed is the same, because it is a general fact that the inverse Fourier transform of a characteristic function is a positive distribution. The marginal distributions that are obtained are simply related to other well-known quantum mechanical quantities. It was shown[2] that for the generic linear combination

$$X = \mu \circledast q + \nu \circledast p\,, \quad (11)$$

where q and p are the position and the momentum, the marginal distribution $w(X, \mu, \nu)$ (normalized in the variable X and depending on two vector-valued real parameters μ and ν) is related to the Wigner function $W(q, p)$. For n degrees of freedom one has

$$w(X, \mu, \nu) = \int \exp[-ik \bullet (X - \mu \circledast q - \nu \circledast p)]\, W(q, p)\, \frac{d^n k\, d^n q\, d^n p}{(4\pi^2)^n}\,. \quad (12)$$

We see that Eq. (12) is formally identical to (3) of the classical case. For a pure state with wave function $\Psi(y)$, the marginal distribution would be[8]

$$w(X,\mu,\nu)=\frac{1}{(2\pi)^n|\nu_1\cdots\nu_n|}\left|\int\Psi(y)\exp i\sum_{j=1}^{n}\left(\frac{\mu_j y_j^2}{2\nu_j}-\frac{y_jX_j}{\nu_j}\right)d^ny\right|^2. \qquad (13)$$

Eq. (12) may be inverted and the Wigner function expressed in terms of the marginal distribution, like in the classical case of Eq. (5)

$$W(q,p)=\left(\frac{1}{2\pi}\right)^n\int w(X,\mu,\nu)\exp\left[i\left(X-\mu\circledast q-\nu\circledast p\right)\bullet\mathbf{1}\right]d^n\mu\, d^n\nu\, d^nX. \qquad (14)$$

Therefore the usual quantum mechanical quantities can be reconstructed from the marginal distributions. These quantities (wave function and Wigner function) have a nature quite different from the classical quantities, however the marginals $w(X,\mu,\nu)$ are in both cases positive distributions with the same physical meaning.

For a system with Hamiltonian

$$H=\frac{p^2}{2}+V(q), \qquad (15)$$

the marginal distribution satisfies the quantum time-evolution equation

$$\frac{\partial w}{\partial t}-\mu\bullet\nabla_\nu w-i\frac{1}{\hbar}\Big\{V\left(-\nabla_X^{-1}\circledast\nabla_\mu-i\frac{1}{2}\hbar\nu\circledast\nabla_X\right)$$
$$-V\left(-\nabla_X^{-1}\circledast\nabla_\mu+i\frac{1}{2}\hbar\nu\circledast\nabla_X\right)\Big\}w=0, \qquad (16)$$

which provides a dynamical characterization of quantum dynamics, alternative to the Schrödinger equation.

The evolution equation (16) can also be written in the form

$$\frac{\partial w}{\partial t}-\mu\bullet\nabla_\nu w-\nabla_x V(\tilde{q})\bullet(\nu\circledast\nabla_X w)$$
$$+\frac{2}{\hbar}\sum_{n=1}^{\infty}(-1)^{n+1}\left(\frac{\hbar}{2}\right)^{2n+1}\frac{\nabla_{i_1\cdots i_{2n+1}}V(\tilde{q})}{(2n+1)!}(\nu\circledast\nabla_X)_{i_1}\cdots(\nu\circledast\nabla_X)_{i_{2n+1}}w=0, \qquad (17)$$

where $\tilde{q}$ stands for the operator

$$\tilde{q}=-\nabla_X^{-1}\circledast\nabla_\mu,$$

and a sum over repeated indices is implied.

In Moyal's[11] formulation of quantum mechanics in phase-space, the transition from the classical to the quantum structure is a deformation of the Poisson algebra[12] with deformation parameter $\hbar$. In the symplectic tomography formulation, that we are describing, classical and quantum mechanics are described by the same set of positive probability distributions $w(X,\mu,\nu)$, the $\hbar$–deformation appearing only in the time-evolution equation (17).

For the propagator

$$w(X,\mu,\nu,t)=\int\Pi(X,\mu,\nu,X',\mu',\nu',t,t_0)\,w(X',\mu',\nu',t_0)\,d^nX'\,d^n\mu'\,d^n\nu', \qquad (18)$$

the equation is

$$\begin{aligned}&\frac{\partial \Pi}{\partial t} - \mu \bullet \nabla_\nu \Pi - \nabla_x V(\widetilde{q}) \bullet (\nu \circledast \nabla_X \Pi) \\ &+\frac{2}{\hbar}\sum_{n=1}^{\infty}(-1)^{n+1}\left(\frac{\hbar}{2}\right)^{2n+1}\frac{\nabla_{i_1\cdots i_{2n+1}}V(\widetilde{q})}{(2n+1)!}(\nu \circledast \nabla_X)_{i_1}\cdots(\nu \circledast \nabla_X)_{i_{2n+1}}\Pi \\ &= 0\,,\end{aligned} \tag{19}$$

with boundary condition

$$\lim_{t\to t_0} \Pi\left(X,\mu,\nu,X',\mu',\nu',t,t_0\right) = \delta^n\left(X-X'\right)\delta^n\left(\mu-\mu'\right)\delta^n\left(\nu-\nu'\right). \tag{20}$$

3 Lyapunov exponents

3.1 Density formulation in classical mechanics

Lyapunov exponents and other ergodic invariants in the classical theory are usually formulated in terms of quantities related to trajectories in phase-space, like tangent maps, refinement of partitions, etc.[13] Here, as a preparation for the formulation of Lyapunov exponents in quantum mechanics, using the marginal distributions $w(X,\mu,\nu)$, we explain briefly how these quantities may, in classical mechanics, be expressed as functionals of phase-space densities rather than in terms of trajectories. For more details we refer to Ref. 14.

A density in phase-space is a non-negative, normalized, integrable function, the space of densities being denoted by D

$$D = \left\{\rho \in L^1 : \rho \geq 0, \|\rho\|_1 = 1\right\}. \tag{21}$$

D is the space of functions that, by the Radon-Nikodym theorem, characterize the measures that are absolutely continuous with respect to the underlying measure in phase-space. However, to define Lyapunov exponents by densities, it is necessary to restrict oneself to a subspace of *admissible densities* defined as follows:

To each $\rho \in D$ we associate a square root, that becomes an element of an L^2 space. We then construct a Gelfand triplet

$$E^* \supset L^2 \supset E\,, \tag{22}$$

where E is the space of functions of rapid decrease topologized by the family of semi-norms $\|x_\alpha \partial_\beta f\|_2$ and E^* is its dual. Because E is an algebra $f \in E$ implies $f^2 \in E$. Therefore for each f such that $\|f\|_2 = 1$, $\rho = f^2$ is an admissible density. The restriction to such a subspace of admissible densities is necessary to be able to define Gateaux derivatives along generalized functions with point support. Gateaux derivatives along derivatives of the delta function play for densities the same role as the tangent map for trajectories. In this setting the Lyapunov exponent is[14]

$$\lambda_v = \lim_{t\to\infty}\frac{1}{t}\log\left\| -v^i D_{\partial_i\delta_x}\left(\int d\mu(y) y P^t \rho(y)\right)\right\|. \tag{23}$$

Here $v \in R^{2n}$, $\|\cdot\|$ is the vector norm and the Gateaux derivative $D_{\partial_i\delta_x}$ operates in the argument of the functional, that is, on the initial density

$$D_{\partial_i\delta_x}F\left(\rho(y)\right) = \lim_{\varepsilon\to 0}\frac{1}{\varepsilon}\left\{F\left(\rho(y)+\varepsilon\partial_i\delta(y-x)\right) - F\left(\rho(y)\right)\right\}. \tag{24}$$

μ is the invariant measure in the support of which the Lyapunov exponent is being defined and P^t is the operator of time evolution for densities

$$P^t \rho(y,0) = \rho(y,t)\,. \tag{25}$$

A simple computation shows that the expression (23) is equivalent to the usual definition of Lyapunov exponent in terms of trajectories and the tangent map

$$\lambda_v = \lim_{t\to\infty} \frac{1}{t} \log \left\| DT_x^t v \right\| \,, \tag{26}$$

where DT_x^t stands for the evolved tangent map applied to the vector v at the phase-space point x. Here and in Eq. (23) x and y are phase space vectors, that is, in the notation of Sect. 2, $x = (q_x, p_x)$ and $y = (q_y, p_y)$.

According to the Oseledec theorem,[15,16] for μ–almost every point x there is a decreasing sequence of vector spaces

$$R^{2n} = E_1(x) \supset E_2(x) \supset \cdots \supset E_r = \{0\}\,,$$

such that, by choosing the vector v in $E_s(x) \setminus E_{s+1}(x)$, the sth Lyapunov exponent is obtained by the above calculation.

A similar construction is possible for the metric entropy.

3.2 Classical and quantum Lyapunov exponents by marginal distributions

Let us now translate the equations of the preceding subsection in the tomographic framework discussed in Section 2. Initial densities are, by the tomographic map, mapped to initial tomographic densities by (3)

$$\rho\,(q,p) \to w\,(X,\mu,\nu,t=0) \equiv w\,(X,\mu,\nu)\;. \tag{27}$$

To compute the Gateaux derivatives notice that the generalized density ($\in E^*$)

$$(v_1 \bullet \nabla_q + v_2 \bullet \nabla_p)\,\{\delta^n\,(q-q_0)\,\delta^n(p-p_0)\} \tag{28}$$

is mapped to the tomographic generalized density $w_\eta (\in E^*)$

$$w_\eta\,(X,\mu,\nu) = ((v_1 \circledast \mu + v_2 \circledast \nu) \bullet \nabla_X)\,\delta^n\,(X - \mu q_0 - \nu p_0)\;. \tag{29}$$

According to Eq. (23), to compute the Lyapunov exponent, one has to obtain the expectation value of a generic phase-space vector on the time-evolved perturbation of the initial density (28). Therefore

$$\lambda_v = \lim_{t\to\infty} \frac{1}{t} \log \left\| \begin{array}{c} \int d^n q d^n p \begin{pmatrix} q \\ p \end{pmatrix} \mathcal{K}\,(q,p,q',p',t)\,(v_1 \bullet \nabla_{q'} + v_2 \bullet \nabla_{p'}) \\ \delta^n\,(q'-q_0)\,\delta^n(p'-p_0) dq' dp' \end{array} \right\| \,, \tag{30}$$

where $\mathcal{K}\,(q,p,q',p',t)$ is the evolution kernel for densities

$$\rho\,(q,p,t) = \int \mathcal{K}\,(q,p,q',p',t)\,\rho\,(q',p')\;d^n q'\,d^n p'\,.$$

Notice that in Eq. (30) the integration is carried over the flat phase-space measure $d^n q d^n p$. The result is equivalent to (23) for an invariant measure absolutely continuous with respect to $d^n q d^n p$. However the information and the dependence of the

Lyapunov exponent on the invariant measure is carried by the choice of the initial point (q_0, p_0). The set of Lyapunov exponents that is obtained by (30) is therefore the one that corresponds to the invariant measure on whose support (q_0, p_0) lies.

Eq. (30) may now be rewritten using marginal distributions

$$\lambda_v = \lim_{t\to\infty} \frac{1}{t} \log \left\| \begin{array}{r} \int d^n X\, d^n\mu\, d^n\nu\, e^{iX\bullet 1} \left(\begin{pmatrix} \nabla_\mu \\ \nabla_\nu \end{pmatrix} \delta^n(\mu)\delta^n(\nu) \right) \times \\ \Pi_{\rm cl}(X, \mu, \nu, X', \mu', \nu', t, 0) \\ ((v_1 \circledast \mu' + v_2 \circledast \nu') \bullet \nabla_{X'})\, \delta^n(X' - \mu' q_0 - \nu' p_0)\, dX'^n d\mu'^n d\nu'^n \end{array} \right\|, \tag{31}$$

where $\Pi_{\rm cl}(X,\ \mu,\ \nu,\ X',\ \mu',\ \nu',\ t_2,\ t_1)$ is the classical propagator defined in (8) - (10).

Because (23) is equivalent to the usual definition of Lyapunov exponent, Eq. (31), being equivalent to (23), is also a correct expression for the classical Lyapunov exponent.

Now the transition to quantum mechanics is straightforward. Marginal distributions in classical and quantum mechanics satisfy formally identical expressions and have the same physical interpretation as probability densities. The only difference lies in the time-evolution which in classical mechanics obeys Eq. (7) and in quantum mechanics the $\hbar$-deformed equation (16). Therefore the Lyapunov exponent in quantum mechanics will also be given by equation (31), with however the classical propagator $\Pi_{\rm cl}$ replaced by the quantum propagator Π for marginal distributions, defined in (18) - (20).

4 Example: One-dimensional systems with time-dependent potentials

We consider here one-dimensional systems with time-dependent potentials defined by the Hamiltonian

$$H = \frac{p^2}{2} + V(q,t)\,. \tag{32}$$

For these systems, the Lyapunov exponent expression (31) is

$$\lambda_v = \lim_{t\to\infty} \frac{1}{t} \log \left\| \int dX\, d\mu\, d\nu e^{iX} \left(\begin{pmatrix} \partial_\mu \\ \partial_\nu \end{pmatrix} \delta(\mu)\delta(\nu) \right) F(X, \mu, \nu, t) \right\|, \tag{33}$$

where $F(X, \mu, \nu, t)$ is the time-evolved perturbation, namely

$$F(X, \mu, \nu, t) = \int \Pi(X, \mu, \nu, X', \mu', \nu', t, 0)\,(v_1\mu' + v_2\nu') \times \\ \delta'(X' - \mu' q_0 - \nu' p_0)\, dX' d\mu' d\nu'\,. \tag{34}$$

Passing to the Fourier transform

$$G(k, \mu, \nu, t) = \frac{1}{2\pi} \int e^{ikX} F(X, \mu, \nu, t)\, dX\,, \tag{35}$$

one obtains

$$\lambda_v = \lim_{t\to\infty} \frac{1}{t} \log \left\| \int d\mu\, d\nu \left(\begin{pmatrix} \partial_\mu \\ \partial_\nu \end{pmatrix} \delta(\mu)\delta(\nu) \right) G(1,\mu,\nu,t) \right\|$$
$$= \lim_{t\to\infty} \frac{1}{t} \log \left\| \begin{pmatrix} G^{(2)}(1,0,0,t) \\ G^{(3)}(1,0,0,t) \end{pmatrix} \right\| , \tag{36}$$

where by $G^{(2)}$ and $G^{(3)}$ we denote the derivatives in the second and third arguments and $G(k,\mu,\nu,t)$ is a solution of the equation

$$\begin{aligned} &\tfrac{\partial G}{\partial t} - \mu \tfrac{\partial G}{\partial \nu} - ik\nu\partial_q V\left(-\tfrac{1}{ik}\tfrac{\partial}{\partial \mu}\right) G \\ &+ \tfrac{2}{\hbar} \textstyle\sum_{n=1}^{\infty} \tfrac{(-1)^{n+1}}{(2n+1)!} \left(ik\tfrac{\nu\hbar}{2}\right)^{2n+1} \tfrac{\partial^{2n+1}}{\partial q^{2n+1}} V\left(-\tfrac{1}{ik}\tfrac{\partial}{\partial \mu}\right) G = 0 , \end{aligned} \tag{37}$$

with initial condition

$$G(k,\mu,\nu,t) = -\frac{ik}{2\pi}(v_1\mu + v_2\nu)\, e^{ik(q_0\mu + p_0\nu)} . \tag{38}$$

Therefore, the computation of the Lyapunov exponents, both classical and quantum, reduces to the study of the large time limit of the solutions of Eq. (37). Also the simple expression (36) shows that, despite its apparently complex form, Eq. (31) is a computationally efficient way to obtain the Lyapunov exponent.

For local quadratic potentials it follows from (37) that the same results are obtained in classical and quantum systems. The same is true for non-local quadratic potentials.[10]

A non-trivial example is provided by the standard map. This is a case where the phenomena of wave function localization is believed to have a taming effect on chaos. The Lyapunov exponent analysis gives a characterization of how this taming effect comes about.

The Hamiltonian is

$$H = \frac{p^2}{2} + \gamma \cos(q) \sum_{n=-\infty}^{\infty} \delta(t - n\tau) , \tag{39}$$

the configuration space being now the circle, $q \in S^1$. This system describes a particle rotating in a ring and subjected to periodic kicks. It has been extensively used in studies of quantum chaos[17–21] and has even been tested experimentally with ultra-cold atoms trapped in a magneto-optic trap.[22]

From (37) the equation to be solved now is

$$\frac{\partial G}{\partial t} - \mu\frac{\partial G}{\partial \nu} - \frac{\gamma}{\hbar}\sum_{n=-\infty}^{\infty} \delta(t - n\tau)\sin\left(\frac{\hbar}{2}\nu\right)\{G(1,\mu+1,\nu,t) - G(1,\mu-1,\nu,t)\} = 0 . \tag{40}$$

where we have specialized to the value $k = 1$ because this is the only k-value needed to compute the Lyapunov exponent (36). Notice that we have used here the same tomographic transformations that were described in Section 2 for functions on the line. This is justified by considering all functions as defined not in S^1 but in the suspension of S^1.

From (40) one sees that between any two kicks the function propagates freely, namely

$$G(1,\mu,\nu,t_0) \to G\left(1,\mu,\nu,t_{1_-}\right) = G(1,\mu,\nu+\mu\tau,t_0)\ , \tag{41}$$

and at the time of the kick a quantity is added that is proportional to a finite difference (in μ).

$$\begin{aligned} G\left(1,\mu,\nu,t_{1_+}\right) &= G\left(1,\mu,\nu,t_{1_-}\right) + \\ &\frac{\gamma}{2} f(\nu)\left\{G\left(1,\mu+1,\nu,t_{1_-}\right) - G\left(1,\mu-1,\nu,t_{1_-}\right)\right\}, \end{aligned} \tag{42}$$

where, for the classical case

$$f(\nu) = \nu\,, \tag{43}$$

and for the quantum case

$$f(\nu) = \frac{2}{\hbar}\sin\left(\frac{\hbar}{2}\nu\right)\ . \tag{44}$$

To compute the Lyapunov exponent we need the evolution of the derivatives (in μ and ν) of G at $\mu = \nu = 0$. From (41) and (42) one obtains the following iteration for the derivatives

$$\begin{aligned} G^{(2)}(1,0,0,t+1) &= G^{(2)}(1,0,0,t) + G^{(3)}(1,0,0,t)\ , \\ G^{(3)}(1,0,0,t+1) &= G^{(2)}(1,0,0,t) + \tfrac{\gamma}{2}\left(G(1,1,\tau,t) - G(1,-1,-\tau,t)\right)\ . \end{aligned} \tag{45}$$

Let us consider first the classical case ($\hbar = 0$, $f(\nu) = \nu$ and $\gamma > 0$). Let also $\tau = 1$ and $q_0 = p_0 = 0$ in the initial condition (38). Then, one obtains the following recursion for the derivatives of G at $\mu = \nu = 0$.

$$\begin{aligned} G^{(2)}(1,0,0,n+1) &= G^{(2)}(1,0,0,n) + G^{(3)}(1,0,0,n)\ , \\ G^{(3)}(1,0,0,n+1) &= \gamma G^{(2)}(1,0,0,n) + (1+\gamma)\, G^{(3)}(1,0,0,n)\ , \end{aligned} \tag{46}$$

which has the solution

$$\begin{aligned} G^{(2)}(1,0,0,n) &= A_n(z)v_1 + B_n(z)v_2\,, \\ G^{(3)}(1,0,0,n) &= C_n(z)v_1 + D_n(z)v_2\,, \end{aligned} \tag{47}$$

with $z = 2+\gamma$ and

$$\begin{aligned} A_n(z) &= U_{n-1}(\tfrac{z}{2}) - U_{n-2}(\tfrac{z}{2})\,, \\ B_n(z) &= \tfrac{1}{z-2}C_n(z)\,, \\ C_n(z) &= U_n(\tfrac{z}{2}) - 2U_{n-1}(\tfrac{z}{2}) + U_{n-2}(\tfrac{z}{2})\,, \\ D_n(z) &= U_n(\tfrac{z}{2}) - U_{n-1}(\tfrac{z}{2})\,, \end{aligned} \tag{48}$$

where $U_n(z) = \frac{\sin\left((n+1)\cos^{-1} z\right)}{\sin(\cos^{-1} z)}$ is a Chebyshev polynomial.

For the Lyapunov exponent one obtains in this case

$$\lambda = \ln\left|1 + \frac{1}{2}\gamma + \sqrt{\frac{1}{4}\gamma^2 + \gamma}\right|\,, \tag{49}$$

a result similar to the case of harmonic kicks on the line.[10] One sees that as long as $\gamma > 0$ the exponent λ in Eq. (49) is always positive. This results from the choice made for the phase space point ($p_0 = q_0 = 0$) where the marginal distribution receives the singular perturbation (29). If instead we had chosen ($p_0 = 0$ and $q_0 = \pi$) in the initial condition (38), one sees easily by a change of coordinates in the Hamiltonian that this is equivalent to replace γ by $-\gamma$. Then the Lyapunov exponent λ in Eq.(49) is positive only for $\gamma > 4$. As discussed at length in the next section, this only means that it is the phase space point (p_0, q_0) that defines the measure for which the Lyapunov exponent is computed. Hence, for the measure that supports the hyperbolic point ($p_0 = 0$, $q_0 = 0$) the exponent is always positive, whereas for sufficiently small $\gamma > 0$ the exponent for the measure that supports the elliptic point ($p_0 = 0$, $q_0 = \pi$) is negative.

For the quantum case ($\hbar \neq 0$) let us consider an initial condition $G(1, \mu, \nu, 0) = \mu + \nu$ (corresponding to $p_0 = 0$, $q_0 = 0$, $v_1 = v_2 = 1$) and $\tau = 1$. According to Eq. (45), all one needs to compute the Lyapunov exponent is the time evolution of $G(1, 1, 1, t)$. For this purpose we set up a matrix recursion for the evolution equations (41-42). Define the following matrices

$$M_0 = \begin{pmatrix} 1 & 1 & 0 \\ 0 & 1 & 0 \\ 0 & 0 & 1 \end{pmatrix}, \quad M_+ = \begin{pmatrix} 1 & 1 & 0 \\ 0 & 1 & 0 \\ 1 & 1 & 1 \end{pmatrix}, \quad M_- = \begin{pmatrix} 1 & 1 & 0 \\ 0 & 1 & 0 \\ -1 & -1 & -1 \end{pmatrix}, \tag{50}$$

and vectors $\begin{pmatrix} \alpha \\ \beta \\ \gamma \end{pmatrix}$ where α counts the number of μ's, β the number of ν's and γ is a simple number. Then with Tr denoting the sum of the elements in a vector and

$$x_0 = \begin{pmatrix} 1 \\ 1 \\ 0 \end{pmatrix}, y_0 = \begin{pmatrix} 0 \\ 1 \\ 0 \end{pmatrix}, \tag{51}$$

the initial condition is $G(1, 1, \tau, 0) = \mathrm{Tr}(x_0)$ and the function $f(\nu) = f(\mathrm{Tr}(y_0))$.

On arbitrary functions of 3-dimensional vectors, the operators K_0, K_+, K_- act on the arguments by the matrices M_0, M_+, M_-

$$K_i g(x) = g(M_i x) \,. \tag{52}$$

Then

$$G(1, 1, 1, n) = \mathrm{Tr}\left\{ \left(K_0 + \frac{\gamma}{2} f(y_0) \left[K_+ - K_- \right] \right)^n x_0 \right\}, \tag{53}$$

where it is understood that the power of the operator is fully expanded before the Tr operation is applied to each one of the vector arguments. When this expansion is made, one obtains an expression of the form

$$G(1, 1, 1, n) = n + 2 + \mathrm{Tr}\left\{ \sum_{k=1}^{n} \left(\frac{\gamma}{2}\right)^k \sum_{i=1}^{\frac{2^{k-1} n!}{k!(n-k)!}} c_i f\left(\prod M_{i_1} y_0\right) \cdots f\left(\prod M_{i_k} y_0\right) \right\}. \tag{54}$$

The products of M matrices in the arguments of $f(\cdot)$ contain a variable number of factors, from 1 to k. However for each term, a different combination of products will appear. For $\hbar \neq 0$ the function $f(\nu)$ is proportional to a sine and, if $\frac{\hbar \tau}{4\pi}$ is irrational, the coefficient of each γ^k behaves like a sum of random variables of zero mean. Therefore each coefficient averages to zero and

$$G(1,1,1,n) \sim n+2\,. \tag{55}$$

Large fluctuations are however to be expected in view of the large number of terms in the sums for large n. From (45) the result (55) now implies

$$G^{(2)}(1,0,0,n) \sim 1+n(1-\gamma)+\tfrac{\gamma}{2}\left(\tfrac{n(n+1)}{2}+\tfrac{n(n+1)(2n+1)}{6}\right),$$
$$G^{(3)}(1,0,0,n) \sim 1+\gamma\left(\tfrac{(n+1)(n+2)}{2}-1\right).$$

For large n, $\log G^{(2)}(1,0,0,n) \sim 3\log n$ and $\log G^{(2)}(1,0,0,n) \sim 2\log n$ and the Lyapunov exponent vanishes.

The situation we have been studying ($p_0 = q_0 = 0$ in the initial perturbation) corresponds to the (hyperbolic) case where the classical Lyapunov exponent is positive for any γ. We see here clearly the taming effect of quantum mechanics on classical chaos and its dynamical origin. It results from the replacement in the evolution equation of the linear function $f(\nu) = \nu$ by $f(\nu) = \frac{2}{\hbar}\sin\left(\frac{\hbar}{2}\nu\right)$. This in turn is a consequence of the replacement of the classical Boltzman equation by the quantum evolution equation (17), or in algebraic terms, by the replacement of the ordinary product by the Moyal-Vey product in the non-commutative quantum phase space.

In this model, the origin of the taming effect of quantum mechanics on classical chaos, is traced back to the existence, in the $\hbar$−deformed equation (37), of infinitely many terms in the series which add up to a bounded function in ν. How general this mechanism is, for other quantum systems, is an open question. In any case the taming effect of quantum mechanics, obtained here for the standard map, is more accurate than previous discussions of the same system, because it refers to the behavior of the Lyapunov exponent rather than to indirect chaos symptoms, like the energy growth or diffusion behavior.

References

1. E. Wigner, *Phys. Rev.* **40**, 749 (1932).
2. S. Mancini, V. I. Man'ko, and P. Tombesi, *Quantum Semiclass. Opt.* **7**, 615 (1995).
3. G. M. D'Ariano, S. Mancini, V. I. Man'ko, and P. Tombesi, *Quantum Semiclass. Opt.* **8**, 1017 (1996).
4. S. Mancini, V. I. Man'ko, and P. Tombesi, *Phys. Lett.* A **213**, 1 (1996).
5. S. Mancini, V. I. Man'ko, and P. Tombesi, *Found. Phys.* **27**, 801 (1997).
6. Olga Man'ko and V. I. Man'ko, *J. Russ. Laser Research* **18**, 407 (1997).
7. Olga Man'ko and V. I. Man'ko, *J. Russ. Laser Research* **20**, 67 (1999).
8. V. I. Man'ko and R. Vilela Mendes, *Phys. Lett.* A **263**, 53 (1999).
9. V. I. Man'ko, *Conventional quantum mechanics without wave function and density matrix*, E-print quant-ph/9902079, in *Proceedings of the XXXI Latin American School of Physics* Eds. S. Hacyan, R. Jáuregui, and R. Lopez-Peña, (American Institute of Physics, New York, 1999), pp. 191.

10. V. I. Man'ko and R. Vilela Mendes, quant-ph/0002049, to appear in *Physica D.*
11. J. E. Moyal, *Proc. Cambridge Philos. Soc.* **45**, 99 (1949).
12. F. Bayen, M. Flato, M. Fronsdal, A. Lichnerowicz and D. Sternheimer, *Ann. Phys. (NY)* **111**, 61 (1977).
13. R. Mañe, *Ergodic theory and differentiable dynamics*(Springer, Berlin, 1983).
14. R. Vilela Mendes, *Entropy and quantum characteristic exponents. Steps towards a quantum Pesin theory,* in *Chaos - The interplay between stochastic and deterministic behavior,* Lecture Notes in Physics 457, Eds. P. Garbaczewski,M. Wolf and A. Weron (Springer, Berlin, 1995), p. 473.
15. V. I. Oseledec, *Trans. Moscow Math. Soc.* **19**, 197 (1968).
16. M. S. Raghunatan, *Israel Jour. Math.* **32**, 356 (1979).
17. F. M. Izrailev and D. L. Shepelyanski, *Theor. Math. Phys.* **43**, 553 (1980).
18. D. L. Shepelyanski, *Physica* D **8**, 208 (1983).
19. G. Casati, B. V. Chirikov, I. Guarnieri and D. L. Shepelyanski, *Phys. Rev. Lett.* **56**, 2437 (1986).
20. G. Casati, *Chaos* **6**, 391 (1996).
21. M. Zaslavski, *Chaos* **6**, 184 (1996).
22. M. G. Raizen, *Comments At. Mol. Phys.* **34**, 321 (1999).

AUTHOR INDEX

SUBJECT INDEX

www.ingramcontent.com/pod-product-compliance
Lightning Source LLC
LaVergne TN
LVHW070514160926
848142LV00010B/46
9789810247300